THE USES OF DIVERSITY

Race, Inequality, and Health

Edited by Samuel Kelton Roberts Jr. and Michael Yudell

The Race, Inequality, and Health series explores how forms of racialization have created a wide range of phenomena, from producing inequities in health and healthcare to inspiring social movements around health. The goal of this series is to publish field-defining works across history, the social sciences, the biological sciences, and public health that deepen our understanding of how claims about race and race difference have affected health and society.

Eram Alam, Dorothy Roberts, and Natalie Shibley, editors, *Ordering the Human: The Global Spread of Racial Science*

Sebastián Gil-Riaño, *The Remnants of Race Science: UNESCO and Economic Development in the Global South*

Rob DeSalle and Ian Tattersall, *Troublesome Science: The Misuse of Genetics and Genomics in Understanding Race*

Michael Yudell, *Race Unmasked: Biology and Race in the Twentieth Century*

The Uses of Diversity

*How Race Has Become Entangled
in Law, Politics, and Biology*

Jonathan Kahn

Columbia University Press   New York

Columbia University Press
Publishers Since 1893
New York   Chichester, West Sussex

Copyright © 2025 Columbia University Press
All rights reserved

Library of Congress Cataloging-in-Publication Data

Names: Kahn, Jonathan, author.
Title: The uses of diversity : managing race at the intersections of law,
   politics, and biology / Jonathan Kahn.
Description: New York : Columbia University Press, [2025] | Series: Race,
   inequality, and health | Includes bibliographical references and index.
Identifiers: LCCN 2024050933 (print) | LCCN 2024050934 (ebook) |
   ISBN 9780231220149 (hardback) | ISBN 9780231220132 (trade paperback) |
   ISBN 9780231563086 (ebook)
Subjects: LCSH: Race. | Race—Political aspects. | Equality.
Classification: LCC HT1503.A3 K34 2025 (print) | LCC HT1503.A3 (ebook) |
   DDC 305.8—dc23/eng/20250214

Cover design: Noah Arlow

GPSR Authorized Representative: Easy Access System Europe, Mustamäe tee 50,
10621 Tallinn, Estonia, gpsr.requests@easproject.com

Contents

Acknowledgments vii

INTRODUCTION
Managing the Entanglement 1

CHAPTER ONE
The Roots of Modern Diversities 19

CHAPTER TWO
Modern Diversities Taking Shape 37

CHAPTER THREE
Forensic Diversity 67

CHAPTER FOUR
Diversity Affirmed in the New Millenium 91

CHAPTER FIVE
Diversity Machines:
Race, Biology, and the Sweet Spot of Racialized Time 113

CHAPTER SIX
Getting Bodies: When Diversity Didn't Matter ... 139

CHAPTER SEVEN
Bringing Diversity Back In ... 158

CHAPTER EIGHT
Genetic Entanglements of Sociolegal Diversity in the 2010s ... 176

CHAPTER NINE
Diversity and the Frames of Representation ... 211

CHAPTER TEN
Political Valences of Contemporary Genetic Diversity ... 237

EPILOGUE
Diversity's Pandemic Distractions:
A Case Study of the Contemporary Uses of Diversity ... 262

Notes ... 319
Index ... 405

Acknowledgments

About fifteen (maybe even twenty) years ago, I was at a meeting discussing issues of race and genetics, when a friend and colleague expressed exasperation at the ongoing, seemingly never-ending, need to keep reminding people to be more careful when they are using racial categories in relation to biological concepts. "They just keep biologizing race! It never ends!" And so it doesn't. Since then, I have often repeated this story to friends and colleagues who are now much younger than I am, because they continue to share the same frustration. Working in this area over the years I have seen diversity as a distinctive field where this dynamic has and continues to play out in highly contested and problematic ways. Trying to manage the productive use of race at the intersections of law, politics, and biology sometimes feels Sisyphean. It demands constant and persistent attention. Conflations of race and biology take new forms, driven by different agendas, some relatively benign, others decidedly not. Progress can be and has been made, so if not entirely Sisyphean, perhaps the better metaphor is Whac-A-Mole™. As I say at the end of chapter 8, eternal vigilance is the price of adapting to new challenges to racial justice. It can be tiring, at times, but I have found great sustenance and support among a wide range of scholars committed to the endeavor.

I would like to thank panelists and audience members for feedback on material related to this book presented at the following venues: Northeastern University Humanities Center Fellows Program; Northeastern University Law School Health Law Round Table; the Comparative Health Humanities Symposium, University of Paris; the Annual Health Law Professors Conference; ELSICon; the Law and Society Association; the Governance of Emerging Technologies and Science Conference, Arizona State University; the Jaharis Symposium on Health Law & Intellectual Property, DePaul University School of Law.

The writing of this book was supported by National Library of Medicine Grant G13 LM013533–01. Research was also supported by research grants from the Northeastern University School of Law Dean's Office.

Throughout the text, some material from the following previously published articles may be found: "Diversity's Pandemic Distractions," *Health Matrix: The Journal of Law-Medicine* 32, no. 1 (2022): 149; "The Legal Weaponization of Racialized DNA: A New Genetic Politics of Affirmative Action," *Georgetown Journal of Law and Modern Critical Race Perspectives* 13 (2021): 187; "Precision Medicine and the Resurgence of Race in Genomic Medicine," in *Consuming Genetic Technologies: Ethical and Legal Considerations*, ed. I. Glenn Cohen (Cambridge: Cambridge University Press, 2021), 186–97; "Privatizing Biomedical Citizenship: Risk, Duty, and Potential in the Circle of Pharmaceutical Life," *Minnesota Journal of Law, Science & Technology* 15 (2014): 791; and "Race, Genes, and Justice: A Call to Reform the Presentation of Forensic DNA Evidence in Criminal Trials," *Brooklyn Law Review* 74 (2009): 325.

The support of the members of the RACEGEN listserv has continued to be invaluable. Among the many to whom I am particularly indebted are Sari Altshuler, Ruha Benjamin, Duana Fullwiley, Allan Goodman, Joseph Graves, Regine Jean-Charles, David Jones, Jay Kaufman, Sandra Soo-Jin Lee, Jon Marks, Anne Pollock, Alondra Nelson, Dorothy Roberts, and Patricia Williams. As with my previous books, the people at Columbia University Press have been wonderful to work with. Supportive and responsive, they have been exemplary partners in these endeavors. This book, as ever, is for Karen-Sue and Emma.

THE USES OF DIVERSITY

Introduction

Managing the Entanglement

Everybody, it seems, has something to say about "diversity" these days. An entire multibillion-dollar industry has grown around the concept of "Diversity, Equity, and Inclusion" (DEI). There are untold numbers of books and articles examining the subject from multiple perspectives debating its purported merits or deficiencies. Billionaires go on social media tirades about the subject.[1] Politicians try to get the very word banned from government discourse.[2] My concern here is not to add to this cacophony but to consider diversity from another angle—as a distinctive site where understandings of race as biological or social have periodically become entangled in highly problematic ways that threaten to undermine initiatives to address racial injustice and reinforce dangerously misguided understandings of racial hierarchy. Diversity serves a special role here precisely because, over the past half century, it has become a central organizing concept not only in the domains of law, politics, and commerce but also in the biological sciences, particularly genomics. Thus, even as diversity has played an increasingly prominent role in contentious political debates of recent years, it has emerged as a powerful trope framing major undertakings in the biosciences, where, for example, enlisting "diverse" cohorts of research subjects into massive genomic databases has become a *sine qua non* of realizing the "promise" of personalized medicine.

Diversity has become a prime site for slippage across these domains, with actors from all areas using the concept of diversity in ways that blur the distinction between social understandings of race and biological understandings of genetic variation. These entanglements change and manifest differently over time and in different contexts. Managing these entanglements is critical to ensuring that the concept of diversity is not deployed in ways that produce misguided constructions of race as biological or misapplied understandings of biology in legal or political contexts to undermine initiatives to understand and address persistent issues of racial inequality and injustice. Such invocations of diversity, both social and biological, have deep roots but their most immediate foundations are to be found in the early 1970s, which is where this story begins in chapter 1.

Race has always been a tricky concept. As a term of taxonomic classification in the biological sciences, race "defines an informal subdivision of subspecies which are physically and genetically different." Under this definition, the human species cannot be divided into subspecies.[3] As anthropologist Adam Van Arsdale put it, "The biological construction of race is invalid not because it is impossible but because evolutionary forces have actively worked against such patterns in our evolutionary past."[4] Biology, then, is not in itself what makes race tricky. What makes it tricky—or least, very significant among the many things that make race tricky, has been the entanglement of biological with social constructions of race in ways that justify historical oppression and reinforce existing racial hierarchies. For most Americans, race invokes the categories of the census: White, Black, Asian, Native American, and perhaps Native Hawai'ian or Other Pacific Islander, as well as the ethnic category of Hispanic.[5] These primary racial categories map all too neatly onto earlier racist, racial taxonomies derived from the likes of eighteenth-century naturalist, Carl Linnaeus, who delineated four continentally based subspecies of *Europaeus, Asiaticus, Africanus,* and *Americanus,* each with associated stereotyped characteristics that reinforced a racial hierarchy with White Europeans at the top.[6] The entanglement of race and biology runs deep.

The rise of modern genetics was supposed to have resolved the issue. Among those who study intersections of race and genetics, it is by now almost a cliché to invoke the June 2000 White House ceremony announcing the completion of the first draft of the human genome as a marker of either lost promise or hubristic naïveté with respect to science's ability to resolve social questions about the nature and meaning of race. Standing with Francis Collins, then the director of the National Human Genome Research Institute (NHGRI) and the key leader of the publicly oriented Human Genome Project (HGP) consortium of researchers, and J. Craig Venter, PhD founder of Celera Genomics, a private firm that was, in effect, competing with the HGP to produce the first draft of a human genome, President Bill Clinton declared: "I believe one of the great truths to emerge from this triumphant expedition inside the human genome is that in genetic terms, all human beings, regardless of race, are more than 99.9 percent the same." Following President Clinton, geneticist Craig Venter asserted that this accomplishment illustrated "that the concept of race has no genetic or scientific basis."[7] On its face, such declarations would seem to neatly and nicely make it clear that race has no genetic basis, surely laying the foundations for a definitive disentanglement of biological concepts of genetic diversity from social concepts of diversity. But this has not happened—not in the biological sciences, nor in the worlds of law and politics.

The evolution of these entanglements over time is the primary focus of this book. In particular, I explore how, over the past half century, the concept of diversity has been a distinctive site for producing understandings of race that have crossed lines between the worlds of the biological sciences and those of law, politics, and commerce. Crossing such lines does not always or of necessity produce knots, but it often does, and when it does so it has tended to reify race as genetic in highly problematic and often dangerous ways. Exploring and untangling these knots is my project here. My goal is both to tell a cautionary tale, urging a greater humility among those deploying these concepts in what is often a cavalier or overconfident manner, and to provide some modest guidance on how best to manage these concepts going forward.

For my part, I generally agree with sociologists Michael Omi and Howard Winant who approach race as "an unstable and 'decentered' complex of social meanings constantly being transformed by political struggle."[8] More specifically, evolutionary biologist Joseph Graves and biological anthropologist Alan Goodman define race as "a worldview and social classification that divides humans into groups based on their appearance and assumed ancestry, and that has been used to establish social hierarchies," but note that "what we call *biological race* does not exist in our species."[9] While I assert, along with these authors, that race is not a genetic category, it nonetheless can have biological consequences resulting, for example, in health disparities. As epidemiologist Nancy Krieger puts it, race-based health differences can be understood as "biological expressions of race-relations."[10]

If race is tricky, diversity is protean—it seems to be ever-changing, meaning all things to all people as convenient. Promoters of DEI have deployed it as a representation of all things good and productive in the world of business and academia; yet, certainly in recent years, it has been imbued with more sinister connotations and meanings by icons of the alt-right, such as Ron DeSantis and Christopher Rufo. But diversity is more than a concept upon which we project our fears and desires. It is also itself *productive*: it shapes how we construct and interpret social relations, racial hierarchies, law, and power. It has also provided the basis for invoking state and corporate power to shape institutions central to modern American life. The history of affirmative action attests to this, with the 1978 case of *Regents of the University of California v. Bakke* articulating an idea of the value of diversity in education that became the foundation upon which a vast structure of diversity related law, politics, and business enterprises would be built.[11]

As in the social realm, diversity has also come to suffuse the world of genomics and related biotechnological enterprises, both corporate and state sponsored. Here too it has taken on many meanings, sometimes contradictory, and served many purposes. And it has also been productive—framing and guiding much of the enterprise of modern genomics in powerful ways. I was particularly struck by this in 2015, when President Obama announced the creation of a major new program to be known as the Precision Medicine Initiative (PMI) in his State of the Union address. Obama's stated goal of the PMI was "to bring us closer to curing diseases like cancer and diabetes, and to give all of us access to the personalized information we need to keep ourselves and our families healthier."[12] It soon emerged that a central means to this goal would be the creation of a new, massive genetic database, with the aim of enlisting one million citizens into the enterprise of genomic discovery and related promised developments of diagnostics and therapies. Initially dubbed the "PMI Cohort" it currently operates as the *All of Us* program. As of January 2024, it had registered more than 1,151,000 volunteers.[13]

Following up on his State of the Union address, President Obama authorized the creation of a PMI working group to "articulate the vision for the PMI research cohort."[14] Diversity was central to this vision. But diversity in this context represented many different things. As the working group's report itself noted:

> The U.S. is a diverse society. Americans comprise diverse social, racial/ethnic, and ancestral populations living in a variety of geographies, social environments, and economic circumstances. The U.S. population includes people with extreme wealth and people living in abject poverty and with varying access to social, educational, and other resources....
>
> Racial and ethnic diversity do not just reflect genetic ancestry; they are social constructs rooted in cultural identity and shaped by historic and current events, which influence an individual's behavior, place of residence, and life opportunities.[15]

Here we had diversity presented in terms not only of race and ethnicity, but also time (ancestry), space (geography), and class (social and economic circumstances). Diversity in this initial framing was itself diverse. This is all well and good, and certainly the writers of the report were sensitive to the problems of entangling social and genetic construction of race. Nonetheless in the very next sentence, the working group reduced diversity primarily to genes and race when it asserted that "because genetic variation correlates with self-identified race, a racially/ethnically diverse cohort will provide unprecedented opportunities to examine the complex relationship of ancestral influences, environmental exposures, and social factors."[16]

To support the seemingly straightforward assertion that "genetic variation correlates with self-identified race," the report cited a 2003 article by Burchard et al. from the *New England Journal of Medicine* that was actually part of a forum where the article was paired with another that directly contested the idea of using race as a proxy for genetic variation.[17] In answer to the question, "Has genomics provided evidence that race can act as a surrogate for genetic constitution in medicine or public health?" this latter article's unequivocal answer was no: "Race, at the continental level, has not been shown to provide a useful categorization of genetic information about the response to drugs, diagnosis, or causes of disease."[18] This is not to say that one article was right and the other wrong—merely that it was and remained at the time of the PMI working group report, a highly contested issue. Thus the report, in its framing of the primary organizing concept for a massive genetic database, laid the foundations for further entanglements of race and biology through the concept of diversity.

In the world of law, Sanford Levinson has observed that diversity belongs in a category of "essentially contested concepts," that are, "extremely important but, nonetheless, without truly definite meaning."[19] In his 2003 book, *Diversity: The Invention of a Concept*, Peter Wood, an anthropologist and later president of the conservative National Society of Scholars, argued, with a more obvious focus on affirmative action, that diversity "is above all a political doctrine asserting that *some* social categories deserve compensatory privileges in light of the prejudicial ways in which members of these categories have been treated in the past and the disadvantages they continue to face."[20]

In, *Diversity in America: Keeping Government at a Difference*, another book from 2003, legal scholar Peter Schuck provided a taxonomy of diversity that included such organizing concepts as: "diversity-in-fact and diversity as ideal"; "normative and descriptive diversity"; "ascriptive and consensual diversity"; "intergroup and intragroup diversity"; and "enclave and larger-scale diversity."[21] For each of these he provided brief descriptions, but notably did not articulate a taxonomic grouping based on biology, per se. Nonetheless, he did immediately follow up his taxonomy of different *types* of diversity with a discussion of genetics as a possible *source* of diversity, asserting:

> Much human diversity reflects, first, the almost infinite *genotypic variation* found in our species. Even with the recent coding of the human genome, scientists do not yet comprehend the full nature, extent, and behavioral significance of this variation. What is already known, however, indicates that genotypic

variation accounts for a significant share of interpersonal variations in intelligence, psychology, behavior, and a growing list of other variables.[22]

Notably, Schuck did not provide any citations to support his assertion regarding the genetic basis for interpersonal behavioral variations, but at least he was not making any assertions here regarding genetic basis for inter*group* variations—as, we shall see, have others—most notoriously, Richard Herrnstein and Charles Murray in their 1994 book, *The Bell Curve*.[23]

Schuck's critique of diversity was published the same year as the major Supreme Court case, *Grutter v. Bollinger,* affirming *Bakke*'s articulation of diversity as an interest sufficiently compelling to justify affirmative action in higher education.[24] It was also just three years after the White House ceremony celebrating the completion of the first draft of the human genome.

When one thinks about taxonomies of diversity, it becomes evident diversity is not simply a way of organizing preexisting group qualities. Technologies of diversity, whether legal, regulatory (like the categories we use in the census), or scientific (like genome-wide association studies or genetic ancestry tracing), produce the groups about which we are talking. That is, diversity is not simply something to be managed after the fact; it is also produced, sometimes through the very actions that are intended to manage it.

I am a lawyer with a PhD in American history currently holding a joint appointment between a law school and a department of biology. Over the years I have held appointments in departments of history, political science, social studies, and bioethics. My wife is a cultural anthropologist working on issues of genomics as a cultural object, through whom I have gained a deep appreciation of anthropology and STS (science and technology studies). I relate this simply to provide some background for my particularly eclectic approach to many of the issues addressed in this book.

For many years now, I have been exploring various facets of how law, politics, and science have interacted, each shaping and being shaped by the other, to produce particular understandings and uses of race in American social and commercial contexts. In my book, *Race in a Bottle*,

I examined how legal and commercial incentives drove the racial characterization of a heart failure drug, BiDil, in a manner that dangerously threatened to biologize race.[25] In my next book, *Race on the Brain*, I considered how the elevation, by well-meaning liberals, of implicit bias into a master narrative of race-relations threatened to biologize, not race, but racism. While working on *Race on the Brain*, I was struck by the prominence of the rhetoric of diversity in major biomedical initiatives. By 2019, the National Human Genome Research Institute was highlighting new studies, "Putting Diversity Front and Center."[26] Many of these discussions focused not only on genetic diversity (as one might expect), but also on social diversity. These initiatives touted not only the need to get more diverse genomes to propel research forward, but also the importance of promoting a diverse workforce, in part to help recruit diverse populations, who presumably would provide the ultimately sought after diverse genomes. This inevitably led to characterizations, first of the workforce, then of bodies, and finally of genomes, in terms of racial and ethnic categories. The entanglement was palpable.

All this was unfolding while the case of *Students for Fair Admissions v. Harvard* was working its way up to the Supreme Court.[27] This case would dismantle the diversity rationale for affirmative action in university admissions articulated nearly fifty years ago in *Bakke*, but it is still unclear what its enduring practical effects will be, both in and beyond academia. The majority opinion in that case, thankfully, did not entangle social and biological understandings of race. This in itself was notable because one would think that the biologists of the National Institutes of Health (NIH) who were more familiar with the underlying ideas about the discordance of race and genetic variation would have had a better handle on this than conservative Supreme Court justices.

One might have thought that the biologization of race, with its eugenic overtones, was primarily driven by voices from the Right. While this has often been the case, we also see well well-meaning liberals implicitly and sometimes explicitly biologizing race, often in attempts to address perceived inequities in the domains of biomedicine and public health.[28] Some conservatives, on the other hand, have embraced the idea that race is a social, not a biological construction, in order to challenge affirmative action and related programs as based on nonscientific, and therefore legally invalid, categories. The lesson here is that the entanglement of race and biology within the frame of diversity does not always serve a single political script or produce reliably conservative or liberal results. Rather, in unknotting that entanglement we gain insights into how diversity can be used in a variety of ways to dismantle or reinforce (and sometimes reinvent) racial hierarchies. And so, I undertook the writing of *The Uses of Diversity* to explore the immediate, modern roots of this entanglement and trace it up to the present.

Chapter 1 beings in the early 1970s, when, more by coincidence than design, two foundational framings for discussions of diversity were articulated by Harvard professors with offices just a few blocks apart from each other in Cambridge. The first came in a 1972 article by evolutionary biologist, Richard Lewontin titled, "The Apportionment of Human Diversity,"[29] which became a primary referent for subsequent analyses of the relation between racial categories and genetic variation.[30] The following year, soon after Lewontin had moved from the University of Chicago to Harvard, law professor and recently fired Watergate Special Prosecutor Archibald Cox began working on a Supreme Court brief in the case of *DeFunis v. Odegaard*, in which he first articulated the idea that diversity was a sufficiently compelling state interest to justify affirmative action in university admission programs.[31] In these early iterations, the biological and legal framings of diversity were relatively distinct and unentangled. Indeed, I could find no evidence of any exchange of ideas on the topic between the Marxist, Lewontin and the scion of the establishment, Cox, although both shared an interest in racial justice.

The chapter explores the context out of which each framing emerged and how they manifested in their early iterations. Both were contending with the likes of psychologists Richard Herrnstein (yet another Harvard professor) and Arthur Jensen, who were very much trying to mix race and biology in order to challenge the idea of affirmative action, and indeed most programs aimed at ameliorating racial injustice. In this context, Lewontin's critique of biologized conceptions of racial difference provided critical breathing room for Cox's arguments about the social value of diversity in educational contexts that was ultimately taken up by Justice Powell in his critical opinion in *Bakke*. I then consider the immediate aftermath of *Bakke*, including the creation of the Office of Minority Health under President Reagan in 1985, and the origins of the Human Genome Project in the late 1980s. During this period, despite the agitation of scholars such as Herrnstein and Jensen, diversity as a biological concept remained relatively distinct from diversity as a social concept.

Chapter 2 starts exploring some early examples of their reentanglement in the later 1980s and 1990s. It opens with a consideration of factors propelling the emergence of the modern diversity management industry, including 1987's *Workforce 2000* report on diversifying trends in the American workforce, and the work of business diversity pioneers such a R. Roosevelt Thomas who advocated for a shift from Justice Powell's frame of affirmative action in *Bakke* to one more explicitly engaging with the concept of diversity itself as driving force in corporate and academic settings. Among the key factors for Thomas was diversity's forward-looking temporal frame and an emphasis on realizing so-called "natural" talents as opposed to the backward-looking "artificiality" of affirmative action.

I then juxtapose this with the emergence of the Human Genome Project (HGP) in the 1990s and in particular with a related enterprise, the Human Genome Diversity Project (HGDP), with an explicit focus on mapping global genetic variation, that problematically reinforced the reification of isolated population groups as genetic in its sampling strategy. But like Thomas and the growing diversity management industry, it was concerned with how diversity could drive the production of valuable outcomes.

These worlds collided and intertwined in the controversy surrounding Richard Herrnstein and Charles Murray's 1996 book, *The Bell Curve*, which, while declaring its enthusiasm for diversity, argued that group-based biological differences across the races accounted for differing levels of achievement and distorted new knowledge from genetic science to revive their old arguments against affirmative action and other programs aimed at ameliorating the effects of racial injustice. Meanwhile, more direct attacks on affirmative action were occurring in the world of law and policy as California adopted Proposition 209 in 1996, effectively banning the use of affirmative action in employment or education in California.[32] Similarly, in the 1996 case of *Hopwood v. Texas*, the U.S. Court of Appeals for the Fifth Circuit struck down affirmative action outright, and dismissed the relevance of Justice Powell's embrace of diversity in *Bakke*, declaring that "there has been no indication from the Supreme Court, other than Justice Powell's lonely opinion in *Bakke*, that the state's interest in diversity constitutes a compelling justification for governmental race-based discrimination."[33]

Yet, even while diversity was facing a backlash in these legal and political realms, the world of biomedicine was doubling down on it, as 1994 saw the adoption of the NIH Revitalization Act, which mandated the inclusion of women and minorities in federally funded biomedical research; with the Food and Drug Administration following suit with similar mandates for drug development in 1997's Food and Drug Modernization Act.[34] Ironically, as the likes of Herrnstein and Murray were entangling race and genetics with an eye toward undermining progressive racial programs, these initiatives entangled race and genetics with the ostensible goal of redressing race-based health disparities.

Chapter 3 then takes a bit of a detour into the realm of what I call "forensic diversity," the use of race and DNA in criminal justice settings. It begins by examining how race came to enter the construction and presentation of forensic DNA evidence in the early 1990s and considers how and why its use persisted and developed over the following decades. As this story unfolds, it addresses the basic question of what race added as a practical matter to the ability of the finder of fact to make fair and accurate decisions in a criminal proceeding. At the core of the development of forensic DNA was the construction of racially diverse reference databases to try to assess the odds that a DNA sample found at a crime scene matched that of a potential suspect. We come to see how technological advances largely rendered the use of race irrelevant to the calculation of odds ratios necessary to establish a match between a DNA sample left at a crime scene and DNA from a suspect. It considers how, nonetheless, the use of race persisted in forensic contexts, both entangling race with genetics and additionally introducing violent crime into this toxic mix. I suggest that there is an inertial power to race in American society that propelled its continued use long after any original rationale for its introduction may have faded. The chapter thus argues that in most cases such evidence should be excluded as irrelevant, or if deemed relevant it should be held inadmissible because the dangers of infecting the proceedings with racial prejudice outweigh any possible benefit that introducing the race-based statistics could provide.

Chapter 4 focuses on the early 2000s, with the first draft of the human genome being completed around the same time that the Supreme Court was affirming diversity as a constitutionally permissible rationale for affirmative action in higher education. This appeared to present a moment of disentanglement of race and biology in frames of diversity, as the declarations of the essential biological unity of the human race at President Clinton's White House ceremony celebrating the Human Genome Project attested. Yet this period also saw the emergence of a new computer program, STRUCTURE, which would go on to become widely used to characterize human population structure and estimate individual admixture from different populations.[35] Seemingly innocuous on its face, its "structuring" of genetic samples into population clusters would soon be used in ways that made it appear, *contra* Venter, that race (or at least continentally based population groupings) really did have a genetic basis.

While geneticists were employing tools that promised to reentangle race and genetics, something very different was happening in the world of law and policy. As diversity seemed ascendant, fierce critiques emerged from both the Left and the Right. From the Left came concerns that a focus on diversity erased or downplayed our racist past and diluted the distinctive significance of race as just one of any number of attributes, like being from South Dakota or having an aptitude for the bassoon, that academic admissions officers might consider. From some on the Right came a more curious, genetically inflected critique. Embracing the progressive notion that race was a social, not a genetic construct, they performed a sort of cognitive jiujitsu, using the weight of that insight against progressive goals. If race was "merely" social, i.e., *not* genetic, they argued, then it lacked the coherence necessary to form legally legitimate classifications. This new critique, while clearly distinct from and often in tension with earlier arguments from the likes of Herrnstein and Murray that focused on racial differences as genetic, nonetheless served the same purpose—to challenge and undermine programs aimed at challenging existing racial hierarchies and remedying past injustices. This right-wing dual challenge to racial justice within the frame of diversity–both genetic and nongenetic—would continue to inform attacks on affirmative action up to the present day.

Chapter 5 explores what I characterize as "diversity machines," genetic technologies such as ancestry tracing and the STRUCTURE program that literally produce categories used to create and manage understandings of diversity—not only in the world of genealogy and biomedicine but also in the realms of law and politics. As genetic ancestry tracing companies began to take off in the early to mid-2000s they tended to categorize their results in crude continental groupings that were readily transmuted into racial categories. As customers received results telling them they were "x" percent "European," "y" percent "African," and "z" percent "Asian," they readily came to understand these results in racial terms, reinforcing genetic conceptualizations of race.[36] Some tried to use such information to establish descent from enslaved populations as a basis for reparations claims.[37] Others, seeking to game the system of affirmative action in both business and education, tried to use such tests to gain preferences, even though phenotypically and experientially they presented as White.[38] Meanwhile, as geneticists continued to explore the nature of global genetic variation, STRUCTURE provided the basis of new contentious debates over whether, or to what extent, race indeed did reflect underlying patterns of genetic difference.[39] As race gained new prominence in biomedical research, it similarly gained new value in the arenas of biotechnology patents and commercial drug development. While diversity remained central to discussions of ancestry and affirmative action, it faded from prominence in this new arena, perhaps because race itself became the central driving engine for capitalizing on new biomedical advances. When diversity was not needed, race stood alone. But in literally capitalizing on it, researchers and corporations powerfully reentangled social and biological constructions of race.

Chapter 6 explores more fully another major arena at the intersection of biomedicine and public policy where diversity faded from the picture: the drive to develop massive genetic databases beginning in the mid-2000s. It focuses in particular on a multiyear study conducted for the NIH by the Genetics and Public Policy Center (GPPC) at Johns Hopkins University to examine methods of effectively recruiting subjects to a national biobank, and the passage of the Genetic Information Nondiscrimination Act (GINA) in 2008. These initiatives underscore how we can understand the uses of diversity also through considering places where it was relatively absent. Here the focus was simply on getting more bodies—regardless of their ascribed identity—to drive forward biomedical research and commercial drug development. In some ways, this links back to the original promise of the Human Genome Project with its emphasis on the idea that humans were genetically 99.9 percent the same. But it also highlights the contingent nature of the uses of diversity, with it often coming to prominence or fading into the background as needed to serve particular ends. It also shows how large-scale genomic enterprises did not of necessity entail the entanglement of biological and social conceptions of race.

Chapter 7 brings this observation into fuller relief as it charts how diversity returned to biomedical research with a vengeance during the 2010s as many of the promised benefits of genomics failed to materialize and research shifted to the search for "rare" genetic variants that might lead to new breakthroughs. The idea here was that finding such rare variants required more diverse research cohorts; and so was born President Obama's Precision Medicine Initiative, whose *All of Us* program centered diversity in its aim to create another massive genomic database. The shift from "personal" to "precision" medicine in the 2010s led to a renewed focus on group differences that provided a new basis for the entanglement of race and biology. We see here how more subtle and sophisticated characterizations of diversity in terms not only of race but also geography, environment, class, and other nonbiological factors tended in practice to collapse back into relatively crude discussions of racial difference—again attesting to the inertial power of race to remain a dominant frame in any system where it is introduced. This dynamic was reinforced by the further entanglement of biological and social understandings of race in the persistent calls to diversify the biomedical workforce, in part, so as to facilitate the recruitment of more diverse research subjects, with the ultimate goal of getting more diverse genomes. Thus, in one single initiative could multiple meanings of diversity become blurred and distorted.

Chapter 8 explores the uses of such distortions in the fields of law and politics during the 2010s. It opens with the story of Ralph Taylor, a

White insurance contractor from Lynwood, Washington, who tried to use his DNA ancestry results to gain access to state contracts under its Minority Business Enterprise (MBE) program. Taylor's story neatly encapsulates how one's stance toward social and genetic diversity could readily shift depending upon their potential uses. Initially, Taylor sought to use a genetic ancestry test showing a small percentage of African ancestry to gain access to the MBE program. His application was denied, in part, because the relevant regulations characterized qualifying individuals more in terms of their social experience and personal identification with a subordinated racial group than in terms of biology. Taylor then shifted his approach, arguing that because such socially inflected constructions were *not* genetic, they were somehow less real, less coherent, and hence arbitrary and invalid. Along the way, he became the darling of right-wing commentators who had longstanding quarrels with any form of affirmative action.

The chapter moves on to explore in greater depth how this challenge to the racial categories employed in diversity initiatives as arbitrary and capricious was echoed in broader conservative attacks on an array of federal and state initiatives aimed at ameliorating racial inequity. These included challenging both the use of racial categories in the census itself and the characterization of Native American tribes as political rather than racial entities under the Indian Child Welfare Act. Ironically, such challenges from the Right embraced the disentanglement of race from biology as a means to challenge such initiatives as "merely" social, lacking the rigor and precision of "science" and hence legally invalid.

Building on my examination of moves toward diversifying biomedical research, chapter 9 explores more fully the conceptualizations of representation constructed by and embedded in the particular uses of diversity in contemporary biomedical initiatives and enterprises. Diversity, of course, must be diversity of something. It is necessarily a representational concept. Representation has both descriptive and normative components. In the political culture of a representative democracy, it necessarily invokes ideas of authorization and accountability, concepts directly implicating power. In recruiting subjects into biomedical research, it often involved urging people to participate so that their "group" would be represented. This built on early calls involving the idea of civic duty to participate in research. But in seeking diverse genomic databases, it also involved more descriptive, and often statistical, conceptualizations of representation grounded more directly in capturing a specific range of genetic variation—harkening all the way back to the Human Genome Diversity Project of the 1990s. These newer initiatives were interested in defining and exploring genetic diversity for the betterment of humanity, but conflated normative and descriptive concepts of representation so as to give the appearance of voice and accountability to research subjects without the substance.

Chapter 10 engages with various entanglements of race and biology in recent legal and political deployments of diversity. Here we see White supremacists and other avatars of conservative backlash increasingly using findings from new genetic technologies to reinvigorate arguments about the "reality" of race as genetic. Disconcertingly, they invoke the work of scientifically sophisticated (but frankly politically clueless) geneticists to buttress their claims of racial difference, often under the rubric of "human biodiversity," a term misappropriated from the antiracist biological anthropologist Jonathan Marks, who first coined the phrase back in the 1990s. This leads, perhaps unavoidably, to a consideration of the clearly eugenic and frequently racist views of Donald Trump.

Considering whether we have reached the "end of diversity," I consider a curious inversion and repudiation of the trope of diversity by erstwhile Republican presidential contender Vivek Ramaswamy who proudly declared in 2023, "Diversity is not our strength. Our strength is what unites us across that diversity."[40] Such challenges to diversity found their legal counterpart in the 2023 case of *Students for Fair Admissions, Inc. v. President and Fellows of Harvard University*, where the Supreme Court rejected Justice Powell's original diversity rationale for affirmative action, providing a bookend to the era that began with *Bakke*.[41] This opened the floodgates to other right-wing attacks on diversity, most prominently in the form of assaults on the idea (and industry) of DEI which had begun in a backlash against the Black Lives Matter movement and gained steam throughout 2023 and into 2024, focusing with a particular virulence on higher education and playing no small role in the forced resignation of the presidents of the University of Pennsylvania and Harvard University in late 2023.

I conclude with an extended epilogue that provides an in-depth case study of diversity in action in responses to the COVID-19 pandemic. Pandemic diseases have a nasty history of racialization. COVID-19 was no exception. Beyond the obvious racist invocations of the "China virus" or the "Wuhan flu" were subtler racializing dynamics that were often veiled in more benign motives but were nonetheless deeply problematic. The racialization of COVID-19 proceeded along two distinct trajectories, each of which threatened to reinforce inaccurate biologized conceptions of race while diverting attention from the social, legal, and political forces historically structuring race-based health disparities.

First, early in the pandemic as significant racial disparities in disease incidence and mortality became evident, a frame of race-based genetic difference came to the fore as a possible explanation. Second, as vaccine development ramped up there came widespread calls for racially "diversifying" clinical trials for the vaccines being tested. The rationales for such diversification were varied but tended to reinforce genetic frames of racial difference. Most common was the assertion (without substantial evidence) that vaccines might work differently in Black or Brown bodies and so racial diversity in trials was imperative for reasons of safety and efficacy.

The epilogue explores the dynamics of how the concept of diversity racialized responses to COVID-19 and considers their broader implications for understanding and responding to racial disparities in the face of pandemic emergencies and beyond. In the short term, vaccine developers did a decent job of enrolling minorities in their clinical trials and the vaccines have proven to have the same safety and efficacy across races. In the long term, diversity in the biomedical context of pandemic response not only distracted from important structural causes of health injustice, it also focused attention on the genetics of disparities in a manner that has the potential to reinforce pernicious and false ideas of essential biological difference among racial groups.

I argue that an uncritical embrace of the idea of diversity in analyzing and responding to emergent health crises has the potential to distract us from considering deeper historical and structural formations contributing to racial health disparities. I explore the dynamics through which initial responses to racial disparities in COVID-19 became geneticized and then move on to unpack the rationales for such racialization, examine their merits (or lack thereof), and consider their implications for developing an equitable response to pandemic emergencies. The next section examines the subsequent racialization of clinical trials for COVID-19 vaccines through the concept of diversity, and explores how the geneticization of COVID-19 racial disparities laid the foundations for a similar geneticization of race in vaccine development. In failing to clearly distinguish social and biological rationales for diversity, such framings, while generally well-intentioned, were poorly supported and worked in tandem with the geneticization of racial disparities in COVID-19 morbidity and mortality to locate the causes of disparities in the minds and bodies of minoritized populations. This further distracted attention from the historical and structural forces contributing to such disparities. I conclude by recognizing a certain intractability to the problems of using race in biomedical research and practice, particularly in the context of public health emergencies, and offer some modest suggestions for improvement that could have significant practical effects if taken to heart by researchers, clinicians, and policymakers.

CHAPTER ONE

The Roots of Modern Diversities

EMERGING DIVERSITIES: BIOLOGY

Richard Lewontin was an eminent geneticist who rose to prominence in the latter half of the twentieth century both for his groundbreaking scientific work and for his outspoken advocacy against the misuse of biology to rationalize or obscure social injustice.[1] In 1972, while on the faculty at the University of Chicago, he published a paper titled "The Apportionment of Human Diversity," which examined genetic variation based on an analysis of the ABO blood group system. He found that 85 percent of all human genetic variation could be found within so-called "racial groups," with only 7 percent segregating by racial classifications.[2] In other words, he found that there was more genetic variation within such groups than between them. This finding, largely confirmed by subsequent studies,[3] provided a critical foundation for the modern development of the idea that race has no genetic basis but is rather a social construct.[4]

Lewontin framed his exploration of human genetic variation and difference specifically in terms of the concept of "diversity." This is perhaps unremarkable, given that the concept of diversity had long been central to the discipline of evolutionary biology. For example, a century earlier in *The Origin of Species*, Charles Darwin remarked upon such things as "the vast diversity of plants and animals which have been cultivated," and "the number and diversity of inheritable deviations of structure."[5] But in modern human genetics, it was Lewontin who brought the frame of "diversity" to the fore in examining human genetic variation.

Lewontin's work did not take place in a vacuum. Following World War II, the United Nations Educational, Scientific and Cultural Organization (UNESCO) issued two statements on race. The first, in 1950, tried to distinguish ideological from scientific conceptions of race, emphasizing the essential unity of humankind and highlighting the environmental influences on human differences. As Jennifer Reardon has noted, the statement "did not argue that race was meaningless, but rather that scientists thought race was meaningless for particular purposes—namely, for the purpose of evaluating meaningful human traits and creating hierarchical distinctions among human beings."[6] Some geneticists and physical anthropologists pushed back against this statement, which led to a second, more tempered statement being issued in 1954 which, in effect, left open the possibility that in the future as yet unproven findings might show some variation in endowments that varied by race.[7]

The UNESCO statements were followed in 1962 by geneticist Frank Livingstone's highly influential observation that "There are no races, there are only clines," (that is, gradual genetic variation over geographic space), that "can be plotted in the same way that temperature is plotted on a weather map."[8] As anthropologist Adam Van Arsdale observed in 2019, "These two landmark observations—Livingstone's argument for clinal variation and Lewontin's argument on the apportionment of human biological diversity—have remained, in many ways, the cornerstones of the arguments against the biological race concept within biological anthropology."[9] Livingstone framed his discussion in terms of human "variation." Similarly, in 1972, the same year Lewontin published his article, the National Institute of General Medical Sciences (NIGMS) established the Human Genetic Cell Repository at the Coriell Institute, characterizing its major collection of genetic differences as one of "Human Variation."[10] It was Lewontin who centered the frame of "diversity" in relation to human genetics.

Around the same time Lewontin published "The Apportionment of Human Diversity," Associated Press reporter Jean Heller first broke the news to the wider public of the infamous "Tuskegee Study of Untreated Syphilis in the Negro Male,"[11] which had begun on the 1930s largely premised on the idea that race-based biological difference might lead the disease to progress differently in Black bodies than in White.[12] The U.S. Public Health Service oversaw the study, which denied subjects access to penicillin after it was discovered to be an effective treatment for the disease while lying to them about the nature and purpose of the study up until Heller's 1972 exposé. The fallout from the revelations led the to the creation of the Office for Protection from Research Risks at the National Institutes of Health (NIH).[13] The broader public outrage at the abuses born from a genetic or biological conceptualization of race created fertile soil in which the seeds of Lewontin's insights might take root and grow.

The year after publishing his article, Lewontin joined the faculty of Harvard University where he soon became embroiled in a controversy with another Harvard professor, social psychologist Richard Herrnstein.[14] Herrnstein had recently published an article on IQ in the popular magazine *The Atlantic* arguing that the increasingly meritocratic political system of the United States would eventually result in a class structure that mirrored genetically based IQ scores.[15] This avowedly hereditarian view of social stratification built upon an incendiary 1969 article published by psychologist Arthur Jensen that concluded, among other things, that 1960s Great Society programs such as Head Start were doomed to fail, particularly in their goal of helping African American children, because in his estimation racial variation in IQ were grounded in population based genetic differences between Blacks and Whites.[16]

Jensen also invoked "diversity" in making his case. He wrote of the "the biological basis of diversity in human behavioral characteristics," and posited that "if diversity of mental abilities, as of most other human characteristics, is a basic fact of nature, as the evidence indicates, and if the ideal of universal education is to be successfully pursued, it seems a reasonable conclusion that schools and society must provide a range and diversity of educational methods, programs, and goals, and of occupational opportunities, just as wide as the range of human abilities."[17] For Jensen, diversity, as a function of purported genetic difference, provided a justification for racial stratification and a rationale for abandoning policies aimed at addressing disparities in educational achievement.

Lewontin had already responded to Jensen in 1970, critiquing his concept of heritability by asserting that "is it not then likely that the difference [in IQ scores between Blacks and Whites] is genetic? No. It is neither likely nor unlikely. There is no evidence. The fundamental error of Jensen's argument is to confuse heritability of a character within a population with heritability of the difference between two populations."[18] It was precisely this sort of misattribution of socially observed racial difference to genetics that Lewontin's 1972 article on the apportionment of human diversity would be used to address. Regarding the programmatic implications of Jensen's work, Lewontin declared:

> The real issue is what the goals of our society will be. Do we want to foster a society in which the "race of life" is "to get ahead of somebody" and in which "true merit," be it genetically or environmentally determined, will be the criterion of men's earthly reward? Or do we want a society in which every man can aspire to the fullest measure of psychic and material fulfillment that social activity can produce? Professor Jensen has made it fairly clear to me what sort of society he wants. I oppose him.[19]

Lewontin's conceptualization of the apportionment of human genetic diversity removed racial groups from the realm of biological difference and hence ran directly counter to Jensen's deployment of diversity—both scientifically and programmatically.

So, when Herrnstein published his article in *The Atlantic*, Lewontin was having none of it. In 1973, newly arrived at Harvard, Lewontin savaged his new colleague's work and challenged him to a debate. Very much echoing his response to Jensen, Lewontin explained in a letter to Herrnstein:

> My purpose is in some small way to counteract the rubbish that you and others have been producing. I do not accuse you of being primarily racist; I do not think you are. Your purpose is to

convince people that their position in society is biologically determined in large part. [But] since in America black people are disproportionally represented in the lower classes, people will believe that you think black people are genetically inferior. You may say they are special, but you have no proof of that either. You have no particle of evidence one way or the other. You are a political propagandist masquerading as a scientist. I will continue to struggle against you in any way at my command.[20]

Herrnstein actually ended up suing Lewontin for defamation, but the case went nowhere.[21] He would go on to coauthor *The Bell Curve* with Charles Murray in the 1990s, further developing his Jensenist arguments about the heritability of intelligence and its implication for social policy.[22] Murray, in turn, would continue to develop these arguments, fully bringing the concept of diversity to the fore in his 2020 book, *Human Diversity: The Biology of Gender, Race, and Class*.[23]

A few years later, Lewontin found himself in another contretemps, this time with his Harvard colleague E. O. Wilson who published the highly influential book, *Sociobiology: A New Synthesis* in 1975.[24] Wilson's primary research focus had been on the behavior of ants, but in *Sociobiology* he presented a sweeping analysis of social behavior in animals ranging from ants to apes. In a final, admittedly speculative, chapter he extended his analysis to humans, arguing that such human traits as xenophobia, war, and specified sex roles were the product of adaptive evolution.[25] While Wilson did not directly address issues of racial difference, Lewontin saw this as a dangerous step toward crude biological essentialism. Together with Harvard colleagues Stephen J. Gould and Jonathan Beckwith, and other notable academics, he wrote an open letter titled, "Against Sociobiology," to the *New York Review of Books*. The letter situated Wilson's work in the tradition of eugenics and other forms of biological determinism that "consistently tend to provide a genetic justification of the *status quo* and of existing privileges for certain groups according to class, race or sex."[26]

So here at Harvard, in the early 1970s, we have an early articulation of a formative debate about the nature of human genetic diversity that would come to shape discussion about race and social stratification for decades to come. One the one side, Lewontin and his analysis of ABO blood groups showing that genetic variation does not map neatly on to socially constructed racial groups. On the other, Herrnstein, building on Jensen, arguing in effect that the subordinate position of Black people in America might (at least in part) be attributed to genetic differences. In a distinct position was Wilson, not speaking explicitly about race but reinforcing the sorts of biologically deterministic arguments being put forward by Jensen and Herrnstein.

EMERGING DIVERSITIES: LAW

Meanwhile, a few blocks away from Lewontin's office in the venerable Museum of Comparative Zoology, Archibald Cox had newly returned to Harvard Law School after President Nixon fired him as special Watergate prosecutor in the infamous "Saturday Night Massacre" of October 20, 1973.[27] Around the same time that Lewontin was sparring with Herrnstein over genetic diversity and race, Cox was becoming involved in the affirmative action case of *DeFunis v. Odegaard*, where he would develop legal arguments about a very different type of diversity in relation to race—social, experiential, and viewpoint diversity. His concerns, in many respects, were quite similar to Lewontin's: to identify and address the observed reality of social stratification among racial groups. Where Lewontin's scientific work used the concept of *genetic* diversity (or lack thereof) to dismantle arguments that genetics could explain such stratification, Cox's legal work aimed to develop a concept of the value of *social* diversity sufficiently compelling to justify the use of race to dismantle the stratification itself.

Marco DeFunis had been denied admission to the University of Washington's law school in 1971. He sued, claiming the school's formula for evaluating applicants discriminated against him because he was White. The trial court found in DeFunis's favor and issued an injunction compelling the university to admit him to the law school.[28] The case worked its way up to the Supreme Court in 1974 where Cox became involved writing an amicus brief for Harvard University in support of the University of Washington.[29] Ultimately, the Supreme Court dismissed the case as moot because by that time DeFunis was close to graduating from law school.[30]

DeFunis was a critical precursor to the famous 1978 case of *Regents of the University of California v. Bakke.*[31] *Bakke* is commonly understood as a foundational case in the modern jurisprudence of affirmative action.[32] It was there that Justice Powell notably articulated diversity as a compelling state interest sufficient to warrant taking race into account in properly constructed affirmative action programs because it promoted a "beneficial educational pluralism."[33] Powell's opinion became a touchstone because the Supreme Court in *Bakke* was deeply divided. Four liberal justices voted to uphold California's affirmative action policies arguing that benign uses of racial classifications to help racially disadvantaged groups should only be subject to intermediate scrutiny under the Equal Protection Clause of the 14th Amendment—that is, be upheld if the state can show that the policy serves "important governmental objectives and must be substantially related to achievement of those objectives."[34] In contrast, the four conservative justices took the more narrow position that the university's affirmative action policy violated Title VI of the Civil Rights Act which, they argued, demanded that all federally funded programs be color-blind.[35] Powell's lone opinion thus became effectively controlling as the four liberal justices concurred in his holding that racial classifications *could* be used for sufficiently compelling reasons, while the four conservative justices concurred in his finding that in, this particular case, the university's specific affirmative action program was not sufficiently narrowly tailored to serve that end because it had a strict numerical set-aside of sixteen slots in the incoming class of one hundred for minority students.[36]

Cox argued for the University of California in *Bakke* and put forth the diversity rationale for affirmative action that Powell would adopt. But as legal scholar David Oppenheimer has shown, Cox first developed this argument several years earlier in his amicus brief in *DeFunis.*[37] In his search for the roots of the use of diversity in legal concepts, Oppenheimer found generally that before 1978 there was "virtually no discussion of the value of diversity; [rather] the justifications for affirmative action and other civil rights remedies focused on promoting equality and combating discrimination."[38] He noted that the conventional wisdom was that Justice Powell had developed the diversity justification by relying on an amicus brief filed by Harvard in the *Bakke* case in 1978. But, through some persistent historical sleuthing, Oppenheimer uncovered Cox's hitherto unexamined 1974 amicus brief in *DeFunis*, which largely set forth the arguments for the value of diversity that would later come to inform much of Powell's opinion in *Bakke*.[39] Powell would append to his opinion an amicus brief filed jointly by Columbia University, Harvard University, Stanford University, and the University of Pennsylvania which lifted language about the value of diversity verbatim from Cox's earlier *DeFunis* brief.[40]

How then did Cox, and later Powell, talk about diversity? In his *DeFunis* brief, Cox reached back to the nineteenth century and the vision of Harvard's transformative President Charles Eliot, who expounded on the importance of cultivating a student body that represented a diverse range of experience and geographic background. Oppenheimer quotes from a speech Eliot made in 1911, two years after stepping down from his fifty-year term as Harvard's president, stating: "I cannot imagine greater diversity than there is in Harvard College. It is not superficial; it is deep. It is shown in the variety of races, religions, households from richest to poorest, and in the mental gifts and ambitions."[41] Cox expanded this conception arguing that for diversity to be meaningful, a school must enroll more than a small number of individuals from any racial or ethnic group; and this could be done without the use of strict numerical quotas.[42]

Sociologists Anthony Chen and Lisa Stulberg note that in the 1960s this general concern for diversity in higher education had taken on a distinctively racial tone in the face of the civil rights movement and the passage of the 1964 Civil Rights Act, finding that "by 1967, there was a critical mass of institutions that sought to secure the educational benefits of a diverse study body, and racial diversity ranked among the forms of diversity that they valued the most."[43] Scholars such as political scientist John Skrentny or legal scholar Peter Schuck have presented the emergence of the diversity rationale in *Bakke* as largely a response to the violent political unrest of the late 1960s and early 1970s, or as education reporter Peter Schmidt put it, affirmative action was a product of "black rage and white fear."[44] In contrast, Chen and Stulberg argue that their research into primary sources found that "a nontrivial number of admissions officers at selective institutions around the country embraced the educational value of diversity—including the educational value of racial diversity—more than a decade before *Bakke*."[45]

This is not to say that *Bakke* was not a transformational opinion in many respects. Rather it shows that even as Powell's opinion drew heavily on Cox's brief; and Cox, in turn, drew on an existing tradition of and rising solicitude for the value of various forms of diversity in higher education.

Herrnstein and Jensen constructed genetic diversity as a possible explanation for racial stratification. The productive power of this biological concept of diversity served to undermine calls for policies to redress racial inequality. Cox, in contrast, developed a legal framework where historicized conceptions of social diversity could provide a justification for addressing racial stratification. Powell ultimately adopted a somewhat diluted version of Cox's vision of diversity. The productive power of this concept of diversity would serve as a foundation for affirmative action policies for decades to come.

While not explicitly at issue in *DeFunis* or in Powell's opinion in *Bakke*, Lewontin's work on the apportionment of human diversity can be seen as creating critical breathing room for Cox's vision of diversity insofar as it allowed for the marginalization or containment of the arguments put forth by Herrnstein and Jensen. For example, in its *Bakke* amicus brief supporting the University of California's affirmative action policy, the Law School Admission Council asserted:

> To use educational attainment as the measure of ability, while refusing to come to grips with the glaring question of *why* minority applicants rank lower on these measures, or to consider past discrimination as the obvious cause, would involve a tacit but implicit adoption of a premise of genetic inferiority... [Law schools] should not be compelled to follow an approach which denies, *sub silentio*, that unlawful segregation has produced any effect, and which therefore could be explained only on some undemocratic and unscientific assumption that biologically or genetically, members of minority groups are incapable of equal education attainment.[46]

In referring to (and dismissing) the "premise of genetic inferiority" the Law School Admission Council's brief cited Jensen's 1970 article, "Selection of Minority Students in Higher Education." The brief went on to assert, that under an affirmative action program, "The important factor, of course, is not [an applicant's] race in any immutable genetic sense, but the unique experience and cultural ties which are inextricably intertwined with perceived race in this country in his own lifetime."[47] Such clear and direct invocation of the distinction between race as a social versus genetic construct was precisely the sort of stance facilitated by Lewontin's work.[48]

EMERGING DIVERSITIES: POLITICS

The year before *Bakke* and *Hill*, the Office of Management and Budget (OMB) issued Directive 15, which defined the basic racial and ethnic categories to be utilized by the federal government for three reporting purposes: statistical, administrative, and civil rights compliance.[49] Such classifications are essential to measuring, evaluating, and ultimately constructing the sort of diversity at issue in affirmative action programs. These federally mandated standards emerged as a consequence of major governmental programs and legal initiatives instituted since the 1960s. The OMB categories provided the basis for both census information and access to a variety of governmental goods and services that are contingent upon membership in a particular racial or ethnic group.[50] For example, federal users of racial data provided by the census include: the Department of Education, Department of Justice, Department of Labor, Equal Employment Opportunity Commission, Federal Reserve, Department of Health and Human Services, Housing and Urban Development, Department of Agriculture, and the Veterans Administration.[51] As Alice Robbin noted, "groups must be counted in order to make credible claims for political representation, demonstrate discriminatory practices against them, seek and obtain legal remedies, receive governmental assistance for a host of social programs, and evaluate current, as well as develop new public policy."[52] Additionally, such classifications provide the framework for evaluating school desegregation, electoral districting, and other civil rights initiatives.[53]

Familiar to us today as basic census categories, this first iteration of Directive 15 specified the following categories for race: American Indian or Alaskan Native, Asian or Pacific Islander, Black, White; and for ethnicity: Hispanic origin, Not of Hispanic origin. Already attentive to the sorts of conflation of social and genetic constructs identified by Lewontin and others, Directive 15 was careful to specify that these classifications were for "administrative reporting and statistical activities . . . [and] should not be interpreted as being scientific or anthropological in nature."[54] Nonetheless, once a template for classification is introduced into any working system it tends to take on a life of its own. As we shall see, these OMB Directive 15 categories would come to structure not only how social data for affirmative action programs or voting information was collected but also how biological information in clinical trials and genetic research would be organized, interpreted, and reported.[55]

Ultimately, in his 1978 opinion in *Bakke*, Powell both saved race as a viable category for affirmative action policies and devalued it, characterizing it as "only one element in a range of factors a university properly may consider in attaining the goal of a heterogeneous student body."[56] Race, in other words, was merely one metric of possibly desirable types of diversity, a "plus factor," no different from geography, the ability to play the trombone, or a distinctive interest in Sanskrit texts. As Powell put it, "the diversity that furthers a compelling state interest encompasses a far broader array of qualifications and characteristics of which racial or ethnic origin is but a single though important element. Petitioner's special admissions program, focused solely on ethnic diversity, would hinder rather than further attainment of genuine diversity."[57] Whatever "genuine diversity" might be for Powell, it was something more than just racial or ethnic in composition. More specifically, Powell's concept of diversity produced a deracinated conception of race that obscured its connection to historical injustice. It also rendered race-informed policies more palatable to a society already in backlash against the civil rights advances of the 1960s. It did this in part by shifting substantive demands for power into procedural questions of representation.

On the political front, we see the Democratic Party platform of 1972 speak of diversity for the first time, focusing on "celebrat[ing] the magnificence of the diversity within its population, the racial, national, linguistic and religious groups which have contributed so much to the vitality and richness of our national life." It went on to offer "recognition

The Roots of Modern Diversities

and support of the cultural identity and pride of black people are generations overdue. The American Indians, the Spanish-speaking, the Asian Americans—the cultural and linguistic heritage of these groups is too often ignored in schools and communities." In contrast, the Republican Party platform spoke of "our non-public schools, both church-oriented and nonsectarian," and celebrated the "diversity they help to maintain in American education."[58]

In the context of the school desegregation battles of the late 1960s and early 1970s, "non-public schools" meant White flight and segregation academies.[59] This was the Republican diversity of separate but equal. The platform of the George McGovern Democratic Party was unabashedly liberal and multicultural—and doomed to resounding defeat at the polls that fall. Come 1976, the Democratic Party platform made no mention of diversity at all. It would return in 1980, with a muted call to "recognize the value of cultural diversity in education," that nicely echoed Powell's broad characterization of a diversity that subsumed and obscured race. Such was the palatability (and ambiguity) of Powell's articulation of diversity that the Republican platform of 1980, while still embracing the importance of support for nonpublic schools, would also declare that "diversity in education has great value.... Private colleges and universities should be assisted to maintain healthy competition and enrich diversity."[60] By the end of the 1970s, then, we have the articulation of foundational (though hardly uncontested) conceptions of diversity in both genetic and sociolegal domains.

DIVERSITY EXPANDS

By the mid-1980s diversity was entering the biomedical domain in new and powerful ways and with a distinctive focus on race. While Martin Luther King Jr. notably declared in 1966 that "of all the inequality, injustice in health is the most shocking and inhumane,"[61] it was not until the early 1980s that leaders from Black medical schools and other Historically Black Colleges and Universities (HBCUs) presented a report titled, *Blacks and the Health Professions in the 1980s: A National Crisis and a Time for Action*, to Margaret Heckler, then President Reagan's Secretary for Health and Human Services.[62] In 1984, Heckler convened a special task force to examine issues in "Black and Minority Health" that presented a foundational report the following year. The *Report of the Secretary's Task Force on Black & Minority Health* (or the *Heckler Report* as it came to be known) represented the first time the federal government had consolidated minority health issues into a single report and acknowledged the need to confront "persistent health disparities," that had "for so long cast a shadow on the otherwise splendid American track record of ever improving health." While the precipitating report from the HBCUs emphasized the need to take steps to address the lack of minority representation in the health professions, the Heckler Report studiously steered clear of any invocation of affirmative action and never discussed racism or prejudice as possible causes of disparities. Its focus, rather, was on data. The report deemed the "collection of quantitative data on the incidence and prevalence of health problems in these [minority] populations" to be essential to addressing disparities.[63] The then-recently promulgated OMB standards for classifying data by race would play a central role here.[64] Nonetheless, responding directly to the report from the HBCU representatives, the Heckler Report also stressed the need to "examine ways to increase minority representation in preventive medicine, public health, health education, communications, and other health professions."[65]

The Heckler Report also discussed social and cultural diversity, noting:

> America is rich in the diversity of its minorities. There are more than 500 federally recognized American Indian tribes, 23 different countries of origin for Asian/Pacific Islanders, and three major places of origin for Hispanics. This diversity among populations is reflected in language difficulties, in cultural practices and beliefs with respect to illness and health, in differences in their birth rates, in differences in the afflictions which kill them, and in differences in their needs for types of services and the duration of health care.[66]

In exploring this diversity, it never invoked potential biological or genetic differences as explanations for disparities. To the contrary, it specifically declared that "The scientific literature supports a hypothesis that the differences in cancer experience between non-minorities

The Roots of Modern Diversities 31

and Blacks may be largely attributable to social or environmental factors rather than inherent genetic or biologic differences." The Heckler Report noted in particular that "Among the many factors presumed to influence minority health status in the United States today, four social characteristics are believed to be especially significant: (1) demographic profiles, (2) nutritional status and dietary practices, (3) environmental and occupational exposures, and (4) stress and coping patterns."[67] In short, at this point in time, the Department of Health and Human Services was doing a very good job of keeping social and cultural concepts of diversity distinct from biological or genetic ones; and recognizing that the latter were not relevant to explaining health disparities among the former.

In 1986, Heckler followed up by spearheading the creation of an Office of Minority Health in the Department of Health and Human Services (HHS). Two years later, the Centers for Disease Control would found its own Office of the Associate Director for Minority Health; and soon thereafter the National Institutes of Health would establish the Office of Research on Minority Health. By the early 1990s, federal recognition of and concern for racial issues in health were becoming fully institutionalized.[68]

The Heckler Report and subsequent creation of the Office of Minority Health are perhaps all the more remarkable because they occurred during a period of intense backlash against affirmative action under the Reagan administration.[69] During this time, the Justice Department moved from enforcing to challenging affirmative action and school desegregation programs. Reagan slashed the budget of the Equal Employment Opportunity Commission and installed Clarence Thomas (later to be elevated to the Supreme Court by George H. W. Bush) at its head. Thomas had earlier declared that he was "unalterably opposed" to any sort of quota-based affirmative action program. Similarly, to head the Civil Rights Commission, the Reagan White House chose Clarence Pendleton Jr., a fierce opponent of "quotas, proportional representation or the setting aside of government contracts for minority business."[70] Yet, at the same time, here was Margaret Heckler, embracing the need to increase minority representation in the health professions and focusing attention on race-based health disparities. By the early 2000s, health disparities would become more politicized in a manner akin to affirmative action,[71] but as they first emerged as a matter for governmental concern in the 1980s, they appeared to escape such polarizing characterizations.

It was at this time that the first seeds of what was to become the Human Genome Project (HGP) were germinating. In March 1986, the Department of Energy's Office of Health and Environmental Research held a workshop in Santa Fe to assess the feasibility of sequencing the human genome.[72] Its report noted many possible benefits from such a concerted project, both for basic science and clinical applications, ranging from increased understanding of gene expression to the molecular basis of inherited disease. It also highlighted the possible economic benefits arising from the availability of a complete genomic sequence, which would eventually provide "health advantages...sufficient to more than offset the cost of the entire enterprise." Diversity also made an appearance, with the report noting that there was "enthusiasm for sequencing as the ultimate tool for understanding and defining the extent of human genetic diversity."[73] Santa Fe was soon followed by meetings at Cold Spring Harbor and in Washington, D.C., where leading scientists discussed the possibility of such a project. The next year, President Reagan's budget included a $13 million line item initiating the genome project, and 1988 saw the first official expenditures on what would become the multibillion-dollar HGP.[74] The HGP would officially begin on October 1, 1990 with the Department of Energy and the NIH jointly presenting a five-year plan to Congress.[75]

While groundbreaking scientists were conceptualizing the Human Genome Project, the Supreme Court was hearing a case, *Saint Francis College v. Al-Khazraji*, about the scope of the 1866 Civil Rights Act (codified at 42 U.S.C. § 1981), which forbids racial discrimination in many settings, including employment.[76] This case is notable for its extensive discussion in a footnote of the relation between genetic and social concepts of race. At issue, was a claim by Majid Ghaidan Al-Khazraji, a U.S. citizen who had been born in Iraq, that his employer, Saint Francis College had denied him tenure because of his race in violation of § 1981. One of the questions before the Supreme Court was whether Al-Khazraji, who was Arab, could claim the protections of the

Civil Rights Act. The College argued that Al-Khazraji was a Caucasian and therefore could not allege the kind of discrimination that § 1981 forbids. Ultimately, Justice White wrote an opinion holding that for the purposes of a § 1981 claim, an Arab could be considered a racial group distinct from a Caucasian race and therefore bring a claim of racial discrimination under the Civil Rights Act.[77]

Central to this holding was a consideration of that fact that historically, at the time the Civil Rights Act was passed in 1866, Arabs were commonly viewed as distinct from the "Caucasian race."[78] Along the way to this conclusion, Justice White engaged in a rather remarkable excursion into debates about social versus genetic conceptualizations of race in a long footnote that merits quotation in full:

> There is a common popular understanding that there are three major human races—Caucasoid, Mongoloid, and Negroid. Many modern biologists and anthropologists, however, criticize racial classifications as arbitrary and of little use in understanding the variability of human beings. It is said that genetically homogeneous populations do not exist, and traits are not discontinuous between populations; therefore, a population can only be described in terms of relative frequencies of various traits. Clear-cut categories do not exist. The particular traits which have generally been chosen to characterize races have been criticized as having little biological significance. It has been found that differences between individuals of the same race are often greater than the differences between the "average" individuals of different races. These observations and others have led some, but not all, scientists to conclude that racial classifications are for the most part sociopolitical, rather than biological, in nature.[79]

White followed this with a long list of citations, including biologist Stephen J. Gould's then-recent and highly popular *The Mismeasure of Man*, and works by anthropologists Ashley Montague and Sherry Washburn and geneticist Theodosius Dobzhansky.[80]

There are a few notable things about White's language in this footnote. At first read it appears to be a clear and forceful articulation of the idea that race cannot be coherently understood as a genetic category. He even stated, in a manner that almost directly echoes Lewontin's *Apportionment of Human Diversity* that "It has been found that differences between individuals of the same race are often greater than the differences between the "average" individuals of different races." Nonetheless, there are some important caveats here. First, he strategically used the passive voice to assert that "it is said that genetically homogeneous populations do not exist." Second, he concluded with the equivocal assertion that "these observations and others have led some, but not all, scientists to conclude that racial classifications are for the most part sociopolitical, rather than biological, in nature."[81] White, then, acknowledged a broad understanding about the nature of race among contemporary scientists but did not accord it any specifically legal weight.

Furthermore, as legal scholar Khiara Bridges has noted, White also approvingly cited the court's holding that the statute bans discrimination against "an individual 'because he or she is genetically part of an ethnically and physiognomically distinctive subgrouping of homo sapiens.'"[82] Bridges argued that the court's "language suggests that, within its ontology of race and ethnicity, it is genetics that makes ethnic 'subgroupings,' which are nevertheless cognizable as races within law."[83] Nonetheless, White did go on to state that "It is clear from our holding, however, that a distinctive physiognomy is not essential to qualify for § 1981 protection."[84] There are thus tensions in White's opinion between the observation of broad scientific opinion that race is not genetic, and its recognition of the legal legitimacy of construing "ethnicity" (if not race per se) as something that could be understood in reference to genetics.

In *Bakke*, Powell very explicitly talked about the social value of racial diversity but did not consider the nature or definition of race itself. He did not confuse or conflate genetic with social race, but neither did he directly consider their possible relation—although some of the amicus briefs did. As the *definition* of race was not at issue in *Bakke*, the *value* of race was not at issue in *Al-Khazraji*. In *Al-Khazraji*, therefore, White did not mention diversity per se. Nonetheless, it very much *was* a case about race and its construction in legal contexts. White directly engaged the meaning of race under § 1981 and found that the legislative conception of race did not need to be limited to or grounded in a genetic understanding of the classification. Nonetheless, as Bridges notes, for all his thoughtful review of the current scientific consensus, White left the door open to future legal constructions of race as genetic.[85]

CHAPTER TWO

Modern Diversities Taking Shape

THE EMERGENCE OF DIVERSITY MANAGEMENT

Diversity may not have been central to Justice White's opinion about employment discrimination in *Al-Khazraji*, but many trace the origins of the idea of "diversity management" in business to a report issued that same year by the conservative think tank the Hudson Institute titled, *Workforce 2000: Work and Workers for the 21st Century*.[1] The report predicted dramatic changes in the racial, ethnic, and gender composition of the American workforce, declaring that "The cumulative impact of the changing ethnic and racial composition of the labor force will be dramatic. The small net growth of workers will be dominated by women, blacks, and immigrants. White males, thought of only a generation ago as the mainstays of the economy, will comprise only 15 percent of the net additions to the labor force between 1985 and 2000."[2] As anthropologist Peter Wood noted in his 2003 book, *Diversity: The Invention of a Concept*, the Hudson Institute's report created a sense of "crisis urgency, and purpose" about diversity, but as he also observed, this was largely due to a widespread misreading of this statement. The *New York Times*, for example, interpreted it to mean that "Native White males, who now constitute 47 percent of the labor force, will account for only 15 percent of the entrants to the labor force by the year 2000." Wood, however, asked one of the authors of the report about this and received the clarification that "the 15% refers to the NET increase in workforce where 'net' refers to the total increase minus the withdrawals (retires, deaths). We should have had a sentence explaining this appear at least a few times in the book."[3] Friedman and DiTomaso later observed of the confusion that the report "uses the term 'new entrants' for a category that would be more accurately labeled 'net additions'.... Thus, readers thought the report described total change, where the report actually described marginal change."[4]

Wood observed that as the mistaken interpretation of the 15 percent figure proliferated, it fueled a boom in the diversity movement.[5] Edelman et al. similarly argued that "*Workforce 2000* appears to have been the catalyst that precipitated diversity rhetoric . . . by constructing a threat regarding a major change in the demographics of the workforce. The professional management literature picked up that (largely erroneous) threat and made it into a full-blown crisis with rhetoric that justified and necessitated a major change in management style."[6] Powell's 1978 opinion in *Bakke* may have provided the legal underpinnings for affirmative action programs based on the concept of diversity, but it was nearly a decade later that *Workforce 2000* provided the rationale for broadly embracing such programs beyond academia and across the corporate world.

Coming on the heels of the *Workforce 2000* report, the 1989 Supreme Court case of *Richmond v. Croson* provided an additional boost to the diversity movement in management.[7] *Croson* involved a minority set-aside provision requiring general contractors who worked for the city of Richmond, Virginia, to subcontract at least 30 percent of the dollar value of each contract to one or more Minority Business Enterprises. The court struck the program down as a violation of the 14th Amendment's Equal Protection Clause because the asserted state interest of remedying past discrimination was not sufficiently compelling given that the evidence presented showed little previous discrimination in that particular area. This marked the first time that a majority of the Supreme Court clearly held that strict scrutiny should be used in evaluating government-sponsored affirmative action programs.[8] It also marked the demise, for all practical purposes, of using broad remedial rationales to justify affirmative action. It was at this point, notes legal scholar Peter Schuck, that the diversity rationale fully came to the fore as a justification for affirmative policies in both public and private settings.[9] In contrast to the difficulty of showing specific systemic past discrimination in particular settings that required direct remedial race-conscious remedies, diversity presented a forward-looking rationale that did not foreground race nor assign guilt to past actors.[10] In other words, after *Croson*, Powell's diversity rationale took on greater weight for those looking to address issues of racial representation in business and education. This concern took on added urgency in the context of the misreading of the *Workforce 2000* report.

Diversity advocates were well-poised to spring into action. Among the pioneers of diversity management in business was R. Roosevelt Thomas, who had established the American Institute for Managing Diversity at Morehouse College in 1984. Before coming to Morehouse, Thomas had served on the faculty of the Harvard Business School and as dean of the School of Business at Atlanta University. Across the country in California, Price Cobbs, Lennie Copeland, and Lewis Griggs produced and distributed a training film titled *Valuing Diversity* in 1988 that would go on to be widely circulated among human relations managers at many major corporations. That year, Copeland also published a two-part article titled "Valuing Diversity: Making the Most of Cultural Differences at the Workplace," in the journal *Personnel*.[11] The *Valuing Diversity* video came to market just a few months after *Workforce 2000*'s publication and a few months before the Supreme Court's decision in *Richmond v. Croson*.

In 1990, Thomas published an article in the *Harvard Business Review* that in many respects might be considered the *urtext* of modern diversity management studies. Titled, "From Affirmative Action to Affirming Diversity," the article explicitly pivoted from the pre-*Croson* focus on remedial affirmative action programs to the now-pervasive rhetoric of diversity as a future-oriented, efficient, and profitable way to address the challenges of the modern business world. Certainly with *Croson* in mind, Thomas began his article: "Sooner or later, affirmative action will die a natural death." Clearly referencing the *Workforce 2000* report, he went on to frame his article with the observation that "white males will make up only 15% of the increase in the work force over the next ten years. The so-called mainstream is now almost as diverse as the society at large." Thomas concluded by invoking a fundamental business rationale for diversity management: "In a country seeking competitive advantage in a global economy, the goal of managing diversity is to develop our capacity to accept, incorporate, and empower the diverse human talents of the most diverse nation on earth. It's our reality. We need to make it our strength."[12]

THE "NATURALNESS" OF DIVERSITY

Thomas's prescription for business was premised on the idea that things had changed since affirmative action first came to the fore some thirty years earlier. While acknowledging that prejudice still existed, he asserted: "it has suffered some wounds that may eventually prove fatal," and that therefore, "the realities facing us are no longer the realities affirmative action was designed to fix." He applauded past affirmative action efforts but his article was suffused with the idea that affirmative action was somehow "unnatural"—it had "an unnatural focus on one group" and an "artificial nature." In contrast, diversity concentrated on "the ability to manage your company without unnatural advantage" that allowed for a focus on "pure competence and character." Thomas never really explained what made diversity more "natural" than affirmative action, but he implied that the coercive or mandatory character of affirmative action was the problem—it got in the way of natural ability. As he put it, "Affirmative action gets the new fuel into the tank, the new people through the front door. Something else will have to get them into the driver's seat. That something else consists of enabling people, in this case minorities and women, to perform to their potential. This is what we now call managing diversity."[13]

Tropes of naturalness and artificiality play a powerful role in the history of affirmative action. Opponents of affirmative action have, from the start, derided the focus on race as an unjust denigration of individual merit. This concern played a role in Powell's *Bakke* opinion. It was precisely the idea of diversity that allowed him to pivot away from such individualized concerns to the university's broader educational interest in a diverse student body. In this way, race itself took on an attribute of merit, at least insofar as it was meritorious to contribute to such diversity. Race as a basis for remedial action, took on meaning as a marker of historical injustice to a socially marked group. In contrast, race as a basis for diversity took on meaning as an innate attribute of the individual. It became something natural—something, perhaps, that might even be viewed as genetic.

Of course, such was not explicitly the case in Thomas's article. But the focus on naturalness and innate individual ability set a frame for such conceptualization—a frame that was directly taken up that same year by none other than Richard Herrnstein. Building on his earlier work on race and IQ, in 1990 Herrnstein published, "Still an American Dilemma,"[14] an article in which he reviewed and critiqued a 1989 report sponsored by the National Research Council (NRC) titled, "A Common Destiny: Blacks and American Society."[15] Herrnstein framed his analysis by positing two models to understand and address racial inequity. The first, which he characterized as the "discrimination model," argued that "Black status results from American social institutions and the race relations that have developed within that institutional structure." This model, largely adopted in the NRC Report, undergirded the rationale for affirmative action. The second, a "distributional model," looked at racial differences not solely in terms of socially mediated opportunities but was also "willing to consider the possibility that the different outcomes are also the product of differing average endowments of people in the two races." While acknowledging that both models may be "partially correct," (citing Jensen along the way), Herrnstein emphatically argued that the distributional model based on inherent racial differences in intelligence needed to be given greater weight in developing policy. Perhaps what Herrnstein objected to most in the report was that it primarily focused on the impact of discrimination while "obstinately refusing to consider the evidence concerning racial differences at the individual level," i.e., at the level of inherent genetic difference in cognitive capacity.[16]

At first blush, Herrnstein's critique of affirmative action may seem miles away from Thomas's. Thomas's more positive vision of diversity saw it as an opportunity to cultivate each individual's distinctive potential and manage the (inevitable) coming diversification of the workforce. Herrnstein's more constrained vision focused on individual differences and inherent limitations to explain and justify existing social stratification. But both were intent on moving away from the discrimination model with its focus on prejudice and the subordination of groups to one more focused on the distinctive attributes of the individual. Thomas focused on the social aspects of such diversity, Herrnstein on the genetics. But both saw their conception of diversity as more natural and hence more viable or appropriate than a civil rights model that involved the assertion of state power to address structural inequities or remedy past injustice. The frame of diversity thus lent itself to the naturalization and essentialization of racial difference in way that remediation, which simply focused on addressing past wrongs, did not.

THE MODEL OF CULTURAL COMPETENCE

As the diversity management industry was taking off in the early 1990s, similar concepts were taking root in the emerging focus on "cultural competence" in medical education and practice.[17] The model of cultural competence itself has had many iterations over the years. With roots in the cultural and political movements of the 1960s and 1970s, the idea began to gain significant traction after the publication in 1989 of the monograph *Towards a Culturally Competent System of Care*,[18] authored by team of researchers led by Terry Cross, an adjunct professor of social work at Portland State University and an enrolled member of the Seneca Nation.[19] The monograph was framed, like the *Workforce 2000* report, by an emphasis on changing demographics, asserting that "these issues must be addressed, especially when one considers the shift in population predicted by the year 2000. Just as the overall minority population will be increasing disproportionately."[20]

The first attribute of Cross's cultural competence model was to "value diversity." The monograph defined its model broadly "as a set of congruent behaviors, attitudes, and policies that come together in a system, agency, or amongst professionals and enables that system, agency, or those professionals to work effectively in cross-cultural situations."[21] This echoed many of the concerns expressed by advocates of diversity management in business; and indeed, a follow up volume published in 1991 directly referenced such work.[22] Both models characterized diversity primarily in terms of culture and experience, but when diversity moved into the realm of health and medicine it opened the door to more casual conflation of cultural diversity with biological or genetic difference.

Medical practice had long been rife with essentialized conceptualizations of innate biological racial difference.[23] For example, since the 1920s, pulmonary researchers had developed and applied race-based standards for assessing lung function,[24] the notorious Tuskegee Experiments were premised on an assumption that syphilis was somehow a different disease in Black bodies,[25] and Louisiana mandated the segregation of Black from White blood in blood banks until 1972.[26] Highly questionable race-corrected algorithms for an array of medical tests persist to this day.[27] As one later critique of the cultural competence movement put it, "Valuing the importance of diversity without equally valuing the importance of equity and social justice relegates the former to a vacuous proposition."[28] In this regard, cultural competence, like Thomas's call to move beyond affirmative action and Herrnstein's invocation of inherent genetic differences, focused on individual attributes (in this case those related to cultural belief and practices) rather than on structural inequality or the persistence of prejudice as the primary object of concern.

NEW GENOMIC DIVERSITIES

While diversity was coming to the fore as a central concept in law, business, and clinical care it remained very much on the sidelines during the early years of the Human Genome Project. As sociologist Catherine Bliss has observed, "As recently as 1990, genomics was a science disassociated with the problem of race or health disparities."[29] For example, in its March 1990 "Human Genome Program Report," the Department of Energy (DOE) mentioned diversity only in terms of "genetic diversity in capacity for DNA repair in response to ubiquitous DNA-damaging agents."[30] This reflected the DOE's early interest in the HGP because of its implications for understanding the effects of radiation on the human body and had nothing to do with race. The DOE-NIH five-year plan, published in 1990, did not explicitly mention genetic diversity or race at all.[31]

This all changed dramatically in 1991 when a group of leading population geneticists and evolutionary biologists came together in a workshop at Stanford University to discuss and propose a sort of parallel project to the HGP that would sample and archive the world's genetic diversity.[32] The Human Genome Diversity Project (HGDP), as it came to be known, was formally organized at an international meeting in Italy in September 1993.[33] The initial report from the 1991 workshop opened with a statement recalling Lewontin's work on the apportionment of human diversity, noting that "classical premolecular techniques have already proved that a significant component of human genetic variability lies within populations rather than among them." It discussed diversity primarily in terms of "population groups" and did not mention race, although it did refer to ethnic groups.[34]

The HGDP was conceived at a time when the idea of biological diversity was at the forefront of political and social consciousness. In 1992, 168 governments attending the United Nations Conference on Environment and Development (UNCED) would sign the Convention on Biological Diversity reflecting the urgent concern to stem the loss of biodiversity. Building off the ideas of the NRC Forum on Biodiversity and legislation such as the Endangered Species Act, the Convention on Biological Diversity implied that losses in genetic variation due to species extinction could threaten human well-being.[35] That same year, E. O. Wilson published his widely acclaimed book, *The Diversity of Life*, in which he articulated many foundational ideas of biodiversity and reflected on man's destruction of the natural world.[36] The 1992 Democratic Party platform also mentioned biodiversity for the first time—while the Republican Party had actually referenced a commitment to maintain "biological diversity" in its 1988 platform.[37] Yet, as Deborah Litvin has noted, the biodiversity rhetoric of "counting species" tended to "reinforce the essentialism of Linnaean classifications and the tendency to view all members of a class or species as homogenous, eclipsing the role of individual variation in the process of evolution and survival."[38] Similarly, in the field of medicine, sociologist Stephen Epstein has noted that around this same time, in the aftermath of the Heckler Report on racial health disparities, advocates for increasing the diversity of medical research subjects "appeared to be relatively unbothered by the risk . . . that [racial] difference might be conceived of in pejorative terms and used to bolster arguments about social inferiority."[39] But, while medical and scientific professionals might have minimized worries about essentialism, such concerns would come to animate critics of the nascent HGDP.

At the inception of the HGDP, scientists debated several different framings of diversity around which the project might be built. They focused particularly on two quite different sampling models. The first, advocated by biochemist Allan Wilson, conceptualized diversity spatially in terms of a grid that focused on taking samples randomly across large geographic blocks of territory. As bioethicist Hank Greely noted at the time, "Wilson's approach, by selecting individuals instead of groups for study, would have avoided the possibilities that the selection criteria would inadvertently bias the results."[40] Such a grid-based spatial approach also would have lessened the likelihood that identified genetic variation would be essentialized along racial lines. Opposed to this was geneticist Luca Cavalli-Sforza, who wanted an approach that focused on groups, particularly "vanishing indigenous peoples."[41] This latter approach ultimately won out with scientists who organized the project agreeing to focus on "geographically isolated or culturally unique populations, including indigenous peoples"[42] in order to explore questions of relatedness and historical descent.[43] By focusing on groups or populations identified by scientists, this approach was more susceptible to racial essentialization.[44]

Jenny Reardon has convincingly shown how scientists were not merely identifying such groups but, in many cases, actually constructing them through the processes of their scientific investigations.[45] That is, in searching for genetic diversity they constructed social diversity, which, in turn, shaped understandings of genetic diversity. Legal scholar Miranda Oshige McGowan raised a similar point in her critique of the uses of diversity in the context of affirmative action around this same time. In the wake of assaults upon the University of California's affirmative action policies in the mid-1990s, McGowan noted the shortcomings of certain liberal constructions of educational viewpoint diversity that undergirded the rationale for affirmative action:

> By choosing race or ethnicity as the proxy for diverse viewpoints and perspectives, schools may fail to consider the factors that members of these groups see as defining themselves and distinguishing themselves from other groups, which they presumably would bring to bear in campus life. This failure subordinates the individual and group self-conception to the conceptions held by

policy makers and administrators, which are likely to reflect dominant social constructions. Such programs, in other words, may fall victim to the same lack of understanding they are supposed to promote.[46]

Just as the HGDP's focus on discrete groups risked scientists imposing identities on those groups from the outside, so too might well-meaning affirmative action policies reflect dominant social constructions of relevant groups rather than their members' own self-conception. This was, perhaps, most evident in the creation of the census category "Asian" which lumped together a vast array of social, cultural, and geographic variability. Where someone might identify as Japanese American or Indian American, or Thai American, the census, in effect, produced a new panethnic identity of Asian American.[47] In these cases, we see how diversity is not only descriptive, it is productive—creating and imposing particular identities onto purportedly discrete groups.

Critics of the HGDP, such as anthropologist Jonathan Marks and Executive Director of the Indigenous Peoples Council on Biocolonialism Debra Harry, argued that "the genetic study of human diversity is intrinsically interesting and potentially valuable, but that there is no justification for focusing on exotic peoples." Echoing Lewontin, they noted that "simply put, ethnicity does not map well onto human genetic diversity; the vast majority of human genetic variation is known to be within-group variation."[48] Bioethicist Eric Juengst, who also served as an advisor to the HGDP, similarly argued that "we should look for scientific alternatives to the use of identified social groups as templates for population-genomic research." Hearkening back to Livingstone's observation that "there are no races, only clines," Juengst suggested developing approaches akin to those of field biologists that "allow demic boundaries or the representations of population clines to be drawn as the available human genetic data permit." He noted that such a grid-based "random-sampling approach to genetic diversity would bear little resemblance to a map of the world's self-identified autonomous human groups."[49] Human genetic diversity, in other words, was not some natural "thing" out in the world to be discovered. Different sampling strategies and approaches to classification could yield very different constructions of diversity. Even so, leaders of the population-based approach, such as Cavalli-Sforza, were careful to articulate that they were not talking about racial groups—indeed, they very much subscribed to the idea that race was not genetic.[50] Instead, as Greely put it, the HGDP intended "to collect DNA from members of about five hundred of the roughly five to ten thousand human populations in the world."[51]

Among the rationales for diversity was accuracy. Many believed that relying on the HGP alone "would give scientists an incomplete and misleading picture of human genetics if there were no attempt to study human genetic diversity."[52] Greely, who was a member of the North American Regional Committee of the HGDP and the chair of that committee's ethics subcommittee, articulated four additional rationales for diversity: First, fairness, the idea being that "it is fundamentally unfair to the majority of humanity to describe the human genome without including a representative sample of all humans"; second, better understanding of human evolution; third, improving medicine "because it will advance the study of those genetic diseases found largely in non-European populations, and because genetic variation is basic to better understanding a host of diseases found in all peoples"; and fourth, "studying human diversity will help us uncover our shared human history."[53]

Here at the dawn of the modern genomic era, we already have a juxtaposition of political, medical, and scientific rationales for diversity that invites conflation of social and genetic conceptualizations of group difference. The rationale of fairness builds on tropes of representation and equity that suffused the burgeoning diversity management movement. Connecting fairness to increased medical knowledge also recapitulated ideas put forward by R. Roosevelt Thomas and others that diversity would increase productivity. The saving grace, perhaps, was that the leaders of the HGDP explicitly eschewed genetic definitions of race, instead focusing on the idea of "population." Nonetheless, over time the idea of population all too easily blurs back into race, as is seen even in Greely's reference to "non-European populations," which presumes such a thing as a "European population" which readily reverts into the idea of "White."

Moreover, even in terms of these limited rationales, anthropologist Margaret Locke noted that because the project was designed "to collect DNA samples with no information about the local environment, phenotypic data, individual life histories, nutrition, or disease histories to match these tissue samples," it could make no claims to "benefit the health of either individuals or communities."[54] Similarly, biologists Joseph Alper and Jonathan Beckwith questioned the fairness rationale, arguing that if the "major goal were really to improve the health of ethnic minorities and indigenous populations, then a *direct* attempt to define the genetic and, at least as importantly, the environmental factors associated with the particular diseases plaguing these peoples would be required."[55]

The focus on Indigenous populations and the expressed desire to obtain samples before some of them "disappeared" soon gave rise to controversy and contestation.[56] The Rural Advancement Foundation International, a group active in campaigns to prevent the patenting of Indigenous knowledge, alerted the World Council of Indigenous Peoples to be wary of the project,[57] which soon became known as the "vampire project" for its intended practice of swooping in and gathering up blood samples from Indigenous peoples around the globe.[58] As Cavalli-Sforza himself noted years later, "political and ethical difficulties arose in 1994 . . . focused especially on the fear that indigenous people might be exploited by the use of their DNA for commercial purposes ('biopiracy'). . . . From 1994 to 1997, while the NRC committee was organized, met and wrote its report, the HGDP took no major action."[59]

FROM THE HGDP TO *THE BELL CURVE*

As progress on the HGDP began to stall in 1994, discussions of race and genetics in other fora were just beginning to heat up. In September of that year, Canadian psychologist J. Philippe Rushton published *Race, Evolution, and Behavior: A Life History Perspective*, a short treatise offering an evolutionary explanation of a purported racial hierarchy for such characteristics as restraint, nurturing, and intelligence that effectively ranked "Orientals" above "Caucasoids" with "Negroids" consistently at the bottom.[60] Rushton had been developing these ideas during the 1980s, during which time, it was later revealed, he was engaged in an extensive and mutually appreciative correspondence with E. O. Wilson at Harvard.[61] Rushton's work was controversial. It was deemed overtly racist by many and did not gain much purchase in the popular consciousness or even in broader academic circles. The same cannot be said of *The Bell Curve*, written by Richard Herrnstein with Charles Murray and published that same month.[62]

The Bell Curve was about many things and sparked many controversies but at its core it drew on research from psychology and genetics to argue that innate, group-based differences in intelligence, as measured by IQ scores, accounted for much of the inequality in American society. The book drew on earlier publications by Rushton that focused on the sorts of racial differences he discussed at greater length in *Race, Evolution, and Behavior*. As the authors said in an afterword to the 1996 edition of *The Bell Curve*: "Rushton is a serious scholar who has amassed serious data."[63] The book created a firestorm, fueled by the publication of a special issue of the *New Republic* magazine in October 1994, devoted to showcasing its arguments about racial differences in IQ and publishing critical responses.[64] A veritable industry of debate surrounding *The Bell Curve* was spawned over the next few years with volumes of collected documents, essays, and debates being published.[65]

Most of the controversy focused on how the book essentialized racial differences in IQ as innate—i.e., genetic. While some behavioral geneticists actually embraced this argument, many other scientists from the broader genetic community (and many social scientists) aggressively challenged it.[66] Largely in response to Herrnstein's and Murray's dangerous claims about race and genetics, the American Anthropological Association (AAA) issued a *Statement on "Race" and Intelligence,* in December, 1994, declaring that "differentiating species into biologically defined 'races' has proven meaningless and unscientific as a way of explaining variation (whether in intelligence or other traits)" and urging "the academy, our political leaders and our communities to affirm, without distraction by mistaken claims of racially determined intelligence, the common stake in assuring equal opportunity, in respecting diversity and in securing a harmonious quality of life for all people."[67] For the AAA, respect for diversity demanded a rejection of unfounded claims of essential genetic difference among races. In failing to show such respect, *The Bell Curve* threatened to undermine both equality and social harmony.

A central thrust of *The Bell Curve* attacked affirmative action programs. Murray, a political scientist, had long been an opponent of affirmative action. In 1984 he published an article in *The New Republic* titled, "Affirmative Racism: How Preferential Treatment Works Against Blacks," that made no mention of genetics. Genetic evidence did not convince Murray of the futility of affirmative action. Rather, it presented him with an additional, very formidable weapon to add to his already well-established, anti-affirmative action armamentarium. Teaming up with Herrnstein, whose 1990 article "Still an American Dilemma" had raised the issue of innate racial differences in intellectual endowments, gave Murray's old arguments the added force of science—or what critics would call pseudoscience.[68]

Shortly after the book's publication, Michael Lind, then editor of *The National Interest*, observed that "the rehabilitation of would-be scientific race theory on the right—cautious in the work of Murray and Herrnstein, blatant in the writings of Rushton—raises an interesting question: Why now?" He noted that such theories had been around for decades in earlier publications by Jensen, Rushton, and Herrnstein himself. Yet only now had "these theories about race become respectable on the right for the first time since the civil rights revolution." Lind noted the influence of the Pioneer Fund, an organization founded in 1937 with the stated purpose of "race betterment," pouring hundreds of thousands of dollars into supporting neo-hereditarian and eugenic research. Lind ultimately attributed the timing to a shift in American conservatism toward "a new 'culture war' conservatism, obsessed with immigration, race and sex." Along the way, Lind also mentioned E. O. Wilson's foray into sociobiology in the 1970s, noting that "the crude kind of sociobiology that tried to directly connect specific behavioral traits with genes appear to be giving way in the scholarly community to a nuanced consensus view that human potential is flexible but constrained at the margins by heredity. There has even been interesting research done into the connection between language families and genetic groups, by Luigi Cavalli-Sforza and other scholars."[69]

The reference to Cavalli-Sforza is particularly telling because he was, of course, one of the guiding forces behind the creation of the Human Genome Diversity Project. Although aware of such work, perhaps Lind failed to appreciate how, as the HGP and HGDP were taking on greater significance in the popular imagination, DNA itself was becoming a cultural icon, as Dorothy Nelkin and Susan Lindee would argue in their 1995 book, *The DNA Mystique*.[70] Herrnstein's and Murray's arguments may not have been new, but they were gaining traction perhaps because of broader receptivity to genetically inflected arguments in a world increasingly suffused with grand promises of impending genetic breakthroughs. Sensitive precisely to such dangers, the NIH-DOE Joint Working Group on the Ethical, Legal, and Social Implications of Human Genome Research also issued a statement critiquing *The Bell Curve*, expressing grave concern "about the impact of The Bell Curve, and books developing similar themes, because we believe that the legitimate successes of the Human Genome Project in identifying genes associated with human diseases should not be used to foster an environment in which mistaken claims for genetic determination of other human traits gain undeserved credibility."[71]

While discussing racial difference and genetics extensively, *The Bell Curve*'s use of the term diversity was reserved almost exclusively to the social context of affirmative action policy; that is, it kept the concept of diversity distinct from that of genetic "difference." In this regard, as critics noted, it smacked much more of good old-fashioned scientific racism.[72] This perhaps explains how it avoided blurring concepts of social and genetic diversity. Rhetorically it engaged with Powell's articulation of diversity as a social rationale for affirmative action much more than with Lewontin's discussion of human diversity in genetic terms. Thus, for example, Herrnstein and Murray noted that "A student from Montana can add diversity to a college in Connecticut; a good football team can strengthen a college's sense of community and perhaps encourage alumni generosity. Black and Latino students admitted under affirmative action can enrich a campus by adding to its diversity." This was very much in keeping with Powell's conceptualization of diversity as merely a range of valued social attributes in college admissions that might include geography on the same level as race. In the book's conclusion, they declared that "We are enthusiastic about diversity—the rich, unending diversity that free human beings generated as a matter of course, not the imposed diversity of group quotas."[73]

Here we have, perhaps, a nod to the idea that diversity in innate genetic talents generated socially perceived differences; but the focus was still on social, not genetic, diversity. They contrasted implicitly "natural" diversity of individuals with the forced or artificial diversity of "group quotas." In this, the conclusion again strangely echoed R. Roosevelt Thomas's characterization of affirmative action as unnatural just four years earlier.[74] Both seemed to be embracing the value of socially manifested diversity; but where Thomas saw diversity in terms of variety that was valuable in itself, Herrnstein and Murray cast diversity as both an explanation for and justification of social stratification. As they summed up their argument in *The New Republic*, "We argue that the best and indeed only answer to the problem of group differences is an energetic and uncompromising recommitment to individualism. To judge someone except on his or her own merits was historically thought to be un-American, and we urge that it become so again."[75]

This apparent celebration of individualism, however, was disingenuous. The thrust of *The Bell Curve* was about group-based differences—particularly racial differences—in cognitive capacity that might be attributable to genetic variation. As such, it reflected and extended the essentializing tendencies implicit in the HGDP's sampling scheme. The founders of the HGDP, like Thomas (a founder of modern diversity management), thought they were valuing diversity for its own sake as productive both of knowledge and other forms of scientific advancement; but the critics of the HGDP understood that in practice the HGDP would produce ideas of diversity more in line with reinforcing Herrnstein's and Murray's essentializing rationalizations of racial hierarchy.

The year after *The Bell Curve* came out, anthropologist Jon Marks published *Human Biodiversity: Genes, Race, and History*. The first to directly connect the term "biodiversity" with the human species, Marks actually structured this masterful, historically informed account of the changing understanding of relations between race and biology so as to demolish the idea that there was a legitimate or coherent scientific basis for characterizing human races as discrete biological entities. Marks acknowledged the reality of human biological diversity but argued that such diversity had nothing to do with race. He did not directly address *The Bell Curve*, but he did effectively eviscerate Rushton's claims of different brain sizes across races with attendant effects on IQ. In this work, Marks (like Lewontin before him) was endeavoring to decouple diversity from race. "Human biological variation," he argued, "is gradual and continuous. . . . Races are not objective or biological categories. Populations are different from one another, but races are supposed to be large chunks of humanity, and apparently our species doesn't come packaged that way, despite the fact that generations of Euro-Americans have assumed so."[76]

Ironically, soon after the publication of the book, conservative journalist Steve Sailer, a regular contributor to White Nationalist websites such as VDARE and later the darling of the alt-right, appropriated the term "human biodiversity" for purposes directly contrary to those intended by Marks.[77] In the coming years it would be taken up by White supremacists as a more palatable, scientifically inflected way to reentangle race and diversity to serve their racist ends.[78]

INSTITUTIONALIZING BIOMEDICAL DIVERSITY

While controversy was swirling around *The Bell Curve*, President Bill Clinton's administration was busy implementing the NIH Revitalization Act,[79] which had been passed by Congress the previous year to address concerns about representation in federally sponsored medical research. Among other things, the Revitalization Act created the Office of Research on Minority Health within the NIH and required that women and minorities be included as research subjects in NIH-funded research beginning in 1995. Under the dictates of the new law, federally funded researchers would, in effect, have to certify that they had enrolled adequately diverse populations, had made adequate efforts to enroll diverse populations, or could provide a biomedical justification for not enrolling diverse populations.[80] Henceforth, the Revitalization Act's diversity mandate would have the practical effect not only of changing clinical trial enrollment practices but of redirecting researchers' attention toward racial and sex differences while producing prodigious amounts of biomedical data categorized by race and gender over time. In the future, researchers who might not have been interested in such differences, nonetheless would be structurally incentivized to conceptualize biomedical data in terms of race and sex.

As subsequently implemented by the NIH, "minority" came to be characterized in terms of the basic census categories promulgated by the Office of Management and Budget (OMB) in its statistical Directive 15. It specified four categories for race and two for ethnicity: Race: American Indian or Alaskan Native; Asian or Pacific Islander; Black;

White. Ethnicity: Hispanic origin; Not of Hispanic origin.[81] The purpose of promulgating these categories was to provide "a common language to promote uniformity and comparability for data on race and ethnicity for the population groups specified in the Directive."[82] These categories would come to structure and inform federally funded biomedical research (and related drug development) for decades to come.[83] Notably, Directive 15 itself, which was originally promulgated in 1977, opened with the caveat, "these classifications should not be interpreted as being scientific or anthropological in nature."[84] Clearly, there was a sensitivity here to the pitfalls of characterizing race as biological, much in line with Lewontin's work on the apportionment of human diversity published just five years earlier.

Accordingly, the NIH Revitalization Act itself made only passing reference to genetics and contained no discussion at all of the idea of genetic diversity. This is perhaps not surprising given that the Human Genome Project itself was still in its infancy and actual knowledge of medically actionable genetic variations was limited, let alone knowledge of possible correlations between such variations and racially characterized population groups. Nonetheless, some critics such as Director of the National Cancer Institute's Office of Special Populations Research Otis Brawley, expressed concerns that the Revitalization Act could reinforce beliefs that "there are significant biological differences among the races."[85] Sociologist Steven Epstein has noted, "it is striking is that proponents of the biological-differences frame appeared to be relatively unbothered by the risk" identified by the likes of Brawley, that "difference might be conceived of in pejorative terms and used to bolster arguments about social inferiority."[86] But this is precisely what happened when *The Bell Curve* was published the same year the NIH Revitalization Act went into effect. Of course, there was no direct connection between the NIH Revitalization Act and *The Bell Curve*. Nonetheless, they are both of a piece in the historical moment of increasing focus on ideas of difference and representation in biomedicine and population genetics. Such a focus, in itself, was not necessarily either progressive or reactionary. The NIH deployed the idea of difference to rationalize increasing minority representation in clinical trials in order to ameliorate inequity. Herrnstein and Murray deployed the idea of difference in order to rationalize inequity and argue that increasing minority representation via affirmative action was doomed to failure.

The Revitalization Act itself did not use the words "diverse" or "diversity," and used the words "ethnicity" and "race" only once each. The rhetorical focus instead was simply on the inclusion of "women and minorities" in medical research. In this regard, it bears considering how the idea of "inclusion" is not necessarily coextensive or synonymous with diversity. Stephen Epstein has noted in relation to the Revitalization Act that those seeking greater inclusion in biomedical studies effectively constructed analogies between their cause and affirmative action by emphasizing frames of underrepresentation as a key issue for equity in important social spheres. He similarly noted that opponents of the Revitalization Act, such as Republican representatives Newt Gingrich and William Dannemeyer, invoked the language of affirmative action, decrying what they characterized as mandated "quotas" for women and minorities in clinical trials.[87] The Revitalization Act's remedial affirmative action frame was thus distinct from the frame of diversity deployed by the likes of R. Roosevelt Thomas at this time. Moreover, the Revitalization Act itself was never subjected to the same types of legal challenges as affirmative action programs in education, employment, or contracting, which had just been dealt a severe blow in the *Croson* case a few years prior. While the NIH Revitalization Act was working its way through Congress, the affirmative action case of *Adarand v. Pena* was working its way up to the Supreme Court, which in 1995 would issue an opinion effectively extending the strict scrutiny standard to *Croson* to apply to federal as well as state programs.[88] Yet this holding was never used to challenge the Act.

LAW, POLITICS, AND DIVERSITY IN THE 1990s

Adarand involved a challenge to the constitutionality of federal programs designed to provide highway contracts to disadvantaged business enterprises.[89] In the course of reaching its holding, the Supreme Court expressly overruled the 1990 case of *Metro Broadcasting, Inc. v. FCC*, which had held that so-called "benign" uses of racial categories should be subject only to the lower "intermediate scrutiny" standard of review.

Metro Broadcasting was notable for its use of the concept of "broadcast diversity" as an important governmental objective sufficient to justify taking race into consideration in granting broadcast licenses. In coming to this conclusion, the Supreme Court deferred to the Federal Communication Commission's determination that "there is an empirical nexus between minority ownership and broadcasting diversity." In this context, the Supreme Court found the racial diversity of minority owned broadcasting enterprises to be a legitimate proxy for diversity of broadcasting content, and embraced the FCC's 1978 finding that lack of such diversity was "detrimental not only to the minority audience but to all of the viewing and listening public." In its valorization of racially inflected viewpoint diversity, *Metro Broadcasting* was very much in line with Powell's opinion in *Bakke*, which it cited with approval.[90] The Supreme Court here also brought to the fore the idea that viewpoint diversity benefited not only excluded minorities but society as a whole. In this, the Supreme Court in *Metro Broadcasting* echoed the central message of the emerging diversity management industry as exemplified in R. Roosevelt Thomas's assertion that increased diversity would redound to the benefit of all.[91]

Between the Supreme Court's 1990 decision in *Metro Broadcasting* and *Adarand* in 1995, conservative Clarence Thomas had replaced liberal icon Justice Thurgood Marshall, and a new majority of the Supreme Court was having none of the idea that benign ameliorative uses of racial categories merited a more accommodating level of judicial review than malign discriminatory uses. In his dissent, however, Justice Stevens noted that the goal of promoting diversity was not directly at issue in *Adarand*, and he argued that "the proposition that fostering diversity may provide a sufficient interest to justify such a program is *not* inconsistent with the Court's holding today."[92] And so, it seemed diversity had lived to fight another day.

Certainly, the diversity management business kept chugging along. By 1995, a survey of top Fortune 50 corporations found that 70 percent had formal diversity management programs in place, while only 12 percent had no diversity programs at all.[93] Yet, even as the diversity management business continued to gain steam throughout the 1990s, there were also signs of pushback. One critique from the Left argued that increasingly throughout the 1990s, "despite all the positive intentions of sensitivity seminars, diversity audits and training videos, many diversity interventions were shown to have backfired... [leading] to outbursts of antagonism and resentment from those who had been subjected to the scrutiny of difference."[94]

On the Right, California's Proposition 209 presented the most formidable backlash against affirmative action in the political realm. Proposition 209 was a ballot proposition that California voters approved in 1996. It added Section 31 to the California Constitution's Declaration of Rights, which said that the state could not discriminate against or grant preferential treatment on the basis of race, sex, color, ethnicity, or national origin in the operation of public employment, public education, and public contracting. It therefore effectively banned the use of affirmative action involving race-based or sex-based preferences in California. Perhaps reflecting the increasing decoupling of the concept of diversity from affirmative action evident in the earlier work of R. Roosevelt Thomas, none of the formal arguments for or against Proposition 209 that appeared with the ballot used the word diversity. Arguments against Proposition 209 instead focused on specific programs that its adoption would likely terminate, such as outreach and recruitment programs to encourage women and minority applicants for government jobs and contracts; while proponents of Proposition 209 spoke primarily in terms of ending all forms of discrimination, making no distinction between benign or harmful forms—much like the Supreme Court in *Adarand*.[95]

Proposition 209 followed hard on the decision by the regents of the University of California, in July 1995, to end the use of "race, religion, sex, color, ethnicity, or national origin as criteria for admission to the University or to any program of study."[96] The regents clearly rejected what Robert Post has characterized as Justice Powell's "structural and atemporal" idea of diversity (including racial diversity) as central to constructing a vibrant educational experience.[97] Ironically, the regents' resolution nonetheless concluded by declaring, "California's diversity to be an asset" but went on to characterize such diversity in highly decontextualized and individualistic terms, asserting that "because individual members of all of California's diverse races have the intelligence and capacity to succeed at the University of California, this policy will achieve a UC population that reflects this state's diversity through the preparation and empowerment of all students in this state to succeed rather than through a system of artificial preferences."[98] Here, again, we see articulated the idea that affirmative action was somehow artificial, while implying that differences in representation resulting from the absence of affirmative action were more natural or due to innate factors.

Where the regents of the University of California still professed an allegiance to a broad (and perhaps vacuous) concept of diversity, the U.S. Court of Appeals for the Fifth Circuit took a slightly different tack to dismantle the legal status of the concept itself in the 1996 case of *Hopwood v. Texas*. Emphasizing the fact that Powell's swing opinion in *Bakke* had never been formally adopted as binding precedent by a majority of the Supreme Court, the Fifth Circuit struck down the University of Texas's affirmative action program, declaring that "there has been no indication from the Supreme Court, other than Justice Powell's lonely opinion in *Bakke*, that the state's interest in diversity constitutes a compelling justification for governmental race-based discrimination." In the course of its opinion, the court reduced race to phenotype and biology, arguing that that "the use of race, in and of itself, to choose students simply achieves a student body that looks different. Such a criterion is no more rational on its own terms than would be choices based upon the physical size or blood type of applicants." The court went on to note, however, that "while the use of race per se is proscribed, state-supported schools may reasonably consider a host of factors—some of which may have some correlation with race—in making admissions decisions.... A university may properly favor one applicant over another because of his ability to play the cello, make a downfield tackle, or understand chaos theory."[99]

The court's reference to correlation further indicates its essentialized construction of race. Such social attributes as cello playing constituted a legitimate type of diversity. Race was legitimate only to the extent it correlated with such attributes. But, for the court, race itself was a biological attribute, not social. Basing admissions on biological diversity (such as race, or blood type) was not only unconstitutional, it was irrational. The court thus provided a reductive and essentialized version of race, which it then excised from what the state might legally consider as part of a legitimate conceptualization of diversity. Or rather, it constructed the state's interest in a biologized racial diversity as insufficiently compelling to overcome the demands of strict scrutiny review mandated by *Croson* and *Adarand*.

Here we see an indirect realization of Khiara Bridge's concerns that the Supreme Court in *Al-Khazraji* implied that "it is genetics that makes ethnic 'subgroupings,' which are nevertheless cognizable as races within law."[100] If race could be reduced to biology or genetics in the same way as blood type, then the Fifth Circuit would perhaps have had a point. The *Hopwood* opinion nonetheless used implicit characterizations of race as biological to distinguish between rational and legally cognizable forms of diversity (cello or football playing) from irrational and legally suspect forms of diversity (race or blood type). The linking of race to blood type is particularly striking here considering the historical salience of the "one drop" rule that was used to make legal distinctions throughout the Jim Crow South;[101] not to mention ironic, given that Lewontin's 1972 article on "The Apportionment of Human Diversity" analyzed blood groups to show race was not genetic.[102] The *Hopwood* court did not dispense with diversity altogether but rather reconfigured it so as to legally decouple race from the increasing power of this now broadly accepted concept. For the *Hopwood* court, diversity was fine, so long as it did not include race. Where *The Bell Curve* posited connections between race and biology in order to assert the futility of affirmative action, the court in *Hopwood* did so to undermine the legal legitimacy of racial diversity as a state interest.

The survival and persistence of the trope of diversity, even in the face of right-wing backlash, was also evident in the fact that the same year that Proposition 209 passed in California, the Republican Party was embracing the concept of diversity in its party platform and Republican Speaker of the House Newt Gingrich was declaring on national television that "Diversity is our strength!"[103] The Democrats similarly valorized diversity in their party platform, declaring that, America's diversity "enriched" us and would make us "stronger than ever."[104] Diversity was becoming as American as apple pie, perhaps because of its apparent ability to mean all things to all people. But the regents' policy, Proposition 209, and the *Hopwood* decision, coming in the wake of the firestorm of controversy about race, biology, and intelligence provoked by *The Bell Curve*, marked a shift toward conservatives attempting to differentiate between legitimate and illegitimate deployments of diversity—and, in *Hopwood*, using biology to help make its case.

GENETIC DIVERSITY BEYOND THE HGDP

While California voters were adopting Proposition 209 and the Fifth Circuit Court of Appeals was handing down its decision on *Hopwood*, a joint NIH-DOE working group on the Ethical, Legal, and Social Implications (ELSI) of the Human Genome Project issued its first report. The report itself did not discuss diversity or race per se but focused instead on recommendations for implementing "an appropriate mechanism . . . to ensure that a rigorous, focused research program exists to build a foundation of knowledge about ELSI issues and to provide an open, independent forum for a full discussion of the issues."[105] The following year, the NIH would establish the National Human Genome Research Institute (NHGRI) and the Committee on Human Genome Diversity of the National Research Council would issue a report titled, "Evaluating Human Genetic Diversity." The committee had been charged to evaluate the proposal to formally establish the Human Genome Diversity Project. The report concluded that "a global assessment of the extent of human genetic variability has substantial scientific merit and warrants support, largely because of the insight that the data collected could provide into the origin and evolution of the human species."[106]

While acknowledging the potential health benefits accruing from the related Human Genome Project, the report cautioned that it was more realistic for the HGDP, with its focus on global genomic variation, to view biomedical investigation as "secondary targets" because the study would likely have "no immediate health benefits for potential subjects." The report reviewed the various sampling strategies that had earlier been debated, giving particular attention to the "grid" approach advocated by Allan Wilson and Luca Cavalli-Sforza's population-based model that focused on groups, opting in the end to recommend something far closer to the latter approach even though it presented a greater risk of reifying racial or ethnic groups as genetic.[107] The following year, however, the NRC recommended against funding the HGDP, citing outstanding and unaddressed social and bioethical issues.[108]

Around this same time, new developments in microarray technology spurred geneticists to focus more on single nucleotide polymorphisms as a means to characterize genetic differences among individuals.[109] Sociologists Rajagopalan and Fujimura have argued that "rather than the 'turn to difference' emerging as a post-Human Genome Project (HGP) phenomenon, interest in individual and group differences was a central, motivating concept in human genetics throughout the twentieth century." They observed that "in 1996, statistical geneticist and a leader within the HGP Eric Lander and statistical geneticist Neil Risch and epidemiologist Kathleen Merikangas formalized a genetic strategy for the study of complex diseases. They argued that the future of genetics lay in 'genetic association studies.'"[110] Here were the roots of what became known as Genome Wide Association Studies (GWAS) that would come to dominate genetic research in the 2000s. The GWAS approach moved the search for genes that cause disease away from family-linkage studies (which, for example, had provided the bases for discovering the gene related to Huntington's disease in 1993)[111] toward population-wide studies with a focus on groups—which could and did come to include racial and ethnic groups.[112]

Difference, of course, is not necessarily the same thing as diversity. At this point, most talk of diversity in genetics was understandably concentrated around the Human Genome *Diversity* Project. As sociologist Catherine Lee has noted, while medical experts, policy makers, and activists grappled with issues of racial inclusion in biomedical research and practice throughout the 1980s and 1990s, most geneticists, including the public health subfield known as genetic epidemiology, "remained absent from debates over racial inclusion in their efforts to understand what later would be seen as racially stratified diseases like cystic fibrosis and sickle cell anemia."[113] Only beginning around 1997 did the concerns for genetic diversity and inclusion move from the population-based mapping concerns of the HGDP to more general applications of genetics biomedicine. This was very much in line with broader drives for the inclusion of women and minorities throughout biomedical research and practice that were gaining significant traction during this period.[114] The HGP began to concentrate more on the idea of difference as a means to uncover genetic variation of potential biomedical significance. The turn toward race-based inclusion was shifting framings of difference from individuals to racial groupings. But as the group-based approach of association studies came increasingly to the fore in the coming decades (and as the HGDP faded from view in the face of political controversy), diversity would reenter biomedically focused genetic research as a critical means to realize the "promise" of genomics.

One central component of the promise was the idea of using genetics to develop targeted drugs tailored to the specific disease profile and genetic makeup of individuals, known early on as "pharmacogenetics" and now generally characterized as "pharmacogenomics"; a shift reflecting the move to research based on searching large databases for variants of interest. While pharmacogenetics in some form had been around since the 1930s,[115] Adam Hedgecoe, in his study of the politics of personalized medicine, notes that "industry's intense interest in pharmacogenetics and genomic technologies as a whole ... dat[es] from around 1997."[116] This coincided with the NRC report on *Evaluating Human Genetic Diversity*, but also was occurring at a time when biomedical researchers and the industry more generally were beginning to respond to the mandates for racial and ethnic inclusion of the NIH Revitalization Act.[117]

Additionally in 1997, Congress passed the Food and Drug Administration Modernization Act (FDAMA), which directed that "the Secretary [of Health and Human Services] shall, in consultation with the Director of the National Institutes of Health and with representatives of the drug manufacturing industry, review and develop guidance, as appropriate, on the inclusion of women and minorities in clinical trials."[118] The FDA itself followed up on this in 1998 by issuing the Demographic Rule, which required submissions for drug approval to "tabulate" participants in clinical trials by age group, gender, and race.[119] The following year the FDA issued a guidance entitled "Population Pharmacokinetics," making recommendations on the use of population pharmacokinetics in the drug-development process to help identify differences in drug safety and efficacy among population subgroups, including race and ethnicity.[120]

Reflecting the new regulatory mandates, a review of articles indexed at the federally managed PubMed database of biomedical journals, using combinations of the terms "race," "ethnicity," "pharmacogenetics," and "pharmacogenomics" (coined in 1998) indicates a steady rise in attention paid to racial categories in biomedical research during this time. Compared to the 1980s, the 1990s saw a nine-fold increase in journal articles using these terms, and again in the 2000s there was a five-fold increase over the 1990s. This all provides a critical backdrop to what would emerge in the coming decades as a commercial and professional interest in extracting profits and prestige from correlations among identified social categories of race and presumed underlying genetic variations of medical significance.[121]

While various biomedically oriented clinical and research institutions were coming to grips with these federal mandates, the Office of Management and Budget had been revisiting the very terms of racial and ethnic classifications that structured these calls for diversification and inclusion. As Melissa Nobles has noted, between 1993 and 1997, the "OMB actively sought public comment through congressional subcommittee hearings ... and by notices posted in the Federal Register. At OMB's request, in 1994, the National Research Council's committee on national statistics conducted a workshop that included federal officials, academics, public policy analysts, corporate representatives, and secondary school educators." Of central concern were demands by a wide range of civil rights groups and activist organizations to add a "multiracial" category to the census. Organizations representing multiracial and multiethnic Americans, Arab Americans, Irish Americans, and German Americans lobbied for the addition of new categories, such as "middle-eastern," or for the disaggregation of the "white" category itself to reflect what they understood to be the range of ethnic diversity within that monolithic classification.[122] As Michael Omi, a participant the NRC workshop, noted of the debate, "race and ethnicity will continue to defy our best efforts to establish coherent definitions over time. The real world is messy with no clear answers. Nothing demonstrates this convolution better than the social construction of racial and ethnic categories."[123] Ultimately, the OMB rejected the demands but did, for the first time, allow people to check more than one category on the census. Additionally, the "ethnic" category of "Hispanic" or "Non-Hispanic" was broken out into a separate section to be answered as separate from racial classifications.[124]

As in the original caveat to the 1977 Directive 15, most references to racial classifications in the discussion of the 1997 revisions were careful to emphasize that they were not talking about biological or genetic categories. The report from the NRC workshop even cited the work of anthropologist Jonathan Marks when noting that "Biological scientists also find no evidence for fixed races as they are popularly perceived. Current biological studies point to a wide variety of genetic characteristics that are shared across supposedly different races as well as very different characteristics that are not shared by groups that are supposedly of the same race."[125]

RACE SHAPING GENETICS, GENETICS SHAPING RACE

In contrast to the group-based focus of the 1990s regulatory mandates, Francis Collins, then-director of the newly created NHGRI (and who would later become director of the entire National Institutes of Health from 2009 to 2021), was writing about how the new knowledge being generated by the HGP would revolutionize medical care of the individual patient. In his 1999 article, "The Medical and Social Consequences of the Human Genome Project," all his talk was of "individualized medicine."[126] This was the promise of pharmacogenomics—getting the right drug, in the right dose, to the right patient.[127] For Collins, "this vision of genetically based, individualized preventive medicine is exciting, and it could make a profound contribution to human health." Collins also acknowledged that the current five-year plan for the HGP (from 1998 to 2003) included, "new areas of research in the ELSI [Ethical, Social, and Legal Implications] program ... such as identifying and addressing issues that link genetics to personal identity and racial or ethnic background and examining the implications of these links for philosophical and religious traditions."[128] Critically, at this early stage in the modern genomic era, Collins was not conceiving of race as any sort of needed proxy or aid to finding genetic variants of medical interest. Race was playing virtually no role in his framing of how new genomic technologies would further medical advancements. Hence there was no discussion of racial diversity as a component of genetic research at this time. The focus on race came through the ELSI program that was exploring the possible implications of genetic breakthroughs for *social* understandings of racial or ethnic identity. In other words, at this point, the focus of NHGRI was not on a concern for how understandings of racial diversity might affect genomic research but rather how genomic research might affect understandings of racial diversity.

Collins's framing stood in stark contrast to what was happening at the same time with clinical studies and drug development. The latter, often driven by commercial considerations of gaining access to developing new products and markets, often implicitly characterized race as an intrinsic genetic characteristic with significant implications for understanding biological mechanisms of drug safety and efficacy. This approach directly entangled social categories of race with biological or genetic framings. Collins, however, was focusing his genetic lens on the individual, while remaining open to something more akin to a "biocultural" conceptualization of race; that is, a social category that shapes, and is in turn shaped by, biological processes and environmental contexts.[129]

This tension would come to inform discussion of diversity, race, and genetics for the next twenty years. A critical dynamic to identify and trace is when the frame shifts from a focus on race's implications for genetics to a concern for genetics' implications for race. The former frame comes to be animated by tropes of "promise": the idea that realizing the promise of genomics depended upon increasing diversity throughout the biomedical enterprise. In biomedical realms it often focused on existing, observed racial disparities in health or drug efficacy and posited some sort of race-correlated genetic basis for the difference. It would often manifest in calls to recruit more diverse (primarily racially diverse) research subjects and thereby tended to reinforce essentialized notions of race as genetic. In the political and legal realms, this meant recruiting a more diverse (again, primarily racially diverse) biomedical workforce. As such, the frame was deeply informed by broader discussions of affirmative action and the burgeoning literature and industry of diversity management. While clearly dealing with racial categories as social, in practice, the realm of political concerns becomes readily entangled with serving the interests of biomedical realms: socially diverse researchers are needed to recruit purported genetically diverse research subjects—and the categories of diversity in both cases were characterized by the same racial terminology.

The latter frame tended to be more complex and contested. On the one hand, going back at least to Lewontin's early work, genetics could be cast as showing the underlying unity (i.e., *lack of diversity*) among human races. In this frame, racial diversity became clearly situated as a fundamentally sociocultural phenomenon. That is, here genetics' implications for race were to undergird positions that race could not be understood as a coherent genetic construct. On the other hand, after the completion of the HGP (and to a certain extent following upon the work of the HGDP) some researchers would come to pay increasing attention to what genetics could tell us about the deep history of human ancestry, often casting their findings in terms of racial groupings. In some ways recapitulating the early HGDP debates over sampling strategy, researchers in this area of population genetics would oscillate between more fine-grained, localized characterizations of genetic populations and characterizations that either directly employed or could be readily collapsed back into traditional racial groupings. In this arena the implications of genetics for race could be in direct tension with the legacy of Lewontin, tending to rebuild a new edifice of geneticized understanding of race.

CHAPTER THREE

Forensic Diversity

Difference in the form of genetic diversity among racial groups was also very much at the forefront of another area of intersection between law and science around this time: the forensic use of DNA in criminal justice settings.[1] There are two major steps in using DNA for purposes of forensic identification. First, a sample left at the crime scene by the perpetrator is compared to a sample from a suspect. Second, if there is a "match" then statistics must be used to calculate the frequency of that DNA "profile" in an appropriate reference population.[2] This latter step is required because, although every person's DNA is unique, it is impractical to compare the full, three billion nucleotide base pairs between two samples for forensic purposes.[3] Therefore, two samples will be compared only at a limited set (historically between four and thirteen) of "loci," or specific parts of the genome. For this practice to be effective, it is necessary to find loci that are highly variable between individuals and test only for them.[4] Humans, however, are essentially identical in about 99.5 percent of their DNA.[5] Finding the specific points of variation among individuals, therefore, can be difficult.[6] Addressing this challenge is where racial categories and notions of diversity came to play roles as the technology developed.

In 1985, English geneticist Alec Jeffreys first described a method for developing a DNA profile of a person in a manner that might be used for purposes of forensic identification.[7] Jeffreys's innovation consisted in observing that, in particular regions of the human genome, short segments of DNA are repeated between twenty and one hundred times.[8] These repeat regions became known as "variable number of tandem repeats" or VNTRs. Different VNTR "alleles"—or variations—are composed of different numbers of repeats. In order to examine and visualize the VNTRs, Jeffreys employed a technique known as restriction fragment length polymorphism (RFLP) which used a restriction enzyme to cut the regions of DNA surrounding the VNTR. By looking at VNTRs from several distinct loci on the genome, it was possible to calculate the probability that a particular genetic profile comprised of distinct sets of VNTRs would appear in one or more individuals in a particular population.[9] A standard way to estimate the frequency of a particular profile was to count occurrences in a random sample of an appropriate reference population and then use classical statistical formulas to place upper and lower confidence limits on the evidence.[10] The resulting conclusion of identity or nonidentity between two samples was therefore necessarily probabilistic.[11] In conducting the comparison, investigators came to adopt the "product rule"[12] for determining the "random match probability" (RMP)—the probability of finding the same DNA profile identified in the crime scene sample in a randomly selected, unrelated individual.[13] Any given VNTR may be calculated to occur at a certain frequency in a random population. By the early 1990s, the standard was to test for VNTRs at four independent loci on the genome. The product rule allowed for multiplying each independent genotype frequency together to produce an overall probability of a match at all four loci.[14]

Jeffreys's innovation was first used in a forensic setting in England in 1986.[15] Forensic DNA testing was first used in the United States in 1987.[16] Soon thereafter some commercial laboratories made use of this "fingerprinting" procedure and in 1988 the U.S. Federal Bureau of Investigation implemented forensic DNA techniques.[17] Critical to the acceptance of forensic DNA in courts was the development of standards of technical proficiency and accuracy in generating RMPs.[18] The product rule was one such standard, requiring that each chosen loci be understood as being inherited independently of the others.[19] Also important were basic crime scene management techniques for the identification and handling of DNA samples.[20]

Questions about the reliability of DNA evidence surfaced as early as 1989 in cases such a *People v. Castro* in New York and the Minnesota case of *Schwartz v. State*.[21] Partially in response to these cases, several federal agencies called upon the National Research Council (NRC), an arm of the National Academies of Science (NAS), to study and recommend guidelines for the production and use of DNA evidence.[22] The NRC created a Committee on DNA Technology in Forensic Science, which issued a report in 1992.[23] It is in the context of the production of this report that diversity—in the form of racially characterized variations in allele frequencies—first entered the story of forensic DNA analysis.

The committee covered an array of issues relating to the forensic use of DNA technologies. Among its most controversial findings were those relating to reference populations and the appropriate methodology for calculating RMPs.[24] In order to calculate the odds of any particular VNTR allele appearing at a given locus on the genome, one must have an appropriate reference population.[25] The product rule depends on the assumption of statistical independence of the alleles tested—that is, they do not tend to occur in groups.[26]

Generally speaking, the more "related" a person is to a particular population group the higher the odds are of finding shared alleles—or alternatively stated, the less independence there is among alleles.[27] Siblings would likely share more DNA than cousins; cousins more than others in the same isolated village; members of the same isolated village more than others in the same region; and so forth. Higher odds favor a suspect or defendant because they indicate a greater likelihood that some other person may have left the DNA sample found at a particular crime scene.[28] The choice of reference population, therefore, can play a critical role in shaping the weight and authority of DNA evidence.[29] The choice, however, is not always straightforward. Some of the earliest and most contentious controversies involving the use of DNA technology in forensic science involved choosing the appropriate population against which a suspect's DNA should be compared and defining just how the suspect may be related to this population.[30] Concepts of race played a central role in these debates and continue to frame the way forensic scientists, law enforcement, and the bar produce and interpret DNA evidence to this day.[31] In this respect, debates among forensic scientists echoed those occurring around this same time among geneticists who were debating sampling strategies for the HGDP.

The basic issue was whether, or to what extent, racial or ethnic categories should be used to characterize reference populations against which particular DNA samples could be compared to generate RMPs. The use of such categories could be particularly problematic in the arena of forensic DNA analysis because racial groups, especially those delineated in the U.S. census, are, as the caveat to Directive 15 made clear, fundamentally *social* not *biological* categories.[32] Nonetheless, to the extent that certain population geneticists understood particular racial groups as sharing a common genetic ancestry—usually by using race as a crude surrogate for geographic or continental ancestry—members of those groups could be viewed as more related to each other (like an extended family) than to individuals from other groups.[33] This problematic understanding of relatedness could then affect the calculation of RMPs. Generally speaking, the more fine-grained the characterization of a particular reference population the higher the odds of a random match—again, higher odds favoring the suspect/defendant.[34] In the early years of forensic DNA analysis (when typically, only four VNTR loci were tested), there were concerns that using a general, undifferentiated population database would produce inappropriately low RMPs.[35] The decision to use race in constructing and categorizing reference populations was introduced into forensic DNA analysis in the belief that it would improve the precision of the calculations that generate RMPs.[36]

In early 1991, two pairs of eminent geneticists squared off against each other in the pages of *Science*, a highly influential scientific journal, to debate the problem of using racial categories in forensic DNA analysis. On one side we see, once again, Richard Lewontin together with Daniel Hartl of the University of Washington. On the other side were Ranajit Chakraborty of the University of Texas and Kenneth Kidd of Yale University. Their dispute did not revolve around the question of *whether* to use race but rather *how much* race to use in constructing reference population databases from which to calculate match probabilities.[37]

Lewontin and Hartl questioned the then current practice of calculating allele frequencies in the racial categories used in the census such as "Caucasian," "Black," and "Hispanic," to provide the basis for calculating RMPs. They argued that such groupings were too broad and that substantial "genetic substructuring" occurred *within* the broad racial groupings that should be taken into account in calculating match probabilities. Using the broad racial groupings could produce RMPs with substantially lower odds than those that might be produced using more fine-grained, ethnically identified subpopulations.[38] These concerns grew logically out of Lewontin's earlier article, "The Apportionment of Human Diversity," which showed how genetic variation *within* socially identified racial groupings was actually greater than variation observed *between* such groups.[39] Thus, Lewontin and Hartl observed:

> Among genes that are polymorphic in European national or ethnic groups, the magnitude of the differences in allele frequency among subpopulations differs from one gene to the next.... For example, there are striking geographical clines of allele frequency across Europe for the ABO blood groups: the frequency of the B allele is 5 to 10% in Britain and Ireland, increases across Eastern Europe, and reaches 25 to 30% in the Soviet Union; the frequency of the O allele is 70 to 80% in Sardinians, Irish, and Scottish populations but lower in Eastern European populations. These clines reflect the migrations and political history of Europe over the last few thousand years.[40]

Problems were even greater for the "heterogeneous assemblage" known as "Hispanic," which was perhaps "the worst case for calculating reliable probabilities." Consequently, they concluded that using reference databases organized by the broad racial groupings "Caucasian," "Black," and "Hispanic" was "unjustified."[41] In a sense, Lewontin and Hartl were arguing that humans were both not diverse enough and yet too diverse to use such racial groupings. That is, precisely because genetic variation was primarily local and clinal, any meaningful construction of reference population databases would need to be

Forensic Diversity

similarly fine-grained. Broad, racially characterized reference populations would obscure the true nature of the apportionment of human genetic diversity.

Chakraborty and Kidd argued that Lewontin and Hartl exaggerated both the extent of ethnic substructuring in America and its significance for calculating match probabilities. While conceding that some substructuring existed, they argued that its effects upon frequency estimates generated by using the broader racial databases was "trivial." Chakraborty and Kidd did not deny that using finer-grained ethnic reference populations might produce more precise allele frequency estimates. Rather, their point was that such an approach was unnecessary—and unnecessarily burdensome. Current technology and understandings of population genetics, they asserted, justified the use of broad racial and ethnic categories, which were, additionally, far more practical and currently available.[42] Race was at the center of this early debate. But, again, for these eminent scientists, it was not a question of *whether* to use race but *how*, or more specifically *how much* (i.e., how fine-grained) race to use.

In some respects, this debate echoed the one going on almost concurrently among organizers of the Human Genome Diversity Project. In both cases, geneticists were arguing about the most appropriate or productive means to organize and classify population-based genetic data. In the case of the HGDP the debate centered on contrasts between a grid approach to sampling versus one based on "isolated," largely Indigenous populations. As Chakraborty and Kidd argued that Lewontin and Hartl's approach was unnecessarily burdensome, the HGDP similarly would come to reject the grid approach to global genetic sampling as logistically impractical. The language of the HGDP researchers very explicitly and centrally invoked the idea of diversity. The goal was to provide insight into the range of genetic variation around the globe and better understand its implications for population history and possible medical applications down the road. In the case of forensic DNA, the focus was more on establishing a basis for genetic "identity" (or sameness) rather than diversity, and the goal was to provide a reliable and just foundation for using DNA as evidence in criminal justice settings. Racially characterized population groups became relevant not as a marker of diversity but as a means to provide greater accuracy as to the identity between a DNA sample found at a crime scene and the DNA of a suspect.

This is not to say that these approaches affirmatively lacked scientific merit, merely that it was not relative merit that ultimately drove their adoption. Lewontin and Hartl's focus on getting more finegrained characterization of population groups was perhaps more in accord with the HGDP conceptualization of population diversity. The HGDP's founders, after all, explicitly eschewed the idea of race as a genetic category and chose to focus more on smaller "ethnic" groups—numbering in the hundreds. Chakraborty and Kidd, in contrast, by arguing for the adequacy and utility of racially categorized reference population DNA databases reinforced the frames being presented by the likes of Herrnstein, Murray, and Jensen in their work on race and IQ. In both cases, the approach that ultimately won out did so largely based more on considerations of cost and convenience than on scientific merit.

These debates took place while the NRC committee was conducting its study of DNA technology in forensic science.[43] Its report, issued in 1992, discussed both sides of the issue without specifically taking sides. It did, however, choose "to assume for the sake of discussion that population substructure may exist and to provide a method for estimating population frequencies in a manner that may account for it." The report explicitly mentioned diversity only once, citing Lewontin's earlier work to support its contention that "contrary to common belief based on difference in skin color and hair form, studies have shown that the genetic diversity between subgroups within races is greater than the genetic variation between races." The report further noted that "North American Caucasians, blacks, Hispanics, Asians, and Native Americans are not homogeneous groups."[44] In effect, this approach reflected the concerns expressed by Lewontin and Hartl, recognizing that social categories of race did not map neatly onto discrete genetically definable population groups.

The NRC's 1992 report created problems for prosecutors. By taking cognizance of the difference of scientific opinion regarding the appropriate calculation of allele frequencies and RMPs, it seemed to assert that forensic DNA technologies lacked the sort of scientific consensus needed to support the introduction of such expert evidence.[45] Thus, for example, in the 1992 case of *People v. Barney* the California Court of Appeals cited the NRC report in concluding that disagreement and uncertainty in the scientific community regarding the selection of appropriate reference populations precluded the admission of DNA evidence based on the product rule.[46]

By April 1993, the director of the FBI asked the NAS to conduct a rapid follow-up study to resolve these uncertainties.[47] The NRC then appointed a second committee (NRC II) late in 1994 with a specific mandate to update and clarify discussions of population genetics and statistics as they applied to DNA evidence.[48] Meanwhile, the debate continued in the scientific community. The position advocated by Chakraborty and Kidd received a major boost in 1994 when Eric Lander of the Massachusetts Institute of Technology, previously a vigorous critic of the lack of adequate standards in DNA typing, paired with Bruce Budowle, one of the principal architects of the FBI's DNA-typing program, to write an article in the journal *Nature*, declaring "DNA fingerprinting dispute laid to rest."[49] The article argued that applying the product rule to the frequency estimates for four independent VNTRs generated odds of such magnitude that any technical statistical differences observed between the use of the broad racial databases (as advocated by Chakraborty and Kidd) versus more fine-grained, ethnic subgroup databases (as advocated by Lewontin and Hartl) were "of no practical consequence to the courts."[50]

As Lander and Budowle observed:

> In the vast majority of cases, a jury needs to know only that a particular DNA pattern is very rare to weigh it in the context of a case: the distinction between frequencies of 10^{-4}, 10^{-6} and 10^{-8} is irrelevant in the case of suspects identified by other means.... The most extreme positions range over a mere two orders of magnitude: whether the population frequency of a typical four-locus genotype should be stated, for example, at 10^{-5} or 10^{-7}. The distinction is irrelevant for courtroom use.[51]

Lander and Budowle were not arguing that racial subgroups themselves were not needed or desirable in calculating RMPs. The "distinction" they saw as "irrelevant" was the one between ethnic subgroups, such as "Irish," and larger racial groups, such as "Caucasian." Thus, they were legitimating the then current standard FBI practice of using broad racial groups, such as "Black" and "Caucasian," as reference databases for generating allele frequencies for calculating RMPs. Significantly, Lander and Budowle did not argue for doing away with racial databases altogether in favor of using an undifferentiated general population database. Given the then-current state of forensic technology which generated RMPs from examining VNTRs at only four loci, they deemed racial diversity relevant. They simply did not want too much of it—that is, they did not want law enforcement forced to undertake the burdensome task of developing more elaborate databases that reflected the wide array of genetic population substructuring that actually occurs across the globe. Given the odds generated by testing at four VNTR loci, they deemed the broad racial categories of the census more than adequate for forensic purposes.[52]

Lander and Budowle made a critical distinction between statistical and legal relevance. Although hardly the first to do so,[53] the distinction allowed Lander and Budowle to quiet both the scientific debates and the legal uncertainties swirling around this new and powerful forensic technology. Another critic of the NRC's 1992 report, David Kaye, of the Arizona State University School of Law, made a similar distinction in a 1993 article in the *Harvard Journal of Law & Technology*. Kaye, who would sit on a second NRC committee, wrote that in calculating RMPs, "the real issue ... is not 'statistical significance' but rather practical or substantive significance."[54] The difference was critical for Kaye and others because it provided the basis for validating the then-current law enforcement practices of using broad racial reference population databases. By distinguishing between statistical vs. logical or practical significance, Kaye and others, such as Lander and Budowle,[55] did not refute Lewontin and Hartl so much as bracket off their concerns as irrelevant to the legal applications of forensic DNA technology in courts. Of most immediate significance in terms of the unfolding story of the use of race in forensic DNA technology, was the fact that this distinction played a central role in the NRC's second report, *The Evaluation of Forensic DNA Evidence* (NRCII), issued in 1996.[56]

NRC II: Questions of Race Laid to Rest?

The NRC II report focused primarily on updating and clarifying issues related to population genetics and statistics as they applied to DNA evidence. It argued directly for "using separate databases for different racial groups" even while it acknowledged Lewontin's underlying argument that "the variability among individuals within a population is greater than that between populations." Recognizing the uncertainties inherent in calculating RMPs, the report noted that "the accuracy of the estimate will depend on the genetic model, the actual allele frequencies, and the size of the database." It was confident, however, that "when several loci are used, the probability of a coincidental match is very small." Nonetheless, the report recommended incorporating a ten-fold margin of error in RMP calculation, stating: "if the calculated probability of a random match between the suspect and evidence DNA is 1/(100 million), we can say with confidence that the correct value is very likely between 1/(10 million) and 1/(billion)."[57]

At first glance, such a range may seem rather large, but the report legitimizes it by returning to the distinction between statistical and legal relevance. "The proper concern," it asserted, "is not whether the probability is large or small, but how accurate it is. Probabilities are not untrustworthy simply because they are small. In most cases, given comparable non-DNA evidence, a judge or jury would probably reach the same conclusion if the probability of a random match were one in 100,000 or one in 100 million."[58] In other words, the large range presented earlier in the report was of little practical or legal significance so long as it was *good enough* to guide a judge or jury in their deliberations. It was good enough for two reasons: first, because it was *accurate*—accuracy here was crucially distinguished from *precision* which the large range of probabilities certainly lacked; second, because the lower end of the range still presented odds so vanishingly small as to render it indistinguishable from the upper end of the range *as a practical matter*—that is, the difference was deemed to be insufficient to have any practical effect on the conclusion a judge or jury would reach in using the evidence.

And yet, even accepting this huge range of variance, the report persisted in using race to organize categories in calculating RMPs. Thus, even while acknowledging that "some assert the word race is meaningless" in a genetic context, the report adopted the categories "white (Caucasian), black (African American), Hispanic, east Asian (Oriental), and American Indian (Native American)" as designated "racial groups" as a matter of "convenience, uniformity, and clarity." Once again, experts here justified the entanglement of social concepts of race with genetic concepts of allele frequency primarily as a matter of practical convenience not scientific rigor. It justified this choice by asserting that "there are reproducible differences among the races in the frequencies of DNA profiles used in forensic settings, and these must be taken into account if errors are to be minimized."[59]

It is instructive to note here just where it was that "difference" made a difference in the calculation of RMPs. Difference was deemed insignificant when it manifested as a thousand-fold range for an "accurate" calculation using the product rule to compare a single sample against a single reference population database—that is, the difference between one in 100,000 and one in 100 million made no practical difference for use of the data in a court of law. To be fair, as noted above, the NRC II report recommended calculating RMPs with a margin of error limited to ten-fold in either direction—but this still translated into a variation of one hundred-fold between the lowest and highest estimate. But when race was at issue in the NRC II report, the difference of frequencies among racial reference populations became critical and had to be "taken into account if errors are to be minimized."[60]

The NRC II report employed the concept of diversity in relation to genetics significantly more than the first report. Perhaps reflecting new developments in the emerging HGP and HGDP, it talked of diversity in very particularized terms relating to the specific frequencies of certain alleles independent of racial context.[61] But it also used the term in its critique of how the first NRC report proposed to sample groups to determine gene frequencies for reference populations. The first NRC report proposed the following strategy:

> On the one hand, it is not enough to sample broad populations defined as "races" in the U.S. census (e.g., Hispanics), because of the possibility of substructure. On the other hand, it is not feasible or reasonable to sample every conceivable subpopulation in the

world to obtain a guaranteed upper bound. The committee strongly recommends the following approach: Random samples of 100 persons should be drawn from each of 15–20 populations, each representing a group relatively homogeneous genetically.[62]

This approach appeared to address the sorts of concerns raised by Lewontin and Hartl but clearly placed a greater practical burden on those creating and managing the databases. While the idea of defining fifteen to twenty population groups in the United States that were "relatively homogeneous genetically" might be problematic, it certainly undercut the idea that the racial groups of the census categories could be meaningfully defined in genetic terms.

The first report here did not use the term diversity, but in critiquing it, the second report did. First, it called the sampling "controversial" and then characterized it as "sampling 100 persons from each of 15–20 genetically homogeneous populations spanning the racial and ethnic diversity of groups represented in the United States." It went on to note that this approach "has the advantage that in any particular case it gives the same answer irrespective of the racial group. That is also a disadvantage, for it does not permit the use of well-established differences in frequencies among different races; the method is inflexible and cannot be adjusted to the circumstances of a particular case."[63]

The second report then moved on to discuss racial diversity in a manner that clearly opened the door to reifying racial categories as genetic:

> There is no generally agreed-on vocabulary for treating human diversity.... For convenience, uniformity, and clarity, in this report we designate the major groups in the United States—white (Caucasian), black (African American), Hispanic, east Asian (Oriental), and American Indian (Native American)—as races or racial groups. We recognize that most populations are mixed, that the definitions are to some extent arbitrary, and that they are sometimes more linguistic (e.g., Hispanic) than biological. In fact, people often select their own classification. Nevertheless, there are reproducible differences among the races in the frequencies of

DNA profiles used in forensic settings, and these must be taken into account if errors are to be minimized.[64]

The NRC here used the idea of diversity to reimpose racial categories upon forensic DNA practices largely because it was more practical and convenient for the purposes of law enforcement, which already used racial categories in its day-to-day operations. Precisely because humans were so diverse genetically, the NRC implied, users of forensic DNA were justified in imposing the order of racial organization to make the range of genetic variation manageable. The idea of "reproducible differences" did a lot of work here in justifying the use of racial categories to structure forensic DNA reference databases.

Race enters into people's consciousness in complex and often unanticipated ways. The NRC II report clearly focused on issues of race in response to the questions raised by the debate between Lewontin/Hartl and Chakraborty/Kidd. That debate involved the relation between social groups of race and genetic variation. Both sides recognized that racial categories were crude surrogates for capturing genetic variation and diversity across groups, but Chakraborty and Kidd were, in effect, arguing that race was nonetheless not "too crude"—that is, it was good enough for practical use in law enforcement because of the ability to generate astronomically low RMPs, even allowing for a substantial range of variation.[65]

As a practical matter, the debate cast into doubt the admissibility of DNA forensic evidence in courts; hence the FBI's urging that the issue be revisited by a second NRC committee.[66] The NRC II report, therefore, aimed to quiet the dispute, rendering it irrelevant to the practical application of forensic DNA technologies in law enforcement. Yet, it is unclear why the NRC II report characterized differences between racial reference populations as meaningful "error" while it deemed the hundred- (or even thousand-) fold range of variance within a single reference population to be of no practical significance. This seems largely to be an artifact of the report's focus on addressing the issues raised by Lewontin and Hartl in a manner that would allow forensic DNA testing to proceed unimpeded by concerns of the accuracy of using racial reference populations to calculate RMPs. The report needed to show that RMPs generated by using racial categories were good enough for

Forensic Diversity 79

practical use in courts of law. The utility and/or validity of using a general population database without reference to either race or ethnic subgroups was never really at issue.

In the end, the report issued the following formal recommendation for estimating RMPs: "In general, the calculation of a profile frequency should be made with the product rule. If the race of the person who left the evidence-sample DNA is known, the database for the person's race should be used; if the race is not known, calculations for all the racial groups to which possible suspects belong should be made."[67] The NRC II report thus legitimized the then-standard practice of using race to generate RMPs. In rejecting Lewontin and Hartl's concerns about broad racial databases, it seems also implicitly to have rejected—or at least failed fully to appreciate—Lewontin's cognate concerns about the incoherence of race as a genetic category and the dangers of reifying race as genetic.

The first NRC report came out in 1992, merely three years after *Richmond v. Croson* mandated that strict scrutiny be applied to the use of racial categories in state sponsored actions. The second report came out in 1996, one year after *Adarand v. Pena* extended this requirement to federal programs. Yet at no point in either report was the constitutionality of using racial categories to structure forensic DNA databases considered. There is no clear explanation for this. But perhaps it might be understood in light of Khiara Bridge's critique of the 1987 case of *Saint Francis College v. Al-Khazraji*, where she noted that despite Justice White's thoughtful footnote on the socially constructed nature of race, his "language suggests that, within its ontology of race and ethnicity, it is genetics that makes ethnic 'subgroupings,' which are nevertheless cognizable as races within law."[68] There was still plenty of room in legal discourse for conceiving of race as genetic. As something constructed as biologically "real," race came to be accepted as a basis for organizing biological information (i.e., allele frequencies) in a matter that was implicitly understood not to implicate the sorts of equal protection concerns raised by affirmative action in explicitly social settings. That is, racial diversity as a function of genetics was constructed as natural and hence a justifiable basis for state action; whereas racial diversity as social was constructed as artificial and to be subjected to strict scrutiny.

CODIS AND THE MOVE FROM VNTRS TO STRS

The NRC II report itself was based largely on an assessment of the then-current practice of testing samples at four VNTR loci. Ironically, by 1997, barely a year after the report had been issued, a new technology had emerged to replace four loci VNTR analysis using restriction fragment polymorphism (RFLP) methods of analysis. In 1985, Kary Mullis and members of the Human Genetics group at the Cetus Corporation discovered a technique known as Polymerase Chain Reaction (PCR) which enabled scientists to make millions of copies of a specific sequence of DNA in a matter of hours. The ability to amplify segments of DNA was critical to forensic analysis. PCR is sensitive, rapid, and not limited by the quantity of DNA as are RFLP methods.[69] PCR enabled a shift in focus from VNTRs to sections of DNA known as "Short Tandem Repeats" (STRs). VNTRs are typically 10–100 bases in length. STRs (also known as microsatellites) are regions of DNA only 2–6 base pairs in length. STRs are highly variable across individuals and are easily amplified by PCR, thus making them very effective for purposes of human identification.[70]

Beginning in 1996, the FBI commenced an effort to develop a set of core STR loci to be used as standard referents for the calculations of RMPs in forensic DNA analysis. In November 1997, the FBI settled on thirteen core STR loci which were chosen to be the basis of the CODIS (Combined DNA Index System) national DNA database which was launched in 1998.[71] New technologies allowing for "multiplex" testing of multiple loci at once were soon capable of regularly generating RMPs rarer than one in a trillion.[72] By 2000, the FBI laboratory and many others stopped using RFLP analysis altogether in favor of PCR analysis of the thirteen CODIS STRs. Because of their use in the FBI database, the thirteen CODIS STRs have become a national (indeed international) standard and have come to "dominate the genetic information that has been collected to date on human beings."[73]

CODIS was initially authorized by the DNA Identification Act of 1994 and became operational in 1998.[74] As of August 2023, there were 16,532,335 offender profiles, 5,190,279 arrestee profiles, and 1,282,418 forensic profiles in CODIS.[75] The profiles themselves are not classified by race. Rather they are primarily used, much like a database of fingerprints, to aid in the investigation of crimes by providing matches or "hits" to DNA evidence left at crime scenes.[76] In the context of establishing an initial match using the CODIS database, race is therefore irrelevant.

Nonetheless, race came to pervade the characterization of forensic DNA data generated using the standard thirteen CODIS loci. This is because establishing a match is only the first step in applying forensic DNA technology. Once a match is found, whether using the CODIS database, law enforcement must still take the further step of calculating an RMP for any given DNA profile. It is at this stage that race enters CODIS—and in a more powerful way than ever before. In addition to the basic CODIS database, in the early 2000s the FBI generated a population file to estimate allele frequencies according to specifically identified racial or ethnic groups.[77] This population file was based on a 2001 study led by Bruce Budowle which typed allele frequencies for the thirteen CODIS loci from forty-one population datasets. Budowle classified the results in terms of five "major population groups": "African American, U.S. Caucasian, Hispanics, Far East Asians, and Native Americans."[78] These allele frequencies have since become the standard reference database for calculating racially identified RMPs.[79] For example, in *People v. Wilson*, a California case from 2006, criminologist Nicola Shea referenced the Budowle study when noting that the California Department of Justice "used databases that the Federal Bureau of Investigation published in the Journal of Forensic Sciences reflecting profile frequencies in the Caucasian, Hispanic and African-American populations."[80]

RACE, TECHNOLOGY, AND "CARE OF THE DATA"

Race v. Technology

The care (or lack thereof) taken in presenting and interpreting racial data in professional discussions of forensic DNA stands in marked contrast to the meticulous care taken concerning the more technical aspects of DNA extraction, amplification, and analysis. The discussions of each in a 2005 article by Peter Vallone, Amy Decker, and John Butler, of the National Institute of Standards and Technology's (NIST) Human Identity Project team, are fairly typical.[81] This particular article involved the characterization of allelic frequencies for seventy SNPs (single nucleotide polymorphisms) in DNA samples taken from three racially marked groups: U.S. Caucasian, African American, and Hispanic. The article presented its techniques for racially identifying the DNA samples as follows: "Anonymous liquid blood samples with self-identified ethnicities were purchased from Interstate Blood Bank, Inc. (Memphis, TN) and Millennium Biotech, Inc. (Ft. Lauderdale, FL)."[82] In short, "self-identification" provided the sum total of all care and technique devoted by Vallone et al. to characterizing genetic samples by race. In contrast, their discussion of the more apparently technical aspects of how they manipulated the samples once in the lab ran to nearly 700 words and over two pages of text. The point here is not to assess (or even understand) the intricacies of the technical analysis performed by Vallone et al. on their DNA samples. Rather it is to contrast the extreme care and detail devoted to elaborating the techniques performed in the lab with the casual and perfunctory discussion of how the samples came to be racially marked in the first place.

As scientists, Vallone et al. understandably went into greatest detail with respect to techniques and practices in which they were professionally trained and proficient. This reflects their reasonable understanding that the extraction, amplification, and analysis of DNA takes great care and expertise. The contrasting lack of care taken in characterizing the racial identity of the genetic samples indicates an implicit assumption that such characterizations were obvious, uncomplicated, and took no special expertise. This contrast may be understood more broadly as reflecting a conceptual separation of the world of the "social" from that of the "natural," where the former is understood to contain transparent categories accessible to all while the latter required specialized knowledge and expertise for proper analysis and interpretation. In other words, race was seen as easy and obvious; DNA was difficult and complex.[83]

Social v. Genetic Race

Ironically, this separation of the social from the natural was enabled by the work of geneticists such as Lewontin who, together with a wide array of social scientists, had worked diligently since World War II to reconfigure race from a biological into a social construct.[84] It was precisely because race was widely understood as a social phenomenon that forensic scientists were able to effectively marginalize it from their analysis of the biological construct of DNA. As a result, their "care of the data" extended only to the analysis of DNA samples while wholly overlooking the complexities of using racial categories in relation to genetics.[85]

In effect, forensic scientists simply adopted the broad categories of race and ethnicity used in the U.S. census to organize their genetic data. Given the social and political uses which such standards were designed to serve, it should come as no surprise that Directive 15 explicitly acknowledged that the categories it provided were social in character, not biological or genetic.[86] Using these same categories in the context of genetic research, however, presented issues of a different order. As Lee et al. note, "research utilizing race serves to 'naturalize' the boundaries dividing human populations, making it appear that the differences found reflect laws of nature. In fact, the use of race and ethnicity in biomedical research is problematic because it is caught in a tautology, both informed by, and reproducing, 'racialized truths.'"[87] This dynamic reinforces what sociologist Michael Omi has characterized as an "interesting dilemma" facing scientists in the United States: "One the one hand," Omi asserts, "scientists routinely use racial categories in their research. . . . On the other hand, many scientists feel that racial classifications are meaningless and unscientific."[88]

The Obvious Solution: A Nonracial, General Population Database

Race was originally introduced into the calculation of RMPs in the early years of forensic DNA analysis in hopes of providing more refined statistical calculations. The rationale was grounded in the reasonable observation that there is a modicum of genetic variation across certain human populations. Capturing this variation might provide more accurate RMPs. Greater accuracy was important in the early years of forensic DNA analysis when generating RMPs using only four VNTR loci. With such limited data, the variation of RMPs generated using different reference populations could be of forensic significance.

With the advent of multiplex assays testing for the thirteen standard CODIS loci, forensic scientists were able to routinely generate RMPs with denominators far in excess of the entire world's population.[89] Under such circumstances the concerns originally expressed by Lewontin and Hartl that using broader racial categories would not produce accurate enough RMPs fade into irrelevance. When dealing with odds in the hundreds of billions or trillions, the more fine-grained characterizations of genetic variation among ethnic subgroups called for in their original 1991 article in *Science* simply were not necessary as a practical matter.[90]

The issue then shifted from *how much* race to use, to *whether* to use race at all. As made evident by the range of odds generated in cases such as *People v. Wilson* (one of 96 billion Caucasians, one of 180 billion Hispanics, and one of 340 billion African Americans),[91] the use of a nonracially marked general reference population would still generate RMPs whose reciprocals would exceed the world population many fold, rendering any differences between RMPs generated by using race-specific reference populations vs. a general population devoid of forensic significance. Thus, it was no longer necessary even to use the broad racial reference populations advocated by Kidd and Chakraborty back in 1991.[92]

The possibility of abandoning racial reference populations in favor of a general population database was broached in a 2000 report by the National Institute of Justice's National Commission on the Future of DNA Evidence, which noted that "With enough loci it may be possible to have a single database for all the major groups in the United States."[93] In other words, given the ability to generate RMPs in the trillions, it was no longer necessary to have diverse reference databases organized by race. The commission, in effect, called for a unified database—a database without diversity—echoing President Clinton's declaration that same year upon the completion of the first draft of the Human Genome Project that "in genetic terms all human beings, regardless of race, are more than 99.9% the same."[94]

The question remains, why did legal authorities continue to demand diverse reference populations and employ racial categories to generate RMPs?

The Inertial Power of Race

There is no easy answer to this question. I would like to suggest that there is an inertial power to race in American society that propels its continued use long after any original rationale for its introduction may have faded. In particular, I would like to consider three possible dynamics contributing to the persistent use of race in the presentation of forensic DNA evidence even after current technology has obviated the need for race-specific databases: 1) the persistent conceptualization of race as genetic; 2) the confusion of statistical with forensic significance; and 3) the deep-seated American identification of violent crime and race.

First, with respect to genetics: in spite of decades of efforts on the part of social and natural scientists to sever the ties between race and biology, large segments of American society continue to conceptualize race primarily in genetic terms.[95] As anthropologist Sandra Lee noted in 2006:

> The current trajectory of genomic research is increasingly focused on the 0.01 percent genetic difference that is believed to separate one individual from another. The search for functional genetic variability is increasingly taken up in populations that are identified by conventional notions of "race." This trajectory is the result of a confluence of factors, including a growing infrastructure of research materials that are racially categorized through the creation of biobanks. Such sorting practices reflect the ongoing conflict over the meaning of "race" in science and medicine. In the emerging era of the new genetics, in which super-computer technology has given way to an explosion of human genetic data, biobanks that utilize taxonomies of race in the classification, storage and distribution of DNA samples become "racializing technologies" that promote notions of racial biology in research protocols designed to discover group difference.[96]

Sociologist Troy Duster further argued that "new claims that DNA analysis of crime scene data will assist criminal investigations" have led to a "molecular reinscription of race in the biological sciences."[97] The same technology underlying the creation of racialized forensic DNA databases was also being used for drug development and to market new genetic ancestry tracing services.[98] There thus emerged both structural and commercial incentives to continue to use race in relation to genetics. This dynamic undergirded the inertial power of race in forensic DNA analysis by providing a broader context in which race was understood somehow to be naturally or logically connected to genetics. It was further reinforced by the tendency of forensic DNA experts to take race as an obvious, unproblematic category that did not require the same care and analysis as genetic data.

Second, the technical ability to generate statistically significant variation in RMPs across racial databases led to the unquestioned assumption that such variation was also legally significant. Using the thirteen CODIS loci, forensic experts around the world characterized allele frequencies for numerous ethnically and racially marked populations.[99] Modest frequency variation at each individual locus, when multiplied across loci by the product rule, could lead to apparently significant variations in RMPs across races. Thus, in cases such as *People v. Wilson*, the variation in RMPs across race-specific databases might appear, at first blush, to be important. In that case, RMPs varied from one in 96 billion Caucasians, to one in 180 billion Hispanics, and one in 340 billion African Americans.[100] According to the databases, Wilson's genetic profile was more the three times as likely to occur in a Caucasian as an African-American—an apparently significant difference. But in the forensic context, this statistically significant difference has no real practical importance. When the world's population was under seven billion, the difference between an RMP of one in 96 billion and an RMP of one in 340 billion provided no meaningful distinction for a finder of fact. Both were astronomically low probabilities. Nonetheless, the experts' ability to generate statistically significant differences across races propelled the continued use of racial databases; even when these differences were of no practical legal significance.

Race persisted largely because it had become normative; an unquestioned, standardized practice that persisted long after the rationale for it had faded. In *Wilson*, the State of California argued for the legitimacy of using race-specific RMPs primarily on the grounds that such was the "standard practice" and the "generally accepted method for generating match probability statistics," and that "typically" the state and federal labs used "three major U.S. population databases: African-American, Caucasian, and Hispanic."[101]

Third, there is the unfortunate but well-documented tendency in the United States to identify race and violent crime. In their book, *Whitewashing Race*, Michael Brown et al. discuss a cultural shift that began in the 1960s when the image of "the brave little girl walking up to the schoolhouse door in the face of jeering white crowds was replaced by fearsome young black men coming down the street ready to take your wallet or your life."[102] In the context of the rising racialization of crime on the United States, Rothenberg and Wang observe that "from 1990 to 2004, blacks were five times more likely than whites to be incarcerated, and in 2000, blacks and Latinos comprised 63% of incarcerated adults, even though together they represented only 25% of the total population."[103] Similarly, while examining the impact of DNA technology on the criminal justice system, Simon Cole concluded that "At the endpoint of this system is a carceral system that embodies gross race and class disparities, even if differential rates of offending are taken into account: two thirds of people in prison are racial and ethnic minorities, one in eight black males in their twenties are in prison or jail, three-quarters of persons in prison for drugs are people of color."[104] Considering the dynamics that have produced such inequalities, Brown et al. review an array of historical, legal, and sociological data on race and crime in the United States. Discussing a classic observational study of police responses to juveniles in a Midwestern city in the 1960s, they note that police justified their different treatment of black youths on "epidemiological lines," concentrating on "those youths whom they believed were more likely to commit delinquent acts." They argue, however, that "the results of this 'actuarial' reasoning . . . is to exacerbate the very differences that are invoked to justify the racially targeted practices in the first place. This in turn helps to cement the public's image, and the police's image, of the gun-toting gangster or drug dealer as black or Latino. And this confirms the validity of the police focus on youth in the same kind of vicious circle ... described a generation ago."[105] The association of crime and race produced more racialized crime. As Dorothy Roberts has noted, the resulting mass incarceration is "iatrogenic"—by damaging social networks, distorting social norms, and destroying social citizenship, the disproportionate incarceration of minorities has produced a vicious cycle of crime and repression that further reinforces the identification of race and crime in the public mind.[106]

Debates over whether or how to use reference databases categorized by racial descriptors have been present almost since the inception of forensic DNA identification techniques. In a way, they echo some of the early debates at the same time concerning how best to sample global genetic variation for the Human Genome Diversity Project. As discussed earlier, these debates centered on using a geographical grid approach versus identifying distinct, purportedly "isolated" local populations that were subsequently ethnically marked. In both cases, racially and ethnically marked databases were chosen in large part because of the practical utility and relative ease of characterizing diversity in terms of race instead of geography. Yet even as the practical need for racially marked forensic databases faded away, race did not. Early on, forensic (and other) scientists repeatedly made the point that once a particular odds threshold was passed, any difference among profile frequencies was of little or no practical significance. As Yale geneticist Kenneth Kidd noted in the 1999 case, *People v. Soto*, "any difference in estimates over one in a million was pragmatically meaningless."[107] And then, as recently as 2020, we have an article in the journal *Legal Medicine* arguing that "insufficient genetic differentiation observed among the US racial populations as well as inconsequential differences between race-specific and race-neutral RMPs undermine the value of using race in the context of forensic DNA analysis and support the argument that forensic databases should be race-neutral."[108]

Yet the use of race in forensic DNA contexts has continued. Taken together, the persistent conceptualization of race as genetic, the confusion of statistical with forensic significance, and the deep-seated American identification of violent crime and race may be understood to frame and facilitate the inertial power of race to perpetuate itself as a salient category of forensic DNA analysis, long after its practical legal utility had passed.

CHAPTER FOUR

Diversity Affirmed in the New Millenium

Six months after President Clinton's triumphal press conference declaring that the completion of the first draft of the human genome showed that "in genetic terms, all human beings, regardless of race, are more than 99.9 percent the same,"[1] a federal district court in Michigan would hear arguments in a case brought by Barbara Grutter, a forty-three-year-old White woman, against the University of Michigan Law School alleging that its affirmative action policies unconstitutionally discriminated against her, resulting in the denial of her admission.[2] In 2001, the district court had found in Grutter's favor, squarely rejecting Justice Powell's approach to affirmative action in *Bakke*. Like the Fifth Circuit Court in *Hopwood*, it found that "Justice Powell's discussion of the diversity rationale is not among the governing standards to be gleaned from Bakke," because no other member of the Supreme Court formally joined in "his view that the attainment of a diverse student class is a compelling state interest which can justify the consideration of race in university admissions." In reaching its conclusion the district court distinguished between "viewpoint diversity," which it presented as having perhaps some important educational benefits, and "racial diversity," seeing the connection between them as "tenuous as best."[3]

Over the next two years, the case worked its way up to the Supreme Court. Along the way the Court of Appeals for the Sixth Circuit overturned the district court, finding the diversity rationale sufficiently compelling to justify the law school's admissions program.[4] Then, almost two years to the day after the White House ceremony announcing the completion of the first draft of the Human Genome Project report, Justice Sandra Day O'Connor delivered an opinion affirming the Sixth Circuit Court's decision and formally adopting Justice Powell's diversity rationale as a precedential holding of the Supreme Court.[5] O'Connor clearly declared in her conclusion, "the Equal Protection Clause does not prohibit the Law School's narrowly tailored use of race in admissions decisions to further a compelling interest in obtaining the educational benefits that flow from a diverse student body."[6]

As the pronouncements at the White House ceremony were widely understood to resolve any outstanding doubts about the relation of race to genetics, so too, it was widely assumed that the definitive statement of the majority opinion in *Grutter* would lay to rest ongoing debates about the place of diversity in affirmative action programs. Nothing could have been further from how events subsequently played out.

In genetics, the drive to map human genetic diversity on a global scale, as envisioned in the HGDP, continued unabated but took new forms. Ironically, it increasingly became informed with the rhetoric and categories of social diversity drawn from the domain of affirmative action. As the Supreme Court affirmed the legitimacy of race as a factor in affirmative action, actors in the realm of biomedical research, from bench scientists, to clinicians, to multinational corporations, would increasingly draw on the language of racial diversity to frame, guide, and structure their research and categorize their results.

In the social and legal realm, the triumph of the diversity rationale in the Supreme Court would produce a spate of critiques from across the political spectrum. While most of these would not engage with the issue of the relation of race to genetics, notably some prominent challenges from conservatives would draw directly on the findings like those of the Human Genome Project to argue that precisely because race was *not* genetic, it was too arbitrary a category to be used in legal contexts. In the realm of biomedicine, arguments from the Left urged the disentanglement of biological from social conceptions of diversity as a means to resist the sort of essentialization of race that undergirded subordination. In contrast, this conservative twist on disentanglement would come to inform a powerful strand of opposition to affirmative action and related programs in the legal realm aimed at addressing a long history of racial injustice.

FROM THE WHITE HOUSE CEREMONY TO THE SUPREME COURT: POWELL ASCENDANT?

In the run up to the 2000 presidential election, the Democratic Party platform highlighted its nominee's long-standing engagement with issues of science and technology, noting that "Al Gore, while supporting the completion of the Human Genome Project, has championed legislation to ban genetic discrimination." Such concern logically followed Clinton's and Venter's pronouncements about the genetic unity of the human race. But at this point, even as Barbara Grutter's case was working its way up to the Supreme Court, the platform's statements about diversity, per se, were largely anodyne and diffuse. It did note generally that "Al Gore has strongly opposed efforts to roll back affirmative action programs," and extolled the Democratic Party's embrace of "diversity of views as a source of strength." But it had little else to say on the matter, other than, "America will become much more diverse in the coming century."[7]

The Republican platform similarly invoked the idea that "diversity is a source of strength," almost like a reflexive mantra, but in its larger framing it implicitly placed diversity in tension with unity. It emphasized that, while "our country's ethnic diversity within a shared national culture is unique in all the world ... we must also strengthen the ties that bind us to one another. Foremost among those is the flag.... Another sign of our unity is the role of English as our common language." The Republicans of 2000 might embrace diversity but only if it were contained within a larger hegemonic structure. The platform placed "America's strong and diverse private sector" on a par with ethnic diversity as a source of strength. There was, unsurprisingly, no concern that the diversity of the private sector might somehow challenge the unity of our "shared national culture." Unlike ethnic diversity, the Republican Party saw diverse markets as an unalloyed good that needed no control or containment in the name of national interests. Nonetheless, when it came to genetics, the Republicans shared the Democrats' "concern [with] genetic discrimination, now that genetic testing will become a routine part of medical health care."[8] Neither party, however, directly connected genetic discrimination with issues of race or diversity, instead casting it more in terms of potential threats to medical privacy.

In 2001, Celeste Condit, one of the early analysts of the rhetoric and ideology of modern genetics, coauthored an article arguing that an emerging focus on treating "the biological components of socially constructed racial groupings as differences in the frequencies of appearance of some genetic elements among geographically dispersed populations" offered a new "de-essentialized account of human genetic variation." The authors expressed a hope that the new scientific account of human diversity "should help shift the stasis of these debates away from scientific proofs and questions about innate human abilities." The authors argued that such an account, however, would not, on its own, provide a "solution to issues of affirmative action and the broader discourse of equality in national life." They noted, for example, that Herrnstein and Murray themselves argued that the conclusions in *The Bell Curve* were "true whether differences among human groups are environmentally or biologically induced." In short, Condit et al. argued, "evidence about human variation that contemporary molecular biology has produced cannot close off the broader debate" about social policy and legal issues revolving around issues of diversity. They did not assert that the new genetic knowledge had no relevance or use in addressing these issues, rather, they concluded: "Science can be a useful handmaiden to rhetoric and public affairs, even if it cannot be their arbiter. Our choice to support affirmative action or other racially significant social policies can thus be informed by the scientific evidence we can amass about our biological natures, but it will also depend heavily on what notions of social justice we wish to forward."[9] In other words, despite the triumphalist declarations at the White House ceremony, science alone would not on its own resolve social questions of the nature and value of racial diversity, but its findings could be used constructively to reframe such discussions and delegitimize essentialized understandings of race and innate human characteristics. The challenge, therefore, would be to engage consciously and deliberately in such work.

By the time Barbara Grutter's lawyers were arguing her case before the Michigan Federal District Court, more than 75 percent of Fortune 1000 companies had instituted diversity initiatives.[10] In a 1999 law review article, Sanford Levinson observed that "'Diversity' is ... a ubiquitous topic of contemporary discourse. It has joined 'family values' and 'good medical care' as something that everyone is for."[11] Whether affirmative action would survive in academia, it clearly had firmly taken hold in the business world. Nonetheless, even as Grutter and others were challenging the legal basis of affirmative action in public institutions, critics from both the Right and the Left emerged during the later 1990s, challenging the emerging hegemony of diversity as an organizing concept in corporate personnel management practices.

In 1997's *The Diversity Machine*, Frederick Lynch charted the rise of a large network of diversity management professionals during the 1990s, which had so come to dominate the administrative and rhetorical landscape that by 1996, even conservative Republic Congressional Leader Newt Gingrich was echoing the mantra of "Diversity is our strength!" Charting the rise of the diversity management industry from the 1980s, Lynch decried what he saw as the institutionalization of a sort of left-wing "political correctness" that threatened the very fabric of American society "with its underlying ideology of ethnic-gender proportionalism, cultural relativism, and identity politics." (Concerns that any reader of current debates on "wokeism" will surely recognize).[12]

In contrast, in the 2001 article, "Diversity Rhetoric and the Managerialization of Law," Edelman, Fuller, and Mara-Drita, writing from the Left, decried the ways in which the managerial conception of diversity was disassociated from the tradition of civil rights laws by, in effect, diluting the concept with nonlegally protected categories, such as personality traits. More specifically, they argued that by 2000, "diversity rhetoric replaced the legal vision of diversity, which is grounded in moral efforts to right historical wrongs, with a managerial vision of diversity, which is grounded in the notion that organizations must adapt to their environments in order to profit."[13] Far from imposing left-wing political correctness, this article saw diversity rhetoric as domesticating the civil rights tradition into the service of profit-making multinational corporations.

CIVIL RIGHTS AND DIVERSITY RHETORIC

It is worth pausing for a moment to consider this juxtaposition of civil rights and diversity rhetoric. The "legal vision of diversity," implicitly lauded by Edelman, Fuller, and Mara-Drita, itself moved away from the civil rights tradition. Powell's influential opinion in *Bakke*, had refused to consider redress for historical injustice as a sufficiently compelling interest to justify affirmative action. Where civil rights was a *cause*, diversity was a *thing*. There was no "diversity movement" in the same sense as there was a Civil Rights Movement; rather, as Lynch and others noted, there emerged a "diversity industry." Under the civil rights model, difference was grounded in the idea of redress for past racial discrimination. It demanded common effort in the struggle to right past wrongs. Certainly, these were group-based wrongs, but what mattered was not how one group (Blacks) differed per se from another group (Whites) but rather how their present circumstances demanded recognition and remedy of past (and ongoing) unjust treatment. Under the diversity model, difference was based on assigned characteristics that were valued more or less as contributing to increased productivity (whether corporate or intellectual). It demanded managerial interventions to ensure particular "types" were represented in an institution. Thus framed, diversity more readily provided a basis or tendency to essentialize difference; hence, one of the persistent and recurring criticisms of affirmative action (again, from both Left and Right) was that it rendered racial groups as monolithic and expected one member of the group to speak for or represent the whole.[14]

Perhaps another way of thinking about possible differences between the civil rights model and the diversity model is to consider that under the civil rights model an individual might *benefit* because they were a *member* of a historically subordinated group. It is essentially a model of reparations—of using affirmative action (or other policies) to make up for past wrongs to a group. The individual's membership in that group says nothing about them beyond the fact of membership itself. In contrast, under the diversity model an individual might be *valued* because they *represent* a group in some essential way. The logic of Powell's opinion in *Bakke* was profoundly individualistic; it insisted on taking each individual on their own terms. It therefore directed focus to the distinctive attributes inhered in that individual. Here, the individual's identity was meant to speak volumes about other characteristics they were presumed to embody (literally). During the 1990s, the diversity management industry elevated this focus on inherent attributes to an organizational imperative. It valued a diverse workforce not because it righted historical wrongs but because its individual members were presumed to bring something to the table—their bodies, their viewpoints, their experiences—to benefit the bottom line.

The transition from a civil rights model to a diversity model also entailed a subtle shift in the conceptualization of representation that further reinforced the trend toward essentialization. Of course, the idea of representation is foundational to a democratic republic. The Voting Rights Act of 1965 marked a milestone in efforts to secure effective political voice to disenfranchised African Americans (whose experiences were at the forefront of debates over the bill). But the Voting Rights Act was primarily about empowering African Americans to have a voice in choosing their own representatives. It was not about the attributes of those representatives. Individual voters were not understood as "representing" anything in particular. It was not assumed that a "diverse" voter pool was per se somehow superior or produced better results— simply that it was right (and constitutionally mandated). Powell's diversity rationale, however, particularly as taken up and expanded upon by the diversity management business, centered on the idea that the beneficiaries of affirmative action themselves "represented" something—usually a viewpoint or an experience. Under a civil rights model, the individual *gets* representation through the exercise of the franchise. Under a diversity model, the individual *is* representation—that is, they get access to a good, like college admissions, by virtue of the fact that they purportedly represent some form of diversity.

While cases like *Richmond v. Croson* made it appear that considering race was taboo, they really only insisted on color blindness when dealing with race *qua* race—that is, as a historically specific and normatively salient group-based category. Considering race as a deracinated attribute of diverse individuals remained acceptable. This concept of race within diversity received a powerful affirmation in 2003 when the Supreme Court found against Barbara Grutter and affirmed the University of Michigan Law School's admissions program.

This notion of embodied representation of essential difference had also informed the mandates of the NIH Revitalization Act of 1993 with its affirmative action-like requirements for tracking race. Thus, even as Clinton and Venter were declaring the genetic unity of the human species in 2000, calls for increasing the "representation" of "diverse" racial and ethnic groups would increasingly come to inform biomedical research and practice. The affirmation of the diversity rationale in the *Grutter* case would provide both support and further rationales for this trend.

GRUTTER'S DIVERSITY

In 1998, William Bowen and Derek Bok (former presidents, respectively, of Princeton and Harvard Universities) published *The Shape of the River: Long-Term Consequences of Considering Race in College and University Admissions,* a book one reviewer characterized as "the most thoroughly researched analysis of affirmative action's effects in college admissions that has been done."[15] The authors of this widely reviewed and influential study (it would later be cited by Justice O'Connor in her majority opinion in *Grutter v. Bollinger*),[16] conducted an extensive review of students from a number of academically selective colleges and universities. It focused primarily on the performance, in college and after college, of Black and White students admitted to these schools. The book marshalled copious data to show that Black students at elite institutions performed well and shared in the success that attending such institutions typically brings. It found that affirmative action in admissions programs had "led to striking gains in the representation of minorities in the most lucrative and influential occupations." It also argued that White students benefited from the diversity that these students brought to campus. Bowen and Bok also devoted an entire chapter to what they characterized as "Diversity: Perceptions and Realities."

Notably, much of this chapter explored the idea that, particularly for White students, diversity helped to "develop[] the ability to work with, and get along with, people of different races and cultures."[17] This echoed and reinforced (with extensive data and charts) the arguments that were coming out of the corporate diversity management industry about the benefits of diversity.

Diversity Challenged: Evidence on the Impact of Affirmative Action, a collection of essays edited by Gary Orfield and Michal Kurlaender, published in 2001, came to similar conclusions. One essay by Mitchell Chang argued that "more diversity promotes more interaction and that socialization across racial lines, and is associated with more discussion of issues, better retention in college, and higher satisfaction with the college experience"; while another essay found that "increased campus diversity increases economic productivity" for both Black and White students.[18] In his essay, "A Policy Framework for Reconceptualizing the Legal Debate Concerning Affirmative Action in Higher Education," Scott Palmer, deputy assistant secretary for civil rights at the U.S. Department of Education in the Clinton administration, tried to address the *Hopwood* court's assertion that diversity stereotyped people on the basis of their race, by asserting that the "variety of viewpoints that the University seeks to foster does not come from any innate difference between the races themselves, but rather from the varying life experiences of the individual, due in large part to their racial backgrounds."[19] It was this "experiential diversity," Palmer argued, that could enrich the learning environment. University of Michigan President Mary Sue Coleman made a similar argument while the Grutter case was pending before the Supreme Court in 2003, asserting that "recognizing the education benefits of diversity does not mean that we equate a person's race with a particular point of view. It reflects the fact that race still matters in American society, as it influences our perceptions about the world and the people around us."[20]

These statements mark an important recognition of the need to resist the essentializing tendencies of diversity rhetoric. They rightfully rejected a crude biological essentialism but treaded dangerously close to replacing it with a softer form of cultural essentialism. They did not consider the underlying structure of the logic that eschewed a group-based justice framework in favor of a diversity framework focused on the individual. That logic still demanded a focus on the distinctive attributes of racialized individuals instead of addressing past injustices. For example, Palmer argued that "the belief here is not that a person's race controls his/her viewpoint, but rather that a person's race may affect his/her background and life experience and, in turn, his/her perspective on certain issues." That is certainly reasonable so far as it goes, but it still constructs the racialized individual as representing those experiences. It also takes those diverse experiences as *a priori* givens, making them appear *natural*. It obscures the fact that those differences might correlate with race precisely because of a long history of enacting subordinating laws and policies based on views that racial differences were, in fact, innate. Thus, while diversity advocates might have understood this as a clever way to address racial injustice without running afoul of Powell's opinion in *Bakke*, nonetheless it inescapably led back to frameworks that were dependent on the idea of innate racial difference, Palmer's protestations to the contrary notwithstanding.

On June 23, 2003, the Supreme Court issued its opinion in *Grutter v. Bollinger*, affirming the constitutionality of the University of Michigan Law School's affirmative action program for admissions.[21] The same day, in the companion case of *Gratz v. Bollinger*, the court struck down a the same university's undergraduate admissions program because, unlike the law school's, it used a more rigid point system for evaluating potential candidates and hence was not "narrowly tailored" to serve the state's compelling interest in fostering educational diversity.[22] In her opinion for the court in *Grutter*, Justice O'Connor noted that the law school's admissions policy, "aspires to achieve that diversity which has the potential to enrich everyone's education and thus make a law school class stronger than the sum of its parts." The policy does not restrict the types of diversity contributions eligible for "substantial weight" in the admissions process, but instead recognizes "many possible bases for diversity admissions." O'Connor went on to discuss Powell's opinion in *Bakke* at length because it "has served as the touchstone for constitutional analysis of race-conscious admissions policies." In the end, she effectively adopted Powell's approach and endorsed the "view that student body diversity is a compelling state interest that can justify the use of race in university admissions."[23] Legal scholar Ofra Bloch succinctly identified three central rationales in O'Connor's opinion for elevating diversity to a compelling state interest:

> The *first* goal is the utilitarian pedagogical and market-driven objective of preparing students for the workforce... *Second,* the Court acknowledged *anti-stereotyping* benefits of diversity and explained that it "promotes 'cross-racial understanding' " and helps to "break down racial stereotypes,"... The *third* and most dominant objective the Court attributes to diversity is the prospective value in sustaining the American democracy.[24]

Patrick Shin and Mitu Gulati argue that in propounding these rationales, O'Connor went further than Powell, who had primarily asserted that diversity was valuable for increasing the variety of viewpoints and experiences represented in educational contexts."[25] While O'Connor's focus on promoting democratic citizenship was certainly laudable, its promise was inescapably tempered by her gratuitous declaration that "We expect that 25 years from now, the use of racial preferences will no longer be necessary to further the interest approved today."[26] In any event, we certainly see the influence of the 1990s rhetoric of the diversity management industry, promoting the practical value of diversity in the marketplace and diffusing racial tensions.

DIVERSITY'S TRIUMPH AND ITS CRITICS

Looking back from over a decade later, Professor of Sociology Ellen Berrey noted in her book, *The Enigma of Diversity,* that "Following the court's decisions in *Gratz* and *Grutter,* campus leaders made the legal notion of diversity central to the university's new undergraduate admissions policies, the design of their affirmative action initiatives, and their public discourse on the student body. Diversity became a guiding principle for the internal admissions process, and race remained an important factor in how admissions officers did recruitment."[27] By 2001, more than 75 percent of Fortune 1000 companies had instituted diversity initiatives of some form or another.[28] As *Grutter* was being argued, Yale law professor and critic of affirmative action Peter Schuck saw the writing on the wall. In his 2003 book, *Diversity in America: Keeping Government at a Safe Distance,* he argued, or more accurately lamented, that "diversity's rhetorical power and prestige are at their height, and this exalted status is unique in history and in the world." Schuck was writing his critique of diversity while *Grutter* was working its way up to the Supreme Court. His primary argument was that government "should use its bully pulpit to praise diversity in general and even particular diversities. . . . But it should not try to create or promote any *particular* kind of diversity"; and that "law is a singularly poor instrument for performing that function." He wanted "diversities" to emerge more naturally from social interaction, hence the subtitle of his book: "Keeping Government at a Safe Distance."[29]

Adopting, in effect, a color-blind approach to the legal management of race, Schuck argued that "Race is perhaps the worst imaginable category around which to organize group competition and social relations more generally. . . . Whether benignly intended or not, using the category of race—which many affirmative action proponents depict as both socially constructed and immutable—to distribute advantage and disadvantage tends to ossify the fluid, forward-looking political identities that a robust democratic spirit inspires and requires. Justice Blackmun's earnest hope that we could get beyond race by emphasizing it has not been borne out."[30] Schuck here betrayed the same sort of racial exhaustion exhibited by Justice Bradley in his 1883 opinion in *The Civil Rights Cases* (restricting the reach of the 14th Amendment to "state action," i.e., officially sanctioned discrimination) where he declared that "When a man has emerged from slavery, and, by the aid of beneficent legislation, has shaken off the inseparable concomitants of that state, there must be some stage in the progress of his elevation when he takes the rank of a mere citizen and ceases to be the special favorite of the laws." This, just eighteen years after the end of the Civil War.[31] Schuck also anticipated Justice Roberts a decade later when he struck down key portions of the 1965 Voting Rights Act in 2013 because "things [had] changed dramatically" since it was enacted.[32]

Schuck's analysis focused primarily on diversity as used by Powell in his *Bakke* opinion. As Robert Post has noted, O'Connor's opinion in *Grutter* actually approved quite distinct justifications for affirmative action in higher education that Schuck did not fully address.

Specifically, like Shin and Gulati, Post found that *Grutter* held that "state universities could use affirmative action in order (1) to train persons to work in 'an increasingly diverse workforce'; (2) to maintain 'our political and cultural heritage' by making certain that 'knowledge and opportunity... be accessible to all individuals regardless of race or ethnicity'; and (3) to 'cultivate a set of leaders with legitimacy in the eyes of the citizenry' by ensuring that 'the path to leadership be visibly open to talented and qualified individuals of every race and ethnicity.'" He concluded that "These three justifications for affirmative action reach far beyond the diversity rationale of *Bakke*, which was the primary focus of Schuck's attention."[33]

Anthropologist Peter Wood was also writing a critique of diversity as *Grutter* was proceeding but was able to include a postscript addressing it in his 2003 book, *Diversity: The Invention of a Concept*. Wood argued that "the diversity movement... has achieved a substantial record of increased social discord and cultural decline," and he did not mince words in trashing "the intellectual shallowness and legal vapidity of O'Connor's opinion." He noted that the opinion formally elevated Powell's diversity rationale from the status of *dicta* in *Bakke* to a formally recognized constitutional concept with precedential value. He heaped his greatest scorn on the deference the opinion showed to professional educators by accepting more or less at face value their assertion that diversity in education was a compelling state interest.[34]

Critiques of *Grutter* and the diversity model were not limited to the Right. Prominent exponents of critical race theory, such as Derrick Bell and Charles Lawrence III, also expressed deep ambivalence about this purported victory for affirmative action. In a 2003 article titled, "Diversity's Distractions," Bell laid out four basic concerns with the Supreme Court's foregrounding of diversity as a rationale for affirmative action:

> 1) Diversity enables courts and policymakers to avoid addressing directly the barriers of race and class that adversely affect so many applicants; 2) Diversity invites further litigation by offering a distinction without a real difference between those uses of race approved in college admissions programs, and those in other far more important affirmative action policies that the Court has rejected; 3) Diversity serves to give undeserved legitimacy to the

heavy reliance on grades and test scores that privilege well-to-do, mainly white applicants; and 4) The tremendous attention directed at diversity programs diverts concern and resources from the serious barriers of poverty that exclude far more students from entering college than are likely to gain admission under an affirmative action program.[35]

Bell saw the decisions in *Grutter* and *Gratz* as prime examples of his theory of "Interest Convergence" whereby "we could not obtain meaningful relief until policymakers perceived that the relief blacks sought furthered interests or resolved issues of more primary concern."[36]

In this regard, the Supreme Court's focus on diversity reflected concerns manifest throughout diversity management literature of the 1990s that focused on how much it benefited *all* people, not just discriminated against minoritized individuals. Diversity, however, was not just a distraction from issues of past racial injustice, it was also productive. As Bell noted, it legitimized a focus on grades and test scores while also diverting resources from addressing "serious barriers of poverty." Bell concluded, somewhat bitterly, that "Diversity then is less a means of continuing minority admissions programs in the face of widespread opposition than it is a shield behind which college administrators can retain policies of admission that are woefully poor measures of quality, but convenient vehicles for admitting the children of wealth and privilege."[37]

Charles Lawrence III offered a critique of Powell's diversity argument in 2001 as *Grutter* was working its way up to the Supreme Court. Like Bell, he distanced himself from the liberal embrace of diversity because it failed "to challenge the manner in which traditional standards of merit perpetuate race and class privilege," and pushed aside "more radical, substantive defenses of affirmative action which articulate the need to remedy past and ongoing discrimination." Lawrence directly took on the highly influential *The Shape of the River*, then-recently published by Bowen and Bok, asserting that "their defense [of affirmative action] preserves the status quo," and arguing that their case for "diversity" ultimately was "a case for the integration of a privileged class . . . because the liberal defense of affirmative action accepts the reproduction of elites as the primary purpose of selective colleges and universities."[38]

More specifically, Lawrence asserted that the "'diversity' defense articulates its purpose as 'forward-looking' rather than 'backward-looking.' In so doing, it begins with an implicit denial of the defender's participation in or responsibility for past or contemporary racism."[39] Ironically, Schuck criticized diversity because he saw it as backward-looking, but for him, this was because it continued simply to build governmental programs around racial categories derived from past regimes. Perhaps these two characterizations can be reconciled by Schuck's focus on terminology and Lawrence's focus on power and accountability. Schuck simply appears to think that racial categories of "diversity" were too freighted with the ghosts of the past to be useful going forward; whereas for Lawrence, race demanded a recognition of and reckoning with past injustice that the diversity rationale denies.

What Schuck, perhaps, failed to appreciate was that the diversity of *Grutter* was in some critical respects different from Powell's diversity in *Bakke*. Powell was primarily focused on reconciling the need for considering individual applicants *as* individuals with university administrators' assertions that diversity enriched the educational experience for all. Much of the discourse in *Grutter* focused on the *market* value of diversity; that is, the ways in which diversity contributed critically to meet the demands of a modern globalized economy.[40] Certainly, such concerns were central to the many amicus briefs submitted by various corporations, not to mention claims by the Department of Defense that diversity was essential to building a strong military capable of projecting force globally.[41] These rationales echoed those of the diversity business that had emerged 1990s, building on the work of R. Roosevelt Thomas and others. The difference between *Bakke* and *Grutter* can be understood as reflecting a shift from what historian Gary Gerstel has characterized as the New Deal Order (which embraced the idea of government taking affirmative steps to improve social conditions) to the post-1980s Neoliberal Order (which emphasized the idea that government action should be limited to facilitating the free functioning of market forces).[42]

Powell, of course, was hardly a champion of Democratic New Deal policies. He was a Republican corporatist through and through, as evidenced by his notorious "memorandum" written in 1971, shortly before his elevation to the Supreme Court, in which he decried what he saw as the rise of radical Left attacks on the free enterprise system. Powell called for conservatives to start building countervailing cultural institutions along the lines of what became the Heritage Foundation and the Manhattan Institute.[43] Gerstel saw in the memorandum the seeds of the emerging Neoliberal Order. Nonetheless, like the president who appointed him, Powell operated within the tradition of Eisenhower Republicans who had acquiesced to the core components of the New Order.[44] It would be left to O'Connor, writing twenty-five years later after the full flowering of neoliberalism to characterize diversity in terms of market values.

When Justice O'Connor embraced and affirmed Powell's focus on diversity, she expanded upon it, projecting its significance beyond the walls of academia into the global marketplace, emphasizing that the benefits of diversity "not theoretical but real, as major American businesses have made clear that the skills needed in today's increasingly global marketplace can only be developed through exposure to widely diverse people, cultures, ideas, and viewpoints."[45] Diversity, then, was changing even as it appeared to remain the same. It was evolving into a commodity—or at least a commodifiable characteristic of individuals and institutions. Critics of diversity from the Right saw this as distorting market forces. Its champions saw it as enhancing and responding to such forces. But both sides were operating essentially within a neoliberal framework that took the market as the measure of all things.[46]

From this framework emerged a distinctive contemporary iteration of racial capitalism.[47] Discussing affirmative action programs and related diversity initiatives, Nancy Leong has explored how "white individuals and predominantly white institutions use nonwhite people to acquire social and economic value." Leong argues that the sorts of programs ratified by *Grutter* "rel[y] upon and reinforce[] commodification of racial identity, thereby degrading that identity by reducing it to another thing to be bought and sold." This was accomplished largely through marketing strategies that enhanced the standing of particular institutions by highlighting the presence of non-White students (or employees) in promotional materials and campaigns.[48] Beyond academia, perhaps the most iconic example such commercial exploitation of diversity was the "United Colors of Benetton" campaign of the early 1990s. Benetton, a fashion clothing company, went all in on the diversity rhetoric exemplified in the teachings of diversity entrepreneurs such a R. Roosevelt Thomas and

Lewis Griggs,[49] with the high-profile "United Colors of Benetton" advertising campaign, which featured an array of models of different ethnic and racial backgrounds. "The differences between peoples and between individuals are humanity's most valuable resources," declared one promotion, "Diversity is good." As sociologist Celia Lury noted, "Diversity is invoked here by the fashion clothing company Benetton as the ultimate good in a global economy." In this context, she continued, "the production of difference has become a means by which global companies such as Benetton can lay a proprietary claim to goodness itself."[50]

RACE, GENETICS, AND DIVERSITY AFTER *GRUTTER*

Neither O'Connor's majority opinion nor any of the other concurring or dissenting opinions made reference to genetics or any biological science. In its amicus brief, however, the American Association of Law Schools was sensitive to the dangers presented by essentializing arguments in Herrnstein's and Murray's *The Bell Curve*, published just a few years earlier. Addressing possible interpretations of racial differences in academic achievement, the brief asserted:

> The gap is *not* due to different levels of aptitude, as measured by IQ scores, in the respective gene pools of blacks and whites, notwithstanding the much-publicized claim to that effect in Richard Herrnstein & Charles Murray. The Bell Curve: Intelligence and Class Structure in American Life (1995). *See* Richard E. Nisbett, Race, Genetics, and IQ, *in* The Black White Test Score Gap 86, 89 (Christopher Jencks & Meredith Phillips eds., 1998) (finding "almost no support for genetic explanations of the IQ difference between blacks and whites.") Any genetic explanation would be particularly difficult to square with, among other things, two arresting pieces of evidence. First, the gap has no correlation with percentage of African versus European ancestry. *See id.* at 89–91. Second, studies of children fathered by American G.I.'s in Germany find no meaningful IQ difference between children of white fathers and children of black fathers, *see id.* at 91, further confirming that social rather than genetic factors explain the test-score gap.[51]

Diversity Affirmed in the New Millenium

Also part of the brief was an expert report submitted by Marcus Feldman, an eminent evolutionary biologist and mathematician at Stanford University (who had over the years also coauthored articles with Lewontin critical of race essentializing tendencies of work by the likes of Jensen and Herrnstein),[52] declaring:

> The biological concept of race is not tenable; race is a social construct produced by a nation's social and political history.... It follows that sociopolitical and socioeconomic factors are more important than the traditional biological understanding of race in determining outcomes as measured in social science research, such as school performance or in medical research, such as prevalence of some diseases. The importance of this revised understanding of race as a social construct rather than a genetic "fact" lies in the discussions that have permeated the social science literature on the causes of differences in performance and achievement between groups.[53]

A concern for the dangers of reifying racial diversity as genetic was, therefore, not absent from the record in *Grutter*, but given the prominence of passion of the debates over race, genes, and IQ provoked by the publication of *The Bell Curve* just a few years previously, this might be understood as a significant dog that did not bark (or at least only yipped a little). Perhaps the more recent White House ceremony hailing the genetic unity of humankind had eclipsed, at least temporarily, the implications of Herrnstein and Murray's arguments for characterizing diversity in relation to affirmative action. Perhaps those challenging affirmative action thought they did not need to appeal to biology to make their arguments. In any event, when paired with the pronouncements at the 2000 White House ceremony, this may be viewed as a high watermark of the disentanglement of biological from sociopolitical concepts of diversity.

Nonetheless, Peter Schuck took cognizance of genetics in his magisterial 2003 analysis of diversity and related critique of affirmative action. Observing that "diversity is a matter of degree, not an absolute," he went on to assert that "whether one focuses on the 90% or more of DNA that is shared by all humans or instead on the small amount that differs determines whether one emphasizes our genetic similarity or diversity."[54] There are a few things to note about this seemingly innocuous observation. First, it appears in the introduction to his book, so it is part of his framing of the larger project. Second, there is no discussion of *The Bell Curve*, Herrnstein and Murray, or other genetically inflected critiques of affirmative action. On the one hand, this speaks well of Schuck's discernment and avoidance of such spurious arguments. On the other hand, totally avoiding a discussion of them somewhat abdicates a responsibility to consider how they fit into the story of diversity in America. Third, less than three years after President Clinton's prominent statement at the White House ceremony that "all human beings, regardless of race, are more than 99.9 percent the same," Schuck has reduced the similarity to 90 percent.[55] This is not insignificant. By this time, among those familiar with the Human Genome Project, the language of 99.9 percent, or at least 99 percent, was common parlance. One could characterize Schuck's mischaracterization as a matter of 9.9 percent (a fairly large amount) or more strikingly one could characterize the difference between .1 percent and 10 percent as overstating genetic difference by a factor of one hundred. While Schuck himself does not do much with this figure, he nonetheless was laying the groundwork for those who might. He was, in effect, saying that one could indeed choose to focus on this misstated 10 percent difference to explore its implications for understanding, constructing, and acting upon diversity.

Schuck exploited this opening later in the book when asking the question: "What conditions account for the differences in things that we cognitively recognize as different?"[56] He invoked the new genetic knowledge of the Human Genome Project, not to resolve the question of genetic explanations of racial difference, but to keep it as an open question:

> Much human diversity reflects, first, the almost infinite *genotypic variation* found in our species. Even with the recent coding of the human genome, scientists do not yet comprehend the full nature, extent, and behavioral significance of this variation. What is

> already known, however, indicates that genotypic variation accounts for a significant share of interpersonal variations in intelligence, psychology, behavior, and a growing list of other variables; that the environment (usually defined to include whatever is not genetically determined) also accounts for a significant share; but that the relative proportions between genotypic and environmental causes—and more important, the nature of their interactions—remain very obscure and perhaps even unknowable.[57]

Schuck provided no citations for his assertion that "genotypic variation accounts for a significant share of interpersonal variations," perhaps recognizing that this might lead him down the controversial road to Herrnstein and Murray (or Arthur Jensen and Phillipe Rushton). He nonetheless deployed their work here *sub rosa*, not to directly argue that Blacks had lower IQs than Whites, but that "scientists" could not definitively answer this question for us. Moreover, this statement appeared in a section of the book titled "Sources of Diversity," meaning that he was framing the genetic characteristics in terms of their relation to defining diversity. Nonetheless, Schuck also argued that "scientists have long discredited the notion of race that underlies affirmative action policy, and the latest DNA research provides further evidence, were any needed, of its artificiality and incoherence."[58] That is, race as it related to affirmative action was scientifically incoherent, but "interpersonal variation" as understood in the context of a book critiquing race-based concepts of diversity, might well be significantly conditioned by genetics. Schuck here, thus, created space for biologizing racial differences even while laying the foundation for a distinct challenge specifically to the constitutional basis of affirmative action grounded in the idea that there was, in fact, no genetic basis for race.

In his 2003 critique of diversity, Peter Wood also argued that it was "above all a political doctrine asserting that *some* social categories deserve compensatory privileges in light of the prejudicial ways in which members of these categories have been treated in the past and the disadvantages they continue to face." Wood addressed the implications of the science of race and stated that "to admit students by race implies that we know what race is and can recognize racial differences with sufficient accuracy to make these judgments." In assuming that such classification was feasible, he argued, similarly to Schuck, that the Supreme Court "elevated American folk belief over established scientific evidence that, even in 1978, left no real room for the validity of the kinds of racial classifications that Powell had in mind." Echoing Schuck's idea that the racial classifications underlying affirmative action policies were incoherent, Wood similarly considered that the understanding of human genetic diversity articulated by Lewontin in 1972 and affirmed by Clinton in 2000, meant that race itself was an unscientific category and therefore an invalid basis for legal classification. For Wood as well, tropes of nature and artifice were central to his critique of diversity, as where he asserted that "the gulf between the real diversity of the world and the artificial and often imaginary diversity of our social experiments is very large," or that "the concocted *diversity* of contemporary campus life has precisely that element of charmed artificiality, a deadness and inertia beneath whatever lively rhetorical appearance we invoke."[59]

The following year, an article by prominent conservative law professors, Larry Alexander and Maimon Schwarzschild, explicitly took up this argument. They echoed Marcus Feldman's statements in his earlier pro-affirmative action special report, declaring that "human beings are not divided biologically into three, or five, or any number of 'races.'"[60] But where Feldman was looking back to critique the sorts of arguments propounded by Herrnstein and Murray in *The Bell Curve,* Alexander and Schwarzschild were looking forward to consider new ways to deploy science to undermine the rhetorical frame of diversity by accepting the liberal embrace of the idea of human genetic unity and using it to directly attack the *Grutter* decision on the grounds that "racial (and ethnic) classifications are unscientific, arbitrary, and often nearly meaningless."[61] This was a legally significant move because with the Supreme Court holding that diversity was a sufficiently compelling state interest to justify the use of racial categories in narrowly tailored affirmative action programs, one of the few remaining alternatives (at least with that particular Supreme Court) would be to challenge the underlying validity of the categories themselves. Even as Schuck saw diversity reaching its apogee as a framing concept for guiding law and policy, conservatives were developing new genetically inflected arguments to push forward this new challenge. In contrast to the racially essentializing (and subordinating) approach of *The Bell Curve,* this new, subtler attack built upon the disentanglement of race from genetics represented by the statements at the White House ceremony, to argue that, whatever it might be, diversity could not constitutionally be composed of categories based on race because they were not grounded in science.

CHAPTER FIVE

Diversity Machines

Race, Biology, and the Sweet Spot of Racialized Time

In ratifying Justice Powell's opinion, Justice O'Connor also adopted its dismissal of the relevance of history. She refused to recognize past racial injustice as a legitimate rationale for crafting affirmative action programs. To pass constitutional muster, such programs had to be forward-looking, focusing on issues such as preparing students for citizenship and business leadership in an increasingly diverse, globalizing, world market. Considerations of time, however, were not wholly absent from O'Connor's opinion. Noting the progress that had ostensibly been made in the twenty-five years since *Bakke*, she declared (without any justification other than an apparently whiggish faith in inevitable progress): "We expect that 25 years from now, the use of racial preferences will no longer be necessary to further the interest approved today."[1] O'Connor here betrayed a sort of anticipatory racial exhaustion, akin to that exhibited by Schuck that same year, and laying the groundwork for Roberts's declaration a decade later in *Shelby County v. Holder* that "things had changed."[2] Here, she was in effect declaring that by 2028 "things will have changed" such that consideration of race in programs like affirmative action would no longer be warranted. In this respect, O'Connor was declaring a limited temporal shelf life for affirmative action. It could exist in the present, cut off from the past, but could only persist into the future for a limited amount of time. Only in this sweet spot of racialized time would she allow institutions to adopt affirmative action programs.

A similar conceptualization of a sweet spot of racialized time was entering genetic research around this time, as well. In 1997, biological anthropologist Mark Shriver published an article titled, "Ethnic-Affiliation Estimation by Use of Population-Specific DNA Markers." This work grew out of an interest in the same sort of work being performed in the arena of forensic DNA analysis by scientists such as Chakraborty and Kidd. Both involved constructing genetic databases using racially marked "reference populations" to provide estimates of the genetic identity of particular individuals. As sociologist Troy Duster has noted, "Shriver would later revise this framing, and in subsequent papers, refer only to something he would call as Ancestry Informative Markers" (AIMs).[3] This technology, in turn, would provide the basis for the emergence of direct-to-consumer (DTC) genetic ancestry companies such as 23andMe® in the following decade.[4]

Tracing genetic ancestry can be a relatively straightforward affair. In males, the Y chromosome passes relatively unchanged through generations. One can trace a male's paternal line (father to grandfather to great-grandfather, etc.) back generations with relative accuracy. But, of course, the further back you go the less this tells you about a person's overall genetic ancestry. For example, trace back the Y chromosome for five generations and you have genetic information about only one out of thirty-two of an individual male's ancestors at that moment in time. On the maternal side, similar analysis can be undertaken by examining mitochondrial DNA (mtDNA) through which one can trace one's mother's mother's mother, and so on. Tracing ancestry using AIMs is different, and far less straightforward. AIMs are single nucleotide polymorphisms or other genetic markers "that show relatively large (30 to 50%) frequency differences between population samples."[5] Significantly, as Royal et al. note, "not all people from a given population have the AIMs identified with that population, and people from different populations can have the same AIMs."[6] The basic idea is that AIMs analysis can provide an estimate of how much of one's genome derives from any particular ancestral population. The final results employ the related concept of genetic "admixture." Hence the language from DTC ancestry companies telling customers that they are "x percent African," "y percent Asian," and "z percent European" etc.

As with the forensic DNA databases, these frequency estimates depend largely on the construction and classification of the reference populations used to generate the frequencies.[7] Thus, estimates can vary (and have) over time as more samples were gathered and differing sampling and classification techniques applied.[8] Tracing one's paternal or maternal line was a pretty straightforward matter and generally did not involve, or call for, characterizing any given sample by race. Early users of AIMs and related technologies, however, initially relied heavily on classifications based on continental groupings—European, Asian, African—that readily collapsed into old racial categories. Moreover, such results were also premised on the idea that there are somehow "pure" types of African, European, or Asian DNA. That is, to be "x percent," say, "Asian" necessarily implies that there is such a thing as a 100 percent pure Asian genome. This is highly problematic. As one recent article in *Science* has noted:

> The use of the terms admixture and "admixed individuals"—defined as those who have recent ancestry from more than one population, and typically continental ancestry populations—reinforces notions of discrete categories within humanity. This use does not escape the notion of continental ancestry categories but rather compounds the errors of using such categories because these individuals are typically conceptualized as a mixture of otherwise "pure" continental ancestry populations.[9]

This mistaken understanding of purity also depends on an implicit understanding of time. If we go back far enough in time all humans, ancestrally speaking, are 100 percent African. To break down ancestry into continental (or any other) categories necessarily entails drawing a line, not only around geographical but also temporal boundaries. Scientists must make decisions not only about "where" an AIM is from but "when."

To illustrate, consider the hypothetical case of a White South African whose ancestors may have come to Africa over three centuries ago. Ancestry tracing may designate that person to be primarily of "Dutch" or "European" origin. The "match" here is based on an assumption that one's genome cannot "become" African in merely three hundred years; so genetic identity is fixed at some temporal point prior to migration from the Netherlands to South Africa. But then, perhaps we should go on to ask, which temporal point? If we go back one or two more centuries to the 1500s, we find the Netherlands under Spanish rule. Perhaps there was significant admixture with Spanish populations? Spanish populations, in turn, had significant interaction with North African groups. Does this mean the "Dutch" person in South Africa today might really be descended from North African populations if we go back, say, 800 years? Or what if we go back 1500 hundred years or so to the Hun invasions of Europe? Perhaps our "Dutch" South African settler is "really" descended from "Asians." Even assuming some "pure" notion of European "Dutchness," we might choose to find its roots in distinct early migrations from Central Asia that predate the Hun invasions by millennia.[10]

To characterize AIMs in terms of specified geographic locales, such as continents, scientists must also specify a time frame when that locale becomes determinative of genetic identity. When scientists use continental categories that are readily susceptible to, in Duster's term, "molecular reinscription" as racial, they are also choosing a sweet spot of racialized time: too much earlier or later and the continental/racial designation loses coherence. O'Connor's opinion in *Grutter* racialized time prospectively, locating the sweet spot for race's relevance within a limited twenty-five-year time frame. After that time, she assumed (or at least hoped) that race would no longer be relevant. Scientists developing AIMs technology and related concepts of genetic admixture similarly had to fix a specific point in time (admittedly a much longer era) when the racial identity of an AIM could be fixed. For the Dutch South African settler that had to be sometime before their ancestors moved from Europe to Africa but sometime after their deeper ancestors moved from the Middle East or the Asian Steppe to the Netherlands. Perhaps it was a matter of twenty-five decades (or centuries) instead of twenty-five years, but as in *Grutter*, racial identity before and after that specified time frame was elided or effaced. The biggest difference, perhaps, is that scientists using AIMs technology *imposed* race as a function of time on their samples, whereas *Grutter allowed* others to employ race as an attribute of subjects for a specified time.

ANCESTRY, DIVERSITY, AND LAW

By the mid-2000s, there were already at least two dozen companies using AIMs and related ancestry tracing technologies to market "genetic ancestry tests" directly to consumers.[11] While often characterized as "recreational genetics" these tests would come to have a profound impact on many people's understandings of their identity and problematically blur distinctions among ancestry, race, and genetics.[12] Especially during these early years, DTC genetic ancestry companies often characterized their results using overbroad continental categories that all-too-easily read as racial. Moreover, the quality of the results was highly variable and would often change over time as different companies used different reference databases which were often poorly representative of global patterns of genetic variation.[13] Even as the technology improved over the following decade and companies provided more fine-grained categorizations of their results, critics were still complaining about the indeterminacy and variability of results. For example, in 2018 the *New York Times* reported a story on one person's travails with AncestryDNA® and 23andMe® (by then two of the largest players in the field) wherein their ancestry results varied wildly depending on when one company might update its reference data or reconfigure its confidence intervals,[14] because all ancestry estimates are, in effect, statistical guesses.[15]

Whatever else they were (or continue to be), DTC ancestry companies were, in effect, diversity machines—producing new, commoditized, and marketable understandings of racial and ethnic variation as a function of genetics, all legitimized by the patina of science and wrapped up in purported precise percentages—as in the case discussed in the *New York Times* article, where the subject's results said, at one point, that she was 43.4 percent sub-Saharan African, 36.9 percent European, 12.8 percent Western Asian, and 2.7 percent East Asian (or Native American).[16] Ironically, diversity in the context of affirmative action had eschewed numbers—quantifiable representation of specific racial groups were characterized as impermissible "quotas" in both *Bakke* and *Grutter*; but here, in the realm of the market, numbers and the precision they implied, took on commercial value as purported certification of diverse racial identities.[17] Ancestry tracing companies produced diversity not only *between* individuals, but *within* them as they provided statistical breakdowns of their diverse ancestries.

Such precision and the claims to identity it ratified, took on value in an additional realm as some saw it as an opportunity to invoke science to game the system of diversity-based admissions. In 2006, *New York Times* reporter Amy Harmon recounted the story of Alan Moldawer and his adopted twin sons, Matt and Andrew, who "had always thought of themselves as white."[18] As Harmon tells it, when it came time for the boys to apply for college:

> Mr. Moldawer thought it might be worth investigating the origins of their slightly tan-tinted skin, with a new DNA kit that he had heard could determine an individual's genetic ancestry. The results, designating the boys 9 percent Native American and 11 percent northern African, arrived too late for the admissions process. But Mr. Moldawer, a business executive in Silver Spring, Md., says they could be useful in obtaining financial aid.[19]

Harmon went on to note that "given the tests' speculative nature, [this was in the early days of commercial DNA ancestry ventures] it seems unlikely that colleges, governments and other institutions will embrace them."[20]

Some conservatives eagerly seized upon the potential use of DNA to game the system as a new basis for challenging affirmative action in all its forms. Of Harmon's 2006 article, Jonah Goldberg gleefully wrote in *The National Review*: "Can't people see how unbelievably absurd the racial quota game is when you can game the system to advantage kids who were completely unaware of their mixed genetic 'identity.' The system's going to come crashing down soon. Better get your racial spoils while you can."[21]

In contrast to gaming the system or showing its incoherence, Deadria Farmer-Paellmann saw DNA ancestry tests as a means to help right a great historical wrong.[22] Farmer-Paellmann was an African American activist seeking reparations for the harms of slavery. Reparations is a long fought and complicated issue, which has been conceptualized in many ways, from demands for recognition and apology to more policy-driven arguments for general programs to improve the lot of African Americans.²³ Farmer-Paellmann came up with a distinctive approach that was far more specific. In 2000, she had uncovered archival evidence of insurance companies that had written policies for slave owners on the lives of their slaves. In 2002, she filed a claim against FleetBoston Financial Corporation, Aetna, and CSX for the return of lost wages and consequent lost wealth.²⁴ In 2004, this suit was dismissed on several grounds, one of the most central being a lack of standing to bring a claim. Among the reasons for finding no standing was a failure to establish a sufficiently direct link between the plaintiffs and the alleged harm. As the court noted, "Plaintiffs fail to allege any facts in their Complaint that link the specifically named Defendants to the alleged injuries suffered by the Plaintiffs; nor does the Plaintiffs' Complaint allege a connection between any of the named Defendants and any of the Plaintiffs' ancestors."²⁵

To address this problem, Farmer-Paellmann approached geneticist Rick Kittles, whom she had encountered in the 1990s through their work on the African Burial Project in Lower Manhattan. Kittles had recently founded a DNA testing company called African Ancestry with the aim of using new genetic technology to connect African Americans back to their roots in Africa. Together, Kittles and Farmer-Paellmann considered whether genetic ancestry tracing technology might be used address the standing issue by providing a link "to prove your authenticity... to suggest you should receive reparations."²⁶ In a subsequent complaint they, therefore, alleged that "scientific testing in the form of DNA testing has proven beyond doubt the direct relationship between the instant plaintiffs and the instant defendants,"²⁷ and prayed that the court would declare "that the DNA testing conducted by and of each of the plaintiffs is sufficient to establish the standing and/or direct connection between plaintiffs and Defendants."²⁸

Ultimately, the court dismissed this second complaint, again largely on standing grounds.²⁹ With respect to the new genetic evidence, the court noted that "there may well be no perfect method of determining exactly who is a descendant of a slave, and thus a member of the group entitled to receive reparations," and concluded that "genetic mapping, or DNA testing... alone is insufficient to provide a decisive link to a homeland."[30] The court elaborated on the notion of "genetic standing" and its limitations:

> Plaintiffs face insurmountable problems in establishing "to a virtual certainty" that they have suffered concrete, individualized harms at the hands of Defendants. "An essential prerequisite to bringing suit is the plaintiff's ability to establish with precision her relationship to the injury and the defendant." In terms of slavery reparations, the "'traditional' model . . . seeks suit against a defendant or defendants on behalf of a plaintiff class comprised of descendants of slaves." In such situations, plaintiffs "assume [] that a familial relationship between the ancestor victim and the descendant plaintiff—what might be called hereditary or genetic standing—is sufficient to bring suit." An assumption such as this is difficult to implement in practice. "The notion that standing can be inherited (the 'genetic' theory of standing) is . . . legally . . . suspect; and the notion that groups, rather than individuals, have standing to sue, is legally insupportable."[31]

The court here was not entering into the fraught area of adjudicating racial identity. It focused instead on the limitations of genetic science for establishing standing to bring a particular kind of historically informed legal claim.

Of note as well, is the court's reference to "precision." In the context of using race in relation to forensic DNA analysis, courts had been clear that what mattered for legal relevance was not "precision" (remember there could be RMPs that varied by orders of magnitude as high as one hundred) but "accuracy." When used in criminal contexts to incarcerate, "accuracy" was sufficient for establishing relations between race and genetics; but when used to assert claims for reparations accuracy was not enough, "precision" was now also required.

In contrast to apparently bad faith or opportunistic uses of DNA testing to claim tactical advantage, Farmer-Paellmann's appeal to DNA here was consistent with her long-held assertions of racial and ancestral identity. Unlike the Moldawers, she was not using DNA to make a novel claim of membership in a racial group, but for the more focused purpose of establishing direct ancestral ties to enslaved Africans sufficient to establish standing to pursue her case. Her use of DNA testing may have been tactical in the sense of creating a new avenue to pursue her prior claim, but it was fully consistent with it and hence should not be understood as an attempt to use DNA either to game or undermine the system. The diversity mandate of affirmative action was not in play here. Being a historically based claim for reparations, her case ran directly counter to the logic of diversity, which was only forward-looking and denied the relevance of history.

Technically speaking, the claim was not about race per se but about ancestry, a very particular ancestry to specific enslaved individuals. As such, her claim did not formally reify race as genetic, nor did it directly implicate issues of race as a social construct. Nonetheless, in juxtaposing the distinctively racialized ancestry of slave descendants with genetic technology, it inevitably contributed to a frame for perceiving that connections between race and genetics could have legal and policy implications. Thus, it is not surprising that around the same time the court was dismissing Farmer-Paellmann's arguments about genetic ancestry, we had Amy Harmon writing her article in the *New York Times* about the Moldawer twins.

THE HAPMAP TRIES TO GET DIVERSITY RIGHT

In October 2002, the NIH announced the launch of an ambitious new $100 million project "to create the next generation map of the human genome." In contrast to the Human Genome Project, which focused on producing a draft of a model individual human genome, this new initiative, called the International HapMap Project, was focused on charting genetic variation within the human genome. The idea was that by comparing genetic differences among individuals, researchers could develop a tool to detect the genetic contributions to many common diseases.[32] While the message upon the completion of the Human Genome Project emphasized the essential unity of humankind, the HapMap project aimed to drive genetic discovery forward by focusing on difference. In some ways, the HapMap Project picked up where the Human Genome Diversity Project left off, although with a much more explicit focus on using genetic discovery to improve health. It faced similar issues regarding its sampling strategy to try to capture a wide range of genetic variation efficiently.

Perhaps because of the political fallout from the aborted Human Genome Diversity Project, genetic researchers in the early 2000s did not emphasize the concept of "diversity" per se. Rather, they tended to talk more concretely in terms of population groups and ancestry. As diversity receded into the background, the opportunity for more nuanced and fine-grained characterizations of human genetic variation presented itself. The HapMap project very consciously tried to take this road. But as things developed, the terminology of population and ancestry tended to collapse back into racialized categories. Ironically, as the concept of diversity became less prominent during the early 2000s, the concept of race as a genetic construct reemerged centrally in debates about genetic research and biomedical advancement.

The HapMap Project deployed a population-sampling strategy "to maximize the downstream benefits of the Project for all populations—both sampled and un-sampled populations." It defined "population" as "a group of people with a shared ancestry and therefore a shared history and pattern of geographical migration."[33] As with AIMs, the concept of ancestry played a central role in characterizing genetic samples here. With the goal of capturing greater genetic variation, the first phase of the HapMap Project organized its sampling around four geographically dispersed populations: "(1) 90 individuals ... from the Yoruba in Ibadan, Nigeria; (2) 90 individuals ... in Utah, USA; (3) 45 Han Chinese in Beijing, China; [and] (4) 44 Japanese in Tokyo, Japan."[34] The rationale for this sampling strategy was explained as follows:

> The exact pattern of SNP variants within a given haplotype block differs among individuals. Some SNP variants and haplotype patterns are found in some people in just a few populations. However, most populations share common SNP variants and haplotype patterns, most of which were inherited from the common ancestor population. Frequencies of these SNP variants and haplotype patterns may be similar or different among populations. For example, the gene for blood type is variable in all human populations, but some populations have higher frequencies of one blood type, such

as O, while others have higher frequencies of another, such as AB. For this reason, the HapMap consortium needs to include samples from a few geographically separated populations to find the SNP variants that are common in any of the populations.[35]

The HapMap project was well aware of the debates over sampling strategies that occurred during the early days of the HGDP. It acknowledged that "any choice of DNA samples represents a compromise: a single population offers simplicity, but cannot be representative, whereas grid-sampling is representative of the current worldwide population but is neither practical nor captures historical genetic diversity." The HapMap project considered that samples from these populations would be "based on well-known patterns of allele frequencies across populations, reflecting historical genetic diversity."[36] Here the HapMap project notably introduced a temporal component into its characterization of human diversity, again connecting it with the frame of analyzing AIMs.

Sensitive to the dangers of conflating geographically distant groups with race, the HapMap project emphasized: "Because none of the samples was collected to be representative of a larger population such as 'Yoruba,' 'Northern and Western European,' 'Han Chinese,' or 'Japanese,' (let alone all the populations from 'Africa,' 'Europe,' or 'Asia'), we recommend using a specific local identifier (for example, 'Yoruba in Ibadan, Nigeria') to describe samples initially."[37] This unusually careful characterization of data resulted, in part, from deliberate attention to "integrating ethics and science in the International HapMap Project."[38]

Yet, as the data from the HapMap project was taken up and circulated both in the popular press and in scientific journals, these careful classifications were repeatedly conflated with the broad, racially marked continental categories of "African," "Asian," and "European." For example, prominent popular reports of the completion of the first phase of the HapMap project described the samples variously as "people from Africa, the Far East, and western Europe";[39] "people from Asia, Africa and the United States";[40] and "people from Africa, Asia, and the European population."[41] Among scientific journals, an article on linkage disequilibrium in the *American Journal of Human Genetics* typically referred to the HapMap project as sampling "four populations with African, Asian, and European ancestry."[42]

Diversity Machines 123

The International HapMap Consortium (IHC) itself bore a measure of responsibility for such easy conflation. The initial justification for selecting geographically dispersed populations was to capture greater genetic variation.[43] But one could argue that since the greatest genetic diversity is to be found among populations within Africa, a more revealing map might have been developed by sampling geographically dispersed populations within Africa. Alternatively, the IHC could have chosen not to mark the samples according to their points of origin. Moreover, in its actual sampling and classification strategy, the IHC sent conflicting messages regarding the local specificity of the populations. Most prominently, in sampling forty-five Han Chinese and forty-four Japanese, the project design implicitly asked for these two groups to be amalgamated into a larger "Asian" category to parallel the ninety samples each from the Ibadan, Nigeria, and Utah collections. The IHC's otherwise responsible and careful post hoc characterization of the samples thus operated in tension with the message sent by the underlying structure of the project design.[44]

Notably, a major article announcing the project published by the International HapMap Consortium in *Nature* in 2003, did not use the term "diversity" in characterizing its work.[45] Rather, the language was all framed in terms of genetic "variation" across "populations." Driven by the then-dominant Common Disease/Common Variants (CD/CV) hypothesis,[46] (put forward prominently by David Reich and Eric Lander in a 2001 article, "On the Allelic Spectrum of Human Disease,"[47] positing that common disease-causing variants would be found in all human populations which manifest a given disease), the article framed the point of sampling geographically distant populations as finding common patterns of variation that might enable researches to identify which differences were related to disease and drug response.[48]

The logic of the CD/CV hypothesis tended to mitigate against the drive to "diversify" genetic databases to drive the discovery of disease-causing genes. If relevant disease-causing genes were common and found across all populations, then the relative diversity (however defined) of samples in a given database would be less consequential. In contrast, if common diseases were caused by more complex interactions among numerous genes and even rare variants, then DNA sequence variation of any gene-causing disease could encompass a wide range of possibilities, with the most extreme being that each mutation is only found once in any given population. Seeking rare variants that occurred at different frequencies in different populations would demand the development of far more comprehensive and "diverse" reference databases.

Certainly, the International HapMap Project was seeking a species of geographic diversity in its sampling strategy. But as noted above, it was very careful to avoid racial or ethnic descriptors—perhaps because of the uproar still swirling around the abortive HGDP. Subsequent calls for genetic database development during the remainder of the first decade of the twenty-first century also tended to avoid focusing on race per se.

A few years earlier, in 1998, the National Human Genome Research Institute (NHGRI) had created the DNA Polymorphism Discovery Resource (PDR), a database of single nucleotide polymorphisms (SNPs). As anthropologist Sandra Lee noted, "the purpose of creating the PDR was to provide researchers with samples that reflect the diversity of SNP patterns among the U.S. population." But while committed to "diverse" samples, the NHGRI also wanted to disassociate genetic diversity per se from racial and ethnic markers. Therefore, although researchers collected samples from a range of racial and ethnic groups, the NHGRI took the unprecedented step of "packaging samples from different populations into a 'mixed' bundle while at the same time excising racial and ethnic information from individual samples." The goal was to have, in effect, a "colorblind" repository. As Lee noted, "the decision to go colorblind was influenced in large part by the lingering institutional social memory of protest by indigenous groups against the Human Genome Diversity Project."[49] Researchers, however, largely rejected the color-blind approach of the PDR, revealing a persistent belief in the salience of racial categories for genetic research.

Ironically, this was occurring around the same time that the Fifth Circuit Court of Appeals was effectively imposing a color-blind view of the U.S. Constitution in striking down the University of Texas's affirmative action plan in *Hopwood v. Texas*, while California was adopting color blindness for public institutions through Proposition 209. Unlike those color-blind initiatives, however, the intent here was not to roll back racial progress in social arenas but rather to ensure that race did not become reified as genetic in biomedical arenas.

STRUCTURE-ING DIVERSITY

In June 2000, geneticists Jonathan K. Pritchard, Matthew Stephens, and Peter Donnelly published an article titled "Inference of Population Structure Using Multilocus Genotype Data" in the journal *Genetics*, setting forth to "describe a model-based clustering method for using multilocus genotype data to infer population structure and assign individuals to populations."[50] This "model-based clustering method" was a computer program called STRUCTURE which went onto become widely used to characterize human population structure and estimate individual admixture from different populations. The 2000 article itself did not discuss racial or ethnic population groupings per se but two years later Pritchard coauthored a second article with lead author Neil Rosenberg and several other geneticists that applied STRUCTURE to analyze the genomes of 1,056 individuals from fifty-two populations and organize them into clusters. STRUCTURE itself is agnostic as to the number of clusters it will produce (characterized as K), but in the article the authors focused on how STRUCTURE organized the individuals into "six main genetic clusters, five of which correspond to major geographic regions."[51] In contrast to the HapMap project's careful attention not to characterize such diversity along racial lines, this article characterized these regions as Africa, Eurasia, East Asia, Oceania, and America, thus echoing commonly used racial categories of the U.S. census (Black, White, Asian, Native American, Hawaiian/Pacific Islander). But as anthropologist Deborah Bolnick has observed, "it was not possible to determine a single best value for K," meaning it could have been set higher or lower to generate a wide range of possible clusters. She noted that this "is exactly what we would expect given the clinal variation and pattern of isolation by distance found in our species."[52] In other words, STRUCTURE did not show race is genetic, it showed that frequencies of particular genetic alleles vary gradually over geography. It might be possible for a computer program to draw lines at continental boundaries that make it look like the resulting groups are racial in character, but it is also possible for the same program to draw similar lines within and across those same boundaries. As the editors of *Nature Biotechnology* wrote later that same year, "using genetics to define race is like slicing soup. You can cut wherever you want, but the soup stays mixed."[53]

Nearly eighteen years later, Charles Murray, in a follow-up to *The Bell Curve*, tellingly titled *Human Diversity: The Biology of Gender, Race and Class*, would seize on these studies by Pritchard and Rosenberg to try to resuscitate his long-running campaign to undermine the idea that race is a social construct without a biological foundation. For Murray, "the sequencing of the human genome changed everything."[54] By this, Murray meant primarily that STRUCTURE and AIMs technology allowed for a supposedly neutral and objective basis for characterizing racial groups as genetic. From *The Bell Curve* in 1994 to *Human Diversity* in 2020, Murray had been trying to entangle political and biological conceptions of diversity. He did not confuse or conflate them so much as he aimed to prioritize discredited understandings of racial diversity as biological to undermine social and economic policies aimed at ameliorating historical racial injustice. The basic argument was that if *biological* diversity accounted for differences in social standing (particularly in relation to cognitive capacity) then policies constructed around understandings of diversity as *social* or *political* were doomed to fail. Anti-tax crusader Grover Norquist, notably declared in 2001, "I don't want to abolish government. I simply want to reduce it to the size where I can drag it into the bathroom and drown it in the bathtub."[55] Similarly, one might characterize Murray's efforts as using race as a biological concept to undermine the rationales for affirmative action and related programs so that they too could then be drowned in a bathtub.

As flawed as much of *Bakke*'s and *Grutter*'s discussions of racial diversity may have been, they treated it fundamentally as a social phenomenon distinct from biology. Even as the court in *Al-Khazraji* may have left the door open to biologizing race, it nonetheless acknowledged the well-established scientific consensus that race was not genetic. These court opinions kept biological and social conceptions of diversity distinct. Murray was intent on using the work of scientists like Rosenberg to entangle genetic concepts of diversity with social concepts of race. Yet, Marcus Feldman, one of the coauthors of the Rosenberg article clearly had a very different understanding of its significance. As noted, Feldman had earlier coauthored articles with Lewontin critical of race essentializing tendencies of work by the likes of Jensen and Murray's coauthor on *The Bell Curve*, Richard Herrnstein.[56] He also submitted a report in *Grutter* in support of affirmative action, declaring that "The biological concept of race is not tenable" and that social and political factors were far more important than genetics in determining social outcomes.[57] This conclusion runs directly counter to the sorts of arguments Murray would try to use the Rosenberg article to support.

Nonetheless, STRUCTURE can be understood in retrospect as another sort of diversity machine, used by particular scientists (and social commentators) to produce genetic diversity as a function of race. Ironically, STRUCTURE was also presented as, in effect, color-blind; the idea being that the researchers would simply select for "K," the number of clusters they wanted the algorithm to produce. Race would play no ex ante role in shaping the results. Obscured by this were the assumptions, pointed out by Bolnick and others, both regarding where to set "K" and also how the underlying databases to which STRUCTURE would be applied were themselves constructed. The Rosenberg paper that applied STRUCTURE appeared to produce what looked like racially concordant continental genetic groupings.

In a 2022 study looking back on this, and subsequent related analyses that categorized genetic ancestry, an interdisciplinary group of scholars demonstrated how the racial clustering of studies like the Rosenberg paper was largely an artifact of sampling strategies that greatly oversimplified and flattened out the complexity of genetic variation across the globe.[58] The authors noted:

> The most commonly used reference data were created by sampling individuals from a few dozen places spread across the globe. If individuals from these populations are graphed in this manner, distinct clusters roughly representing continental categories are visible.... A prominent early result [the Rosenberg paper] was that genetic ancestry was strongly concordant with continental origins when ascertaining for individuals whose four grandparents were from the recruitment sites. But newly assembled datasets show that if people are sampled differently, such as individuals living in New York City, it becomes clear how impoverished this

view of a structure of distinct clusters is. . . . The clearly separated clusters of reference population individuals, corresponding to different continental groups, merge into a background of continuous genetic variation.[59]

This observation, in many respects, should have been obvious back in 2002. It flowed naturally from Livingstone's 1962 characterization of genetic variation as predominantly clinal. It also reflected precisely the sorts of concerns expressed a decade earlier over different sampling strategies for the Human Genome Diversity Project. Nonetheless, the STRUCTURE program, particularly as deployed by Rosenberg et al., combined with growing interest in AIMs and related ancestry-tracing technologies, led to a new proliferation of studies characterizing genetic diversity in racial terms.

GENETICIZING RACE IN BIOMEDICINE

During the early 2000s, this work on genetics and classification took place against the backdrop of increasing interest in how or whether to use racial categories in biomedical research and practice. Much of the interest in race and genetics centered on the rising concern for the seemingly intractable problem of race-based health disparities that policymakers had been trying to address at least since the Heckler Report in 1985. During this time, a tension developed around what to make of the fact that in some situations, well-documented social and economic factors contributing to health disparities did not seem to account for all of the observed race-based differences in morbidity and mortality.

One approach, arguing for the legitimacy and importance of using racial classifications in genetic studies, was epitomized by an influential article published in the *New England Journal of Medicine* in 2003, by a group of researchers from the University of San Francisco and Stanford, led by Esteban Burchard, and titled, "The Importance of Race and Ethnic Background in Biomedical Research and Clinical Practice."[60] Where the Rosenberg article on STRUCTURE was careful to avoid explicitly using the terms "race" or "ethnicity" in describing their genetic population clusters, the Burchard article placed race front and center in characterizing genetic differences among populations.

The article built on one published the previous year by many of the same authors in the online journal *Genome Biology*, arguing that "genetic differentiation is greatest when defined on a continental basis"; which was precisely the categorization of diversity produced by STRUCTURE in the Rosenberg article. But as Celeste Condit noted at the time, these researchers' characterization of "continents" was highly problematic. For example, Burchard et al. defined Asians as those from eastern Asia including China, Indochina, Japan, the Philippines and Siberia.[61] They excluded the entire landmass of Central Asia from their definition. As Condit noted, "this definition use of the term 'continental' is stretched so far it should be virtually unrecognizable in this case." She observed similar problems with each of the article's groupings, concluding that "none of the groups claimed to be delineated as 'continental' groupings are actually very closely coterminous with a continent as the term is otherwise understood."[62] As happened with the carefully characterized sample groupings from the HapMap project, the authors here set the stage for what I have characterized elsewhere as "racial recursion"—the return of geneticized definitions of race by expanding, contracting, or as Condit characterized it "stretching" nonracial categories, such as continents, to organize genetic data in forms that are readily susceptible to being conflated with racial groupings.[63] Again, as in the case of forensic DNA analysis, slippage derived from a clear lack of care in organizing and characterizing data that was not explicitly biological.

The authors framed their argument in the *New England Journal of Medicine* article as a response to the observation that "with the completion of a rough draft of the human genome, some have suggested that racial classification may not be useful for biomedical studies, since it reflects 'a fairly small number of genes that describe appearance' and 'there is no basis in the genetic code for race.'"[64] Of course, this framing obscured the fact that merely observing that there is no genetic basis for race does not in itself demand that race no longer be used in biomedical studies—merely that it not be used in a way that presumed race to be genetic. Nonetheless, the article did note that some researchers had been arguing for the exclusion of racial and ethnic categories from biomedical research altogether. But this was not really the authors' main concern. Note that the title of the article was not "The importance of race and ethnic *categories* in biomedical research and clinical practice." Rather, it was "The importance of race and ethnic *background* in biomedical research and clinical practice." By "background" they clearly meant *genetic* background. The paper's seemingly modest core contention was that "The evaluation of whether genetic (as well as nongenetic) differences underlie racial disparities is appropriate in cases in which important racial and ethnic differences persist after socioeconomic status and access to care are properly taken into account."[65] The idea here was that if you controlled for nongenetic factors, then any remaining race-based health disparities must be due to some underlying genetic difference. This apparently unexceptionable contention obscured the authors' significantly cramped and limited characterization of nongenetic factors as being encompassed by "socioeconomic status and access to care." While socioeconomic status (SES) and access to care certainly are relevant, so too are such nongenetic factors as diet, exposure to environmental stressors, neighborhood characteristics, or simply the biological impact of experience of racism itself. Leave these out of the equation and it is far easier to posit underlying genetic differences to explain racial health disparities. But easier certainly does not mean more accurate.

This basic move, characterizing social factors contributing to health disparities in a highly restrictive manner so that any remaining disparities appear to be the result of race-based genetic differences, would become a defining attribute of much biomedical research in the following decade. It would also come to drive the call to "diversify" genetic bases in the following decade as a means to address racial inequity, at least rhetorically. More insidiously, this move also created space for opponents of social and policy interventions aimed at addressing racial health disparities to develop gene-based critiques of the growing progressive discourse around disparities. Just as Murray and Herrnstein had deployed genetic arguments about racial difference in *The Bell Curve* to attack diverse social welfare policies and affirmative action, so too did critics in the arena of health policy seek to reframe *disparities* as a function of biological *difference* rooted in racially clustered genetic diversity.[66]

GENETICIZED RACE AND HEALTH POLICY

The major civil rights struggles of the previous several decades had largely focused around issues of desegregation, affirmative action and discrimination in such areas as housing, public accommodations, and employment.[67] To identify and address discrimination in these areas, it was necessary to collect and categorize data by race. The current racial and ethnic categories used in the U.S. census emerged largely in response to needs and pressures created by the civil rights movement and the legislation emerging from it.[68] To track violations of voting rights or employment discrimination claims, it was essential to aggregate data by race. While highly problematic for an array of social and political reasons, the use of racial and ethnic categories in such contexts did not directly implicate them as biological or genetic constructs.[69]

The movement for civil rights continued, for very good reasons, to broaden its focus to encompass a much more explicit concern for health rights. From the creation of the Office of Minority Health in 1985,[70] to the Minority Health and Health Disparities Research and Education Act of 2000,[71] to the proposed Closing the Health Care Gap Act of 2004,[72] major federal initiatives were being undertaken to identify and address racial disparities in health care. As these and related initiatives engaged social, economic, and political influences on disparate health *outcomes*, they implicated racial and ethnic categories as correspondingly social, economic, and political constructs.[73] Such concerns marked a natural progression of civil rights activism from political and economic rights into the realm of health. When, however, racial and ethnic categories were used to guide initiatives to uncover the underlying *causes* of disease they directly implicated these categories as potentially biological and/or genetic concepts. To the extent that otherwise well-intentioned attempts to redress health disparities implicitly or explicitly invoked race as a genetic concept, they ran the risk of fueling what anthropologist Alan Goodman characterized in 2000 as a "comeback" in "racialized notions of biology."[74] This marks a fundamental difference between civil rights activism in the arena of health as opposed to political or economic rights. It gained added force in the aftermath of the completion of the Human Genome Project as increasing attention was being paid to conceptualizations of human genetic diversity and its possible relation to health outcomes.

In 2000, Congress passed the Minority Health and Disparities Research and Education Act, creating a new National Center on Minority Health and Health Disparities. Three years later, the Institute of Medicine (IOM) published a major report on health disparities titled, *Unequal Treatment: Confronting Racial and Ethnic Disparities in Health Care*.[75] The IOM report was a natural outcome of the increasing attention being paid to health disparities. It chronicled an array of health disparities and connected them directly to social and economic issues of equity, access, and racism. In December 2003, not long after the publication of the Burchard et al. article, the Department of Health and Human Services (DHHS) issued a report on health disparities, supposedly based on the IOM Report. The DHHS report, however, dismissed the "implication" that racial differences in care "result in adverse health outcomes." It turned out that top officials had directed DHHS researchers to drop their initial conclusion that racial disparities were "pervasive in our health care system," and to delete or recharacterize findings of "disparity" as mere evidence of health care "differences."[76] For example, an earlier version of the report mentioned the term "disparity" thirty times in the "key findings" section, while the final report mentioned it only twice and left the term undefined.[77] DHHS officials accompanied this push to use the term "difference" to emphasize "the importance of...personal responsibility" for health outcomes.[78] Ultimately, DHHS Secretary Tommy Thompson backtracked when word of the report's manipulation was leaked by concerned DHHS staff.

This approach exemplifies a dynamic identified by anthropologists George Ellison and Ian Reese Jones, whereby, "the 'geneticization' of individual identity shifts responsibility for genetic conditions onto individuals," and the "collective geneticization of social identities shifts responsibility for social inequalities in health on shared values, beliefs and behaviours."[79] As with Murray's arguments against affirmative action, conservatives in the Bush administration deployed frames of biological difference to undermine the report's message that social conditions, amendable to policy interventions, lay at the root of long-observed and pervasive racial health disparities.

Even as O'Connor's opinion in *Grutter* embraced and reconfigured diversity in terms of market value, the Bush administration's response to the IOM report reflected a similarly neoliberal approach to the problem of racial health disparities. The article by Burchard et al. on, "The Importance of Race and Ethnic Background in Biomedical Research and Clinical Practice," dovetailed with Secretary Thompson's attempted revisions to the IOM report.[80] Both directed attention away from social and environmental forces shaping health disparities to focus on underlying differences in people of color—whether genetic (à la Burchard) or behavioral (à la Thompson). Rather than old, New Deal-style political programs to address underlying social conditions, such framings exemplified a neoliberal focus on individual choice and market-based responses.

COMMODIFYING "GENETIC" RACE

This dynamic was most evident in the case of BiDil, the first drug ever approved by the FDA with a race-specific indication—to treat heart failure in a "Black" patient.[81] African Americans suffer a disproportionate burden of morbidity and mortality for an array of cardiovascular conditions—most notably hypertension.[82] Such conditions have an array of contributing factors, environmental, social, and genetic. Addressing facts such as racism-induced stress, environmental insults resulting from segregated housing, lack of access to healthy food options, or safe spaces for exercise all demand the sort of state-centered social policy that was anathema to Secretary Thompson and others in the Bush administration. Developing a drug to address the problem, however, held great appeal. Such a solution involved market incentives and focused on individual choices, corporate marketing, and purported race-based biological difference that located the problem within Black bodies rather than a racist society.

The FDA approved the drug in 2005, but its story began back in the 1980s when University of Minnesota cardiologist Jay Cohn had the idea that two generic drugs then available for treating hypertension, hydralazine and isosorbide dinitrate, if given together might be beneficial for people suffering from heart failure. With support from the U.S. Veterans Administration, Cohn conducted two, small clinical trials during the 1980s to test his theory. The first trial compared patients on the hydralazine/isosorbide dinitrate (H/I) combination against one group of patients taking a drug called prazosin and one group taking a placebo. The initial results looked very promising and so Cohn conducted a second, somewhat larger trial, comparing the H/I combination against a drug called enalapril, which was from a class known as ACE inhibitors that were then being recognized as a new, frontline therapy for heart failure. The results from this trial were somewhat equivocal, with the H/I combination showing no clear advantage over enalapril. Because enalapril and H/I had different mechanisms of action, Cohn thought combining them might be particularly advantageous and so he sought to conduct a third trial to compare enalapril alone versus enalapril plus the H/I combination. By this time, however, the interest and funding from the Veterans Administration had dried up and Cohn could find no interested corporate sponsor because the H/I drugs were generic, i.e., they had no patent protection. Undeterred, in 1987 Cohn applied for and received what is known as a "methods" patent on the method of coadministering the H/I drugs specifically to treat heart failure. With this patent in hand, he was able to convince a biotech company, Medco, not to fund a third clinical trial (very expensive), but to undertake the work to combine the two generics into a single pill (later known as BiDil) and lay the groundwork for bringing the drug to the FDA for formal approval as a therapeutic to treat heart failure based on data from the earlier trials.

By then, numerous articles had been published about the Veterans Administration-sponsored H/I drug trials and awareness about the viability of using the existing generics to treat heart failure was becoming more widely known. As the FDA review panel that first considered BiDil in 1997 would note, the primary innovation of BiDil was one of convenience—both in terms of combining two pills into one and having a new product that a corporation would have an incentive to market to doctors. That same FDA panel ultimately rejected the BiDil application, not because it thought the drug did not work, but because the data Medco used to support its application (having been drawn from the two Veterans Administration trials of the 1980s) was inadequate. These trials were not designed as new drug trials and so did not have the sorts of statistical controls in place that the FDA liked to see. In short, the data from these trials was not bad but it was a mess as far as the FDA biostatisticians were concerned. They suggested conducting a well-controlled follow-up trial, which they were confident would provide a basis for approval. Medco, however, had no interest in funding such an expensive endeavor and got out of the BiDil business, letting the rights to the drug revert to Cohn.

At this point, race had yet to become a part of this story. None of the many published articles on BiDil mentioned race, nor was it brought up in the materials or hearing before the FDA. Rather, from its inception, BiDil had been conceived as a drug for everyone, regardless of race. But, after the FDA denied Medco's application in 1997, something happened. By this point half of the twenty-year life of Cohn's original methods patent had elapsed. Designing and conducting a follow-up trial and presenting the data yet again to the FDA would likely eat up a significant portion of the patent's remaining ten years. Under such conditions, there was no way Cohn could enlist another corporate sponsor to bring BiDil to market. Faced with this impasse, Cohn went back to the original trial data from the 1980s and for the first time broke it out by race. Cohn then published a paper with Peter Carson arguing that the H/I combination in BiDil worked better in African Americans than in Whites. This paper became the basis for a new twenty-year, race-specific patent to BiDil as a method for treating heart failure in African Americans that was issued in 2000. This new patent was identical to the first in many respects, except in its claim to have unexpectedly discovered an even greater efficacy for "Black" patients. Cohn relicensed the intellectual property rights for BiDil to NitroMed, a Boston-area start-up firm with no other products yet on the market. Cohn and Carson's paper also became the basis for reapproaching the FDA, which issued a letter to NitroMed in 2001 stating that, based on the earlier data from the 1980s, it would be amenable to reviewing an amended new drug application for BiDil pending the completion of a confirmatory trial in an African American population. To no one's surprise, this trial, the African American Heart Failure trial, known as A-HeFT, had positive results. What it did not show, however, was that BiDil worked any differently or better in Black patients than in anyone else. This, for the simple reason that the trial enrolled only African Americans and so could not make any claims about its comparative efficacy. At the time and for years afterward, Cohn would freely assert that he believed BiDil worked in people regardless of race, and that he himself readily prescribed it to his White patients. In short, the reason for marketing BiDil as race-specific had much more to do with extending its patent life for commercial benefit than with any underlying medical or biological rationale.

By the time the FDA approved BiDil in 2005, with its race-specific indication, it was, as an institution, far more attuned to general issues of racial representation in drug development—if somewhat oblivious to the dangers of reifying race as genetic in biomedical contexts. Shortly after approving BiDil, the FDA issued a guidance entitled, "Collection of Race and Ethnicity Data in Clinical Trials," that recommended a standardized approach for collecting and reporting race and ethnicity information in clinical trials that produce data for applications to the FDA for drug approval. The guidance explicitly referenced BiDil as a positive example of how to use race in drug development trials.[83]

In a testament to the complexity of the racial politics of biology, many African American advocacy groups, most notably the NAACP and the Congressional Black Caucus (both aided by financial support from NitroMed), strongly supported the approval of BiDil. In many respects, this was quite understandable. Here finally, after centuries of exploitation by the medical establishment, was an intervention designed actually to benefit African Americans. Yet inextricably bound up with its approval was the message that Black people were somehow biologically different from other groups. With such spurious imputations of biological difference invariably come implied hierarchies. The FDA's logic for granting the race-specific approval was simply that because the drug had only been tested in Black people that must be the group for which it was approved. But to this point, most drugs then currently on the market had also been approved based on data from race-specific trials—trials conducted overwhelmingly on White subjects. When the subjects were White, the FDA did not see their race. It was an unmarked norm. A drug tested on White people was considered good enough for everybody because the category "White" was understood as coterminous with the category "human." It was only when the test subjects were Black that the FDA approved a limiting race-specific indication for the drug. In the case of BiDil the unintended yet undeniable message of the FDA approval was, therefore, that Black people were somehow less fully representative of humanity than were White people.

Notably, as the concept of diversity reached perhaps its apogee of social and legal salience in the aftermath of the *Grutter* decision, it was largely absent from these debates. Indeed, with its single-race clinical trial, BiDil's framing was, in some respects, antithetical to the idea of diversity in biomedicine. Perhaps this was due in part to a reluctance among genetic researchers to highlight the term in the aftermath of the political backlash against the Human Genome Diversity Project. But as diversity receded, race more explicitly came to the fore. But bear in mind, in the case of BiDil, it was primarily legal and commercial considerations that drove its racial framing. Even when diversity was not explicitly a part of the picture, the story of BiDil shows how such incentives could provide a strong impetus toward reentangling social and biological constructions of race.

Ironically, in the aftermath of President Clinton's declaration of the genetic unity of the humanity, explicit references to racial difference increasingly came to the fore in discussions of representation and difference in biomedical research and clinical practice during the early 2000s. Immediately following the completion of the first draft of the human genome studies, a social constructionist approach to race and genetics predominated in scientific studies. But as Dorothy Roberts and Oliver Rollins have noted, by 2003 "this trend quickly reversed . . . as the use of biological notions of race increased dramatically."[84] BiDil was perhaps one of the most prominent examples of this trend, but it manifested broadly throughout studies of biomedical research and clinical practice.

CHAPTER SIX

Getting Bodies

When Diversity Didn't Matter

The trope of diversity was also notably absent from one of the other major genomic enterprises of the mid-2000s—the drive to enroll large numbers of volunteers into massive new genetic databases to propel the genomic enterprise to its next stage of discovery and application. At this time, the Common Disease/Common Variant (CD/CV) hypothesis was still dominant. Under this model the main challenge was to identify genetic variants that underlay common diseases, such as diabetes. If those variants were common to all carriers of the phenotype (e.g., diabetes) then simply enrolling massive numbers of human subjects into large-scale population studies (LPSs) was logical. In this context, diversity did not matter, bodies did.[1] The absence of diversity at this stage matters because in the 2010s we will see diversity reenter the scene of recruitment for genomic research with a vengeance. The contrast of later efforts to recruit diverse cohorts into genomic databases with these earlier efforts critically illuminates the different meanings and uses of diversity as a driver of the biomedical enterprise.

Genome Wide Association Studies (GWAS) became prominent during this time as a method to capitalize on such data. GWAS were large-scale studies that genotyped large numbers of samples at thousands or even millions of sites across the genome to find associations between common diseases and specific genetic markers.[2] GWAS held great promise and garnered much attention during the 2000s. They became one of the primary avenues put forward to realize the varied promises of genomics declared at President Clinton's grand White House ceremony in 2000. But as GWAS studies proliferated during the 2000s they were not followed by attendant advances in diagnostics and therapies, although research continued to argue that such studies would sooner or later lead to medical breakthroughs.[3]

As promised genomic miracles failed to materialize in the aftermath of the completion of the first draft of the human genome, genomic research was increasingly perceived as reaching an impasse. Biomedical researchers largely agreed that one critical thing was essential to propel genomics into the future and maintain its legitimacy—more bodies. The Human Genome Project had focused on understanding the structure of the genome. Now efforts shifted to developing a functional understanding of how different genes worked. This required the massive recruitment of subjects to participate in biomedical research. With the urgent focus on recruitment, little attention was paid to diversity. More bodies, wherever they came from, were essential to enter the next stage of promised critical breakthroughs.

GETTING BODIES

The original plan for the first five years of the HGP stated: "The information generated by the human genome project is expected to be the source book for biomedical science in the 21st century and will be of immense benefit to the field of medicine. It will help us to understand and eventually treat many of the more than 4000 genetic diseases that afflict mankind, as well as the many multifactorial diseases in which genetic predisposition plays an important role."[4] The HGP itself formally came to a close in April 2003.[5] The problem was that while a remarkable technical achievement, by 2003 the HGP had yet to make significant progress toward realizing its original great promise of finding new cures for disease. Thus, fast on the heels of the completion of the HGP, researchers, clinicians, and prominent government actors were calling for new initiatives to continue the march toward the promised land of genomic medicine.[6]

In 2004, Francis Collins, then-director of the National Human Genome Research Institute (NHGRI) (later to be elevated to director of the National Institutes of Health (NIH) under President Obama), reported on a December 2003 meeting at the NIH discussing the need for a massive longitudinal study of up to 200,000 people to further develop information needed to treat disease. Collins noted that if the experts agreed on the need for such a population-based cohort study, "then we must collectively seek ways to organize and implement it quickly and efficiently—or face the real possibility that a decade from now the promise of genetic and environmental research for reducing disease burden on a population basis will remain out of reach."[7] Collins was arguing that to truly realize the promise of genetic research, what was really needed was bodies, lots of bodies, at least 200,000 bodies for the one study he was discussing in this paper. Bodies were so important that failure to get them could place the entire genomic enterprise, begun by the HGP, at risk.

Here, at the outset, Collins placed a certain type of risk at the center of his call for genomic innovation: a risk of lost potential and lost hopes. Collins also bound together recruitment, risk, and potential in manner that placed the obligation for realizing continued progress on forces outside of the scientific enterprise of genomics itself into the social and political world where responsibility for the logistics of realizing such a large-scale population cohort study would inevitably fall. What he did not do was center diversity or race. His focus was simply on bodies.

Two years later when further making the case for large-scale prospective cohort studies, Collins and others asserted that:

> The sequencing of the human genome and increased investigation of its function are providing powerful research tools for identifying genetic variants that contribute to common diseases. Recognition is growing, however, that genetic variants alone cannot account for most cases of chronic disease. It is far more likely that environmental and behavioural changes, in interaction with a genetic predisposition, have produced most of the recent increases in chronic disease, and might therefore be the key to reversing this trend.[8]

The HGP was all well and good, but it was conducted primarily in laboratories and involved deriving the complete sequence for only one prototypical genome. Identifying genes alone, it turned out, did not carry us very far down the road toward the promised land of genomic medicine. The thinking now was that to fully understand the complexity of the gene-environment interactions that contributed to most common diseases, scientists would need to study large numbers of genomes in context in order to pick up the many small effects that cumulatively might play a significant role in determining health outcomes.

Getting bodies was complicated. There were already multiple population cohort studies that had been ongoing in the United States for many years. Prominent among these were the Framingham Heart Study and the Jackson Heart Study. The Framingham Study originally recruited 5,209 White men and women between the ages of thirty and sixty-two from the town of Framingham, Massachusetts, and tracked them for decades to study common patterns related to the development of cardiovascular disease. The study enrolled new generations of participants in 1971 and again 2002.[9] The Jackson Study began in 1999 and recruited 5,301 African Americans in and around Jackson, Mississippi. It focused on identifying factors that contribute to the much higher incidence of cardiovascular disease in African American populations.[10]

These significant studies produced much valuable information but operated at a much smaller scale than that contemplated earlier by Collins. Later in 2004, an NIH panel of experts would recommend that a population of 500,000 would be optimal for the sort of large-scale prospective study discussed by Collins.[11] This marked an entirely different scale of research. In ambition it was more akin to one of the national DNA biobanks being developed by such countries as Great Britain, Japan, and Estonia, among others.[12] Accordingly, the NHGRI's plans for a prospective large population-based study (LPS) demanded extensive consideration of both the logistics and ethics of recruitment. Tropes of diversity made only a passing appearance in the recommendation, not in terms of recruitment or representativeness but in expressing the hope to "identify the causes of common diseases across a diversity of ages, geographic regions, and ethnicities."[13]

A focus on recruitment clearly animated the NHGRI's decision, in 2006, to award $2 million to the Genetics and Public Policy Center (GPPC) at Johns Hopkins University to engage in a two-year cooperative Public Consultation Project (PCP) on public attitudes toward participating in a possible large-cohort study of genetic and environmental contributors to health. Dubbed, "Making Every Voice Count: Public Consultation on Genetics, Environment, and Health," the project came to involve focus groups, community leader interviews, and "town halls" in five U.S. cities, as well as a 4,000-person national survey. It also developed educational materials to provide background information for the various targets of engagement.[14] In the years that followed, the GPPC, under the direction of Kathy Hudson, came to play a central role in furthering the federal government's efforts to get bodies for an LPS.

Increasing pressure to enroll participants throughout biomedical research has brought into being what sociologist Steven Epstein has characterized as "a new science . . . that might be called 'recruitmentology.' " While applicable to a broad array of recruitment practices and actors for diverse projects (ranging from small observational studies at academic health centers to large, multinational clinical trials for drug development), Epstein notes that "practitioners of recruitmentology seek to produce and disseminate knowledge about how to successfully recruit and retain participants." In contrast to the science of clinical trials, which evaluates the efficacy of therapies, Epstein posits that the "science of recruitmentology evaluates the efficacy of techniques necessary to get bodies into a trial in the first place, and to keep them there throughout the life of the experiment."[15] The GPPC efforts on behalf of the NHGRI in laying the groundwork for an LPS fall squarely under the rubric of recruitmentology. For the GPPC, identifying "the best recruitment strategies" was a central concern of its PCP.[16]

The NHGRI came to the GPPC because prior to undertaking an initiative as massive as a 500,000 person LPS, it understandably wanted to gauge public attitudes about and willingness to participate in such a project. The PCP involved focus groups and town halls conducted in diverse locations across the country. At sixteen focus groups in six cities, GPPC representatives showed participants a video explaining the proposed LPS, and then discussed whether the study should be done and what factors would influence their willingness to participate. Following the focus groups, the GPPC conducted twenty-seven individual interviews about the proposed study with community leaders in the same locations. The GPPC used the information derived from the focus groups and interviews to shape the subsequent town hall meetings and the design of a national survey.[17]

The GPPC held five town hall meetings in 2008 in Jackson, Mississippi; Kansas City, Missouri; Philadelphia, Pennsylvania; Phoenix, Arizona; and Portland, Oregon (the same cities where it conducted the focus groups). The meetings had audiences ranging from seventy-six to 134 people. The GPPC made efforts to have the participants roughly match the demographics of their local communities. Each meeting began with a presentation by a "senior member" of the GPPC staff who welcomed the participants and explained that the Public Consultation Project was hoping to "gather feedback on a proposed larger-cohort government study of genes, environment, and health."[18] The proceedings were then turned over to be moderated by Jonathan Ortmans of the Public Forum Institute, which describes itself as "an independent, nonpartisan, not-for-profit organization . . . [that] helps policymakers make sound public policy decisions to encourage entrepreneurship, job creation and economic growth."[19]

The moderator framed the event with the following three sets of questions:

1. Do you think the government should create a national biobank? Why or why not?
2. Would you participate in such a biobank? Why or why not?
3. What conditions need to be in place in order for the biobank to happen?[20]

Participants were then shown a nine-minute video discussing genetic variation and its possible contribution to disease. The video also described how the project planned to collect genetic samples and data about medical history, diet, lifestyle, and environmental exposures from up to 500,000 U.S. residents. It informed viewers that researchers, both public and private, would have access to this information to study how genes, environment, and lifestyle contribute to disease.[21] During the course of the video, a female narrator stated:

> No program large enough to do this has ever been done in the United States. But other countries have begun studying

gene-environment interactions. Many people in these counties have given permission to researchers to take genetic samples . . . including Great Britain, Iceland, Estonia, Japan, and Canada. Here in the United States the National Institutes of Health and other federal health agencies have a similar project in mind but it has not yet been approved or funded. NIH would like to get a lot of public input before going ahead to make sure US citizens are comfortable with the project, have say in how it's run and are willing to participate in it.[22]

There are a few noteworthy aspects to this particular framing. First, in referring to existing programs in other countries, the video clearly gave the impression that this was an increasingly common practice, implying that as others had given permission, so too would it be reasonable for these participants to give their permission. Second, in the reference to other *national* projects, as opposed to private ones (for example, projects then being developed at the Mayo Clinic or Vanderbilt University) there was also, perhaps, an implicit call to patriotism (at best) or (less positively) to a jingoistic concern about the United States being left behind in the march of biomedical progress.[23] Third, the stated concerns to make sure citizens were "comfortable, . . . have a say . . . and are willing to participate," while reasonable, may also be problematic. Comfort and having a say are well and good but they seemed clearly geared toward actualizing the third concern—i.e., ensuring a willingness to participate.

The focus groups and town halls certainly allowed for a measure of citizen feedback that could shape how the project was to be carried out. Having a say implied a measure of power, but the town halls did not appear to have provided any ongoing mechanism for citizens to exercise any substantive control over how the project would be carried out. They "have a say" in the recruitment process but not in the substance of the project itself. Thus, "having a say" appears to be part of the preliminary process of making potential subjects "comfortable" so that they will be more willing to participate. This fits squarely within a "subgenre" of recruitmentology identified by Epstein as seeking "to determine the barriers that keep individuals from volunteering."[24]

NHGRI Director Francis Collins also made an appearance in the video to declare that a big study of hundreds of thousands would "really

Getting Bodies 145

give us answers" and serve as "a discovery engine for everything we need to know about medicines in the future."[25] Collins's enthusiasm is understandable. In many respects he was simply following up on the statements made at the White House ceremony announcing the completion of the first draft of the human genome. One problem, perhaps, is that those earlier statements were made at the turn of the millennium and here, eight years later, Collins was still looking to the future and calling for another new venture to keep us on the path toward that ever-receding horizon of biomedical promise.

Collins's remarks also differed from those earlier pronouncements in two distinct ways. First, they were made to propel forward a nascent project rather than celebrate the fruition of an existing one. Second, they were directed at potential research subjects. Such remarks functioned very differently in the context of recruitment. The difference may be understood, in part, by considering the relation between invoking the "promise" of genomics and invoking its "potential."[26] In his exemplary ethnography of the ventures of deCODE Genetics in Iceland and the related creation of Iceland's biobank (alluded to in the PCP presentation), Mike Fortun notes that "the language of promising is a diverse and intricate one, demanding equally diverse and intricate analyses," and that "promising is an ineradicable feature of genomics." This is so, largely because the complexity of contemporary genomics, involving whole genome scans of large populations looking at complex interactions of genes, behavior, and environment was full of uncertainty and contingencies so that results could only be promised, not directly forecast. Making such promises, as Fortun notes, "entails a mixture of a high degree of speculation, an avowed commitment stemming from multiple insecure extrapolations, and bets or gambles placed with a combination of care and risk."[27]

Promises can be influential and do a lot of work in enlisting resources and propelling an enterprise forward. Clinton, Blair, and Venter were largely celebrating the "promise" of genomics in marking the completion of the first draft of the human genome. Generally speaking, once made, a promise does not further call upon the promisee to be realized. Promise and potential may often be mixed—"if you fund us, we will have the potential to cure cancer!"—but when they are, the promise takes on different valences. Potential, with its connotations of latent power, may require ongoing action from the promisee to be realized. It therefore may make different sorts of demands than a simple promise.

Genomic promises are often really about potential—about unlocking the latent power of information stored in the genome to enable us to cure disease. In the context of developing an LPS, diverse actors invoked this potential to make demands on fellow citizens—in particular, to demand their participation in research to cure disease. In this context, a decision not to participate was not simply turning down a proposed bargain as in a promissory contract ("if you give me this, I promise to give you that"), it was thwarting scientific progress itself ("if you don't participate, we cannot realize the potential of genomics to cure disease"). This was evident in the PCP presentation and related endeavors to recruit subjects into a massive, new, federally sponsored LPS and related enterprises.

Collins's statement in the PCP video presented the idea that participating in an LPS would provide researchers with "a discovery engine for everything we need to know about medicines in the future."[28] This was not a general declaration about the importance of some abstract LPS; it was part of a specific appeal to recruit subjects to participate. It was telling them that their bodies were needed to realize this vision. In a similar vein, the GPPC's flyer describing the Public Consultation Project opened with the bolded question to the prospective participant: "Would you volunteer to help solve medical mysteries?"[29] By framing the project as a call actively to "help" it becomes clear that this was not just about promising future medical advance, it was making a claim upon individuals to participate, with the unstated message being that failure to do so placed the realization of the biomedical potential of the LPS at risk.

"Are you," the flyer continued, "and a half-million of your fellow citizens ready and willing to volunteer...? That question, as well as what incentives would encourage study participation and what concerns people might have, is at the heart of the Genetics & Public Policy Center's Public Consultation Project."[30] The invocation of citizenship is particularly striking, connoting, as it does, conceptions of duty and service to a public good. The GPPC was quite clear, however, that it was also looking for "incentives" to encourage participation, indicating an understanding that a sort of quid pro quo might provide a useful complement to claims of civic duty. More fundamentally, this and other statements about the PCP indicate a fundamental tension within its framing and goals.

The GPPC characterized the PCP as an exercise in "deliberative democracy."[31] The concept of deliberative democracy has been the subject of extensive study and discussion. It may not be readily susceptible to one set definition.[32] But if democracy of any sort involves popular power and control, the PCP conferred little of this upon participants. The focus groups, town halls, and survey were less about seeking public input in order to determine *whether* a federally sponsored LPS should go forward, *to what ends* it might be directed, or *how* it might be pursued, than it was about gathering input on how best to frame a pitch to potential subjects in order to more effectively recruit them into participating in the project. The moderator followed a set script that imposed a clear structure for discussion. The preset agenda specified limited periods of discussion for pre-chosen topics—including "Initial Impressions" (fifteen minutes); "Benefits and Burdens" (thirty minutes); "Acceptable and Unacceptable Types of Research" (ten minutes); and "Return of Results" (ten minutes). The final section was titled "Build your own Contract," a twenty-minute period during which participants were to list "elements that should be included in research agreements between the researchers and study participants for the proposed study."[33]

In this context, "deliberation" was not directed to coming to an informed, mutually agreed upon course of action that empowered participants, it was being used as tool to elicit information from subjects about what "incentives" might encourage them to participate in an LPS, which in turn would then be used to elicit even greater amounts of (genetic) information from them (or other prospective participants in an LPS). The structuring frame of such discussions was not "should we do this?" or "how should we do this?" but "what do we need to do to get you to participate?" These questions, of course, are not mutually exclusive, but the process did not foster any questioning of the underlying enterprise, merely a consideration of how to make recruitment to it more effective. It was, as terms such as "focus group" might indicate, more an exercise in marketing research than democracy.

Such an approach was not unique to the GPPC's and NHGRI's plans for large-cohort genomic studies. In his study of similar focus groups conducted as part of the UK Biobank project, Richard Tutton discusses what he terms "the discourse of participation" in such projects, which he suggests "can be seen as an institutional response to public ambivalence toward science and expertise."[34] The idea of seeking "participants" in research, Tutton observes, is a distinctly recent phenomenon. In the past, subjects were viewed as passive or expendable. In the aftermath of the atrocities of World War II, and with the rise of modern bioethics, they became understood as vulnerable and in need of ethical protection. In the context of biobank recruitment, subjects were increasingly being cast as "empowered citizens"—empowered largely with information that enabled them to make free, informed, and rational decisions. Tutton argues, however, that in practice, the discourse of participation was used by the institutions behind UK Biobank to enact a constrained and impoverished model of participation that was "largely confined to providing samples and data to the project, with the likelihood of receiving some general feedback about the progress and key findings of the research in the future."[35]

Nonetheless, alternative approaches have also been tried. Researchers in Canada devoted considerable attention to developing models of deliberative democracy for engaging the public around the creation of the "BC Biolibrary" which was established in 2007 "to support biobanking and a broad range of health research applications that utilize biospecimens in British Columbia, Canada."[36] In contrast to the GPPC's town halls, which conceived of public engagement primarily as a means to overcome barriers to recruitment, researchers at the W. Maurice Young Centre for Applied Ethics, University of British Columbia, clearly stated:

> When talking about public engagement, we are not referring to unidirectional attempts to increase public awareness of certain aspects of science and technology; nor are we referring to the measurement of "public opinion" on certain controversial issues. Rather, we are concerned with mechanisms whereby there can be meaningful and legitimate public input into policy that involves dialogue between relevant publics with scientists, policy makers, and other stakeholders.[37]

To act on this vision, the BC Biolibrary project took a very different approach to public engagement. It opted for one single group of twenty-five Canadians, chosen to represent "the diversity of values, life experiences, and discursive styles of the citizens of British Columbia," and selected also to give "voice to individuals and groups that would otherwise not be heard." The enterprise centered "diversity" and "representation." While framed in terms of "values, experiences, and discursive styles," the idea was to try to capture a sample that would represent concerns and interests of the broader community. The project invoked diversity as a social and political concept. It made no connection between such diversity and an underlying search to capture any sort of genetic diversity in their samples.

In contrast to the PCP's nine-minute introductory video, BC Biolibrary prepared a workbook for participants, outlining key areas of ethical concern including: Collection of Biospecimens; Initial contact/Introducing the Biobank to potential donors; Linking Samples to Personal Information; Consent; and Governance of Biospecimens and Associated Data.[38] In perhaps the most significant difference from the PCP, participants met over the course of four days (instead of the PCP's three hours) to discuss the issues in depth. Discussion facilitators "gave particular attention to ensuring that all voices were heard and no views glossed over in the formulation of final recommendations."[39]

This last concern to respect participants' voices, while seemingly self-evident, gains salience in light of the observations made by anthropologist Karen-Sue Taussig while at one of the actual GPPC town hall events.[40] Taussig spoke of being haunted by the interactions between a participant and the moderator at a town hall in Portland, Oregon. The exchange began following the showing of an introductory video when the participant, whom Taussig called "Sally," asked, "If we participate in this study and you find out we have the breast cancer gene, are you going to tell us?" The moderator deflected the question because it did not fit in with the preset agenda, saying, "That is a really good question and it is one we are going to come to later. Right now we'd like to hear, based on what you know, do you think the study should be done?" About twenty minutes later the moderator turned to the question of return of results but asked the participants to discuss what kind of research they think people should be able to do, or not do, with materials in a biobank. People began generating a list and Sally, whose hand had been raised, said, "So, if you have the breast cancer gene are they going to tell you or not?" The moderator responded, saying, "I'll come to that in a minute," and continued to call on people to add to the list he was generating. More discussion ensued and a while later Sally once again raised her hand and asked, "Couldn't you pay a little extra to get your results?" Taussig noted that at that point, "the moderator turns to look at us with an expression that we read as meaning something like 'can you believe she's asking about this again?' The response was laughter from the rest of the audience. Nonetheless, the moderator did generate some discussion about the issue of getting results back." At the end of the discussion, the moderator asked the participants to vote yes or no about whether researchers should try to return relevant information to biobank participants. Taussig relates that at that point she looked over to Sally's table and saw that she was gone.[41]

The story is haunting, in part, because it seems to conflict so starkly with the GPPCs avowed purpose of "making every voice count" in the public consultation process. It also highlights the degree to which the process was not structured as a true dialogue but as a means to elicit information from the participants. This brings us back to Tutton's observation that discourses of participation and consultation may primarily be serving the ends of recruitment rather than empowerment. Such engagement as that provided by the PCP, however, was only a first step toward recruitment. One of the primary goals of the consultation was to identify concerns and fears that might be acting as barriers to participation. As it turned out, foremost among these were concerns were over loss of privacy and relatedly "the possibility that insurance companies might obtain individuals' genetic information and use it against them." Many participants also expressed a strong desire for research results; that is to have relevant information discovered from their samples returned to them. GPPC researchers found this particularly noteworthy in light of the fact that they had the moderator explain to the participants that "individual research results are not usually returned to study participants because of logistical burdens."[42]

Getting Bodies 151

That the moderator provided no similar intervention regarding the burdens of protecting privacy seems to indicate an implicit understanding on the part of those framing the discussion (i.e., the GPPC) that concerns over privacy were somehow more legitimate and addressable than those involving return of research results (ROR). With respect to privacy concerns, the introductory video made clear that participants' information would be "coded to hide their identities," but made no mention of ROR or other concerns, such as commercialization of research results.[43] The parallel GPPC online survey of 4,659 people similarly found strong interest in ROR and observed: "providing individual research results is a strong motivation to participate; compensating participants $200 may increase participation a similar amount. Incentives, recruitment, and return of results could be tailored to demographics groups' interests." In analyzing the costs and benefits of such measures, the report concluded that "providing even limited individual research results or graduated incentives over time could increase retention and recruitment."[44]

Notably, ROR could impose substantive burdens and responsibilities on researchers involving the actual content of the information they were deriving from the samples. In contrast, protecting privacy primarily imposed procedural burdens concerning management of and access to information. These were real burdens, but they did not materially implicate the way research itself was conducted. That is, ROR involved concerns about what researchers themselves were doing with the data, whereas privacy concerns primarily involved insuring that *other people*, e.g., nonresearchers, not have access to the data. Perhaps more to the point, concerns over privacy were soon to be addressed by legislation that imposed little burden on researchers but promised greatly to alleviate concerns over privacy as a potential barrier to recruitment.

GINA: LAW AS AN ADJUNCT TO RECRUITMENT

The 2008 GPPC report on the town halls highlighted the passage of the Genetic Information Nondiscrimination Act (GINA) in May of that year.[45] The law had two major components: Title I, which prohibited group and individual health insurers from using a person's genetic information in determining eligibility or premiums and from requesting or requiring that a person undergo a genetic test; and Title II, which prohibited employers from using a person's genetic information in making employment decisions, such as hiring, firing, job assignments, or any other terms of employment and from requesting, requiring, or purchasing genetic information about persons or their family members. GINA did not cover life, disability, or long-term care insurance or veterans seeking health care through the Department of Veterans Affairs.[46]

GINA had its roots in legislation introduced thirteen years earlier by Representative Louise Slaughter (D-NY) which garnered bipartisan support but did not pass. Similar legislation was introduced during subsequent congressional sessions as the HGP moved toward its completion in 2003. Between 1996 and 2002 the Senate Committee on Health, Education, Labor and Pensions held five hearings on genetic discrimination, but no progress was made toward enacting specific legislation. In 2003, just as the HGP was coming to a close, Representative Slaughter and Senator Olympia Snowe introduced the first bill with the title, "Genetic Information Non-Discrimination Act." Their efforts continued during subsequent sessions of Congress, but it would take five more years to achieve final passage.[47]

While always animated by concerns to ensure that advances in genetic technologies could be pursued productively without being used to discriminate unfairly against individuals on the basis of their genetic makeup, the need for GINA took on new valences as it moved toward ultimate passage in 2008. Roughly coincident with the rising interest in developing a federally sponsored LPS and the GPPCs efforts to gauge citizen attitudes toward participating in such a study, GINA's advocates began to emphasize more its potential to facilitate recruitment of subjects for biomedical research. Thus, the GPPC town hall report's reference to GINA was not incidental but central to emerging understandings of the significance of the legislation.

In his 2007 testimony on GINA before Congress, then-director of the NHGRI, Francis Collins, neatly exemplified the interweaving of fear as a barrier and a threat:

> We stand on the brink of a revolution in healthcare.... Yet, there is a cloud on the horizon and it is a cloud that has been getting

> darker and more frightening over the course of the last more than 12 years, since I have had the privilege of leading the genome effort and worrying about this issue, and that is that this kind of genetic information, as valuable as it is, might be used against people.... Unless Americans are convinced this information will not be used against them, this era of personalized medicine may never come to pass.[48]

Collins bracketed this statement with the idea of a coming revolution and a new era in genomics. These advances, however, lay on a clouded horizon off in the distance. At first, it seems that the clouds were the threat of discrimination. In fact, however, the way to clear the clouds was not by addressing discrimination per se, but rather by convincing Americans that they would not be discriminated against. These are two very different, though obviously related, things. The real cloud on the horizon, then, or the true threat to realizing the promise of genomics, was the barrier to participation in research erected by fears of discrimination. In her testimony before the same committee, Kathy Hudson was quite explicit about this: "Growing uncertainty and fear threaten the future of genomic medicine."[49]

Diversity was not central to this framing because it was not then understood as relevant to sustaining the claims-making efforts driving genomics forward. Yet, less than five years after the passage of GINA, when fear could no longer be so readily invoked as a barrier to progress, lack of diversity would reemerge as the primary obstacle to genomic progress. Researchers, that is, would find new uses for diversity as both an explanation for the failure to fully realize the promises of genomic medicine and as a frame to drive the expenditure of new efforts and resources on behalf of the genomic enterprise.

The GPPC clearly echoed this framing of GINA as it gauged public attitudes toward participation in large-scale genomic research. Thus, for example, in 2007 Kathy Hudson, then still director of the GPPC (and soon to become Francis Collins's deputy director for Science, Outreach, and Policy at the NIH), emphasized the importance of using law to overcome public fear that threatened to impede genomic progress:

> Without comprehensive legal protections, the public fears genetic discrimination, and that fear has negative effects on both medical research and clinical care. Today, genetics is incorporated into almost all areas of clinical research, and scientists are proposing massive population-based studies that will enable them to identify and distinguish genetic, environmental, and lifestyle-based contributors to disease. *But many potential research participants are deterred by the fear that their information could be used against them by employers or insurers.* . . . The nondiscrimination legislation under consideration *would allow researchers, for the first time, to assure participants* that it is simply against the law for health insurers or employers to use genetic information to discriminate against them.[50]

Here Hudson cast GINA as a recruitment tool that would allow researchers to increase participation by allaying fears about discrimination. Further, a GPPC "Discussion Guide for Clinicians," explicitly stated that discussing GINA "might help your patients feel more comfortable about . . . participating in genetic research."[51]

Ironically, while diversity was largely absent from these initiatives, intense debates were raging about the how or whether to address race in biomedicine more broadly as evidenced in the controversy over BiDil and the 2005 FDA guidance on the use of racial and ethnic categories in drug development. In those contexts, race was taking on salience both in social justice discourse over addressing health disparities and as a commercial opportunity for corporate interests. Such concerns as yet were largely absent from the arena of building massive biobanks.

ADDRESSING THE RECEDING HORIZON OF THE GENOMIC MILLENNIUM

As the sequencing of the first human genome did not in itself bring us to the promised land of genomic medicine, Collins and others identified bodies as the primary barrier to proceeding down this road of potential. Science needed massive numbers of bodies enrolled in LPSs to get the information necessary to achieve genomic breakthroughs. The barriers to recruitment were cast as social. Prominent among social barriers were ignorance, fear, and inertia. The GPPC town halls and surveys were designed to figure out ways to address popular ignorance and fear. Ignorance was to be addressed through education and outreach. Whatever fears were not addressed by education, GINA would resolve by assuring potential recruits that their genetic information could not be used to discriminate against them.

As all this was happening, there was an emerging awareness of a continued failure to realize the full extent of the grand promises of genomics. Brandeis Professor of Biochemistry Gregory Petsko noted at this time that Collins was "beset by people—in the U.S. Congress and from patient advocacy groups—who keep asking him, 'Where are all the cures you promised us?'"[52] A January 2011 *New York Times* article noted that "Dr. Collins has been predicting for years that gene sequencing will lead to a vast array of new treatments, but years of effort and tens of billions of dollars in financing by drug makers in gene-related research has largely been a bust." Collins responded by saying he was, "frustrated to see how many of the discoveries that do look as though they have therapeutic implications are waiting for the pharmaceutical industry to follow through with them."[53] Collins thus located the failure to realize the early potential of the HGP with industry, not science. Even in the face of a historical failure to realize the initial *promise* of the HGP, Collins still invoked the *potential* for science to develop new therapeutic interventions. Collins cast science as the realm of continued potential, demanding more support to be actualized, while he laid the unfulfilled promises of the HGP at the feet of market failure.

Diversity did not play a major role during this initial post-genomic period of building biobanks. The primary focus seemed simply to be on getting bodies—lots of bodies. Barriers to progress were understood as barriers to getting bodies. As these initiatives failed to bring us to the promised land, researchers and policymakers brought diversity back into the picture as a central tenet of genomic progress. Where during the 2000s access to bodies alone was put forth as the key to success, during the 2010s diversity would emerge as the sine qua non of unlocking the potential and realizing the promise of genomics.

As explored further in the next chapter, by 2011 a new conceptualization of risk was emerging—lack of diversity. It came to be seen, in fact, both as an existing barrier to genomic progress and a continuing risk, insofar as failure to adequately diversify clinical trials and biobanks going forward threatened to undermine the promise and potential of genomics. Thus, with the new decade came a new use for diversity: mobilizing continued support, both material and intellectual, for genomics while rationalizing its failure to fully deliver on the grandest of its promises.

CHAPTER SEVEN

Bringing Diversity Back In

The International HapMap Project, with its focus of capturing global genetic variation, was eventually subsumed into the 1000 Genomes Project, which began in 2008 with the goal of developing a comprehensive resource of human genetic variation across worldwide populations.[1] Like the HapMap project, it was more focused on "diversity" than the type of large-scale population studies called for by Collins and explored by the GPPC. By 2015, it had produced a comprehensive, open-access database of genetic variation from a relatively small sample of 2,504 individuals from twenty-six populations across the globe.[2] Shaped by the common disease/common variant hypothesis (CD/CV), the HapMap project had focused on finding only the most common markers of genetic variation, those present in at least five percent of the population. In contrast, the 1000 Genomes Project shifted to focus on rarer variants, occurring in only 1 percent of the population, or so. For this, in contrast to the 270 original HapMap samples, the organizers estimated they would need to fully sequence about 1,000 genomes.[3]

The year 2008 also saw the initiation of the Population Architecture Using Genomics and Epidemiology (PAGE) study. The founding premise of this study was to examine "ancestrally diverse populations within the United States."[4] One of the goals was to assess "the comparability of estimates of genetic risk among racial/ethnic groups."[5] In contrast to the 1000 Genomes Project, this "require[d] large studies comprised of cases and controls defined using identical criteria and sampled ideally from the same study population."[6] For example, one of the earliest publications from the PAGE study (in 2010) "examined genetic associations with 19 validated risk alleles for T2D [type 2 diabetes] in European American, African American, Latino, Japanese American, and Native Hawaiian T2D cases (n=6,142) and controls (n=7,403) from the population-based Multi-ethnic Cohort study (MEC)."[7] But the size of such early cohorts paled in comparison to later PAGE studies such as one published in 2022 that aimed to "identify shared associations for 25 quantitative traits in over 600,000 individuals from seven diverse ancestries."[8]

Even so, the original 2007 "Request for Applications" (RFA) to support grants for institutions under PAGE did not foreground diversity. Its declared purpose focused on providing support "for the investigation, in well-characterized population studies, of genetic variants identified as potentially causally associated with complex diseases in genome-wide association (GWA) and other genetic studies, with the aim of widespread sharing of the resulting population-based descriptive and association data to accelerate the understanding of genes related to complex diseases." Diversity made only a brief appearance well into the document where it simply stated that "applicants are encouraged to include in each study a broad range of participants approximating the diversity of the U.S. population on factors such as age, sex, race/ethnicity, socioeconomic status, U.S. geographic region, and urban/rural residence."[9] Notably, by the time of the second phase of PAGE, initiated in 2013, the "Request for Applications" was immediately framed in its opening sentence by a focus on diversity, stating: "The purpose of this funding opportunity is to provide support for investigators who are affiliated with existing population studies to sample and assess genomic variation from well-phenotyped individuals of diverse ancestry."[10] This change in tone and focus exemplified the larger shift back toward diversity in genomics during this period.

The search for rarer variants and the renewed attention to diversity was perhaps indicative of the limited progress of genomics at the time toward realizing its grand promises. By 2010, influential *New York Times* science reporter, Nicholas Wade was bemoaning the fact that "Ten years after President Bill Clinton announced that the first draft of the human genome was complete, medicine has yet to see any large part of the promised benefits." Wade went on to note that it was not that projects such as HapMap had not produced important information, but, he observed: "with most diseases, the common variants have turned out to explain just a fraction of the genetic risk. It now seems more likely that each common disease is mostly caused by large numbers of rare variants, ones too rare to have been cataloged by the HapMap."[11]

As indicated by Wade's reference to "rare variants," another factor driving renewed interest in diversity was the breakdown of the CD/CV hypothesis. Under this hypothesis if common complex diseases, such as diabetes, had common genetic underpinnings, then finding relevant alleles would naturally arise from analyzing large, undifferentiated arrays of genetic samples. "Common diseases" by definition occur widely across all groups, and presumably "common variants" occur similarly. As an article in *Nature News* noted in 2008, this hypothesis "powered the push into genome-wide association studies," yet by 2008 "that hypothesis [was] being questioned." The article went on to observe that even when GWAS were able to link genes to traits "both the individual and cumulative effects are disappointingly small and nowhere near enough to explain earlier estimates of heritability."[12]

As Professor of Genetics at Duke University David Goldstein noted at the time, "A lot of people are recognizing that screening for common variation has delivered less than we had hoped."[13] By 2013, Director of the Office of Public Health Genomics at the Centers for Disease Control and Prevention Muin Koury recognized:

> GWAS have many limitations, such as their inability to fully explain the genetic/familial risk of common diseases; the inability to assess rare genetic variants; the small effect sizes of most associations; the difficulty in figuring out true causal associations; and the poor ability of findings to predict disease risk. In addition, GWAS have not fully addressed interactions of genes with disease risk factors such as diet, environmental exposures and infectious diseases.[14]

This was not only a problem of scientific progress, it was also a problem of legitimation and accountability. The billions of dollars spent on the HGP and related programs were based on promises to deliver meaningful medical information. As Joseph Nadeau, a geneticist at Case Western Reserve University noted, "The reason for spending so much money was that the bulk of the heritability would be discovered."[15] As one promise faded, new promises based on different premises had to be made if support for the massive endeavors were to continue. This was where diversity came back into play.

As the CD/CV hypothesis failed to pan out, geneticists increasingly focused on the idea that many complex traits were driven by large numbers of rarer genetic variants, each with relatively small effects that cumulatively had significant impacts.[16] To make progress under this hypothesis, geneticists turned to diversity. As the International HapMap Consortium put it in 2010, "despite great progress in identifying genetic variants that influence human disease, most inherited risk remains unexplained. A more complete understanding requires genome-wide studies that fully examine less common alleles in populations with a wide range of ancestry."[17] Echoing these concerns, geneticists Jon McClellan and Mary-Claire King noted that "strong evidence suggests that rare mutations of severe effect are responsible for a substantial portion of complex human disease." They went on to argue that the new focus on rare alleles around this time marked a "major paradigm shift in human genetics." In place of the CV/CD hypothesis, they suggested a "Common Disease-Rare Alleles" hypothesis, observing that "Rare large-effect mutations are now recognized as causes of many different common medical conditions."[18]

Earlier campaigns to get bodies for massive genetic databases were premised on the CV/CD hypothesis, under which diversity was largely irrelevant. But with the shift to the Common Disease-Rare Alleles hypothesis came the idea that rare alleles of large effect might vary in frequency considerably across different populations. To increase the odds of finding such rare alleles one had to increase the diversity of the population databases. One idea here was that the "lack of replication of GWAS among populations could ... be due to differences in genetic architecture."[19] Concerns for genetic architecture hearken back to STRUCTURE and analyses of the distribution of genetic alleles across different marked population groupings. Diversifying reference databases thus became a means to salvage GWAS and sustain the promise of genomics (with its attendant demands for material resources).

Around this same time a similar paradigm shift was occurring in the related field of pharmacogenomics. Pharmacogenomics involves trying to tailor drugs to an individual's genetic profile. It represented one of the most powerfully envisioned applications of the new genetics after the completion of the first draft of the human genome. For every drug, there are some individuals who will respond better, worse, or not at all, depending in part on their genes. Some genes also play a substantial role in regulating how the body metabolizes particular drugs. Determining whether an individual would metabolize a drug more quickly or slowly than another due to his or her genetic profile may help to find the optimal dosage for a particular drug. Echoing some of the hype around GWAS (around the time that BiDil, the race-specific medicine for heart failure, was approved in 2005), one leading researcher at Roche Pharmaceuticals opined: "we're at the beginning right now, and it's [pharmacogenomics] going to absolutely be integrated into the entire drug discovery and development process in the future. Ten years from now, it will just be part of the woodwork"[20]

In the early to mid-2000s, most talk of pharmacogenomics was framed in terms of "personalized medicine"; using genomics to prescribe the "right drug at the right dose for the right person."[21] But by 2008, David Goldstein was arguing that "this talk about personalized risk profiles, using genetics, for most common diseases, and this talk about a whole flood of new drug targets. I think that that's now pretty clearly wishful thinking."[22] This did not mean the pharmacogenomics had failed or even that it was at the same sort of impasses as GWAS. What it did mean is that champions of personalized medicine began to reframe their claims, first in terms of race, then more generally and persistently in terms of diversity.

The shift to race as an interim proxy for personalized medicine was most evident in the case of BiDil. Despite the fact that the clinical trials for BiDil revealed nothing about the genetic mechanisms underlying its efficacy, the approval of the drug was taken by many as evidence of the continued relevance of using race as a genetic category. Even those who were wary of such genetic constructions of race touted BiDil as a sort of interim measure to get to an era of truly individualized pharmacogenomic medicine where the need to use racial categories would fade away.[23] Thus, for example, in the immediate run up to BiDil's approval, there was Lawrence Lesko of the FDA's Center for Drug Evaluation Research asserting that race-based medicine could be a stepping stone to the higher goal of "targeted treatment," i.e., treatment where race need no longer be taken into consideration.[24] Or, as two biomedical researchers put it the following year: "As the science of pharmacogenomics develops more accurate tools to identify the molecular underpinnings of drug response, the need for classification by race will be replaced by more accurate and specific identification of each individual person's likelihood of responding to a particular drug therapy."[25] And again, in 2010, we had NIH Director Francis Collins declaring that "Racial profiling in medicine, even if well-intentioned right now, should recede into the past as a murky, inaccurate, and potentially prejudicial surrogate for the real thing."[26]

As race was meant to fade away, diversity soon emerged to take its place at the forefront of personalized medicine; but along with the shift toward diversity came a recharacterization of "personalized medicine" as "precision medicine." The concept began to emerge around 2010 and came to prominence in 2011 with the publication of a report from the National Academies of Science that distinguished the two terms, noting that while both communicated an idea of individualized medicine, the latter was preferable because it conveyed the idea that genomic medicine would be more accurate for individuals overall, not that each specific individual would get individualized treatment. More problematically, "precision medicine" brought group-based differences back to the fore, and with the notion of group- (or population-) based differences the concept of diversity became central to framing continued efforts to develop pharmacogenomic interventions.[27]

The 2011 report by the National Research Council was called, *Toward Precision Medicine: Building a Knowledge Network for Biomedical Research and a new Taxonomy of Disease*.[28] As indicated from the subtitle, the report was conceived as an initiative to modernize the taxonomy of disease "on the basis molecular information such as causal genetic variants, rather than a symptom-based classification system."[29] The critical rhetorical shift, however, was from "personalized" to "precision" medicine. On its face, this might seem a fairly minor tweak, certainly with few implications for how to use race in biomedicine. However, as Charles Sawyer, the co-chair of the committee issuing the report, later noted, "We consciously spent a considerable amount of time on whether we were talking about personalized medicine or precision medicine.... With the term 'precision medicine,' we are trying to convey a more precise classification of disease into subgroups that in the past have been lumped together because there wasn't a clear way to discriminate between them."[30]

The report itself noted that while, technically, "personalized medicine" did not "literally mean the creation of drugs or medical devices that are unique to a patient," nonetheless, "this term is now widely used, including in advertisements for commercial products, and it is sometimes misinterpreted as implying that unique treatments can be designed for each individual." It posited that the term precision medicine is preferable because it avoids this confusion and would allow a clearer focus on the "the ability to classify individuals into subpopulations that differ in their susceptibility to a particular disease, in the biology and/or prognosis of those diseases they may develop, or in their response to a specific treatment."[31]

An extensive analysis of the rhetorical shift from personal to precision medicine published by Juengst et al. in 2016 explored its implications for reifying race as genetic. The authors cited an interview with one senior editor of a genomics journal who said that "the way I look at personalized medicine is whereby we can stratify patient groups respective of ancestry, ethnicity, into individuals who are more likely to respond using novel technologies."[32] More broadly, among the 143 interviews conducted with an array of parties involved in genomic medicine, Juengst et al. found that the "overwhelming sense among respondents was that genomic medicine would remain at the group level, stratified by empiric genomic disease risk associations and, to lesser degrees, generalizations from racial and ethnic ancestry." And finally, we have the chief medical officer of a gene sequencing company declaring: "We can't get personal in medicine. I can measure what's happening on an individual, but deciding what happens with them, they have to be in a subgroup." Individualized medicine here had given way to group medicine; and racial and ethnic categories remained central, perhaps even essential to this enterprise.[33]

The broader embrace of the rhetorical shift from a focus on individuals to groups was evident at the federal level in a 2016 blog post from Muin Khoury, the founding director of the CDC's Office of Public Health Genomics, where he stated:

> A strong rationale for the shift towards precision medicine was laid out in a 2011 report from the National Research Council (NRC).... The report expressed concern that the use of the term "personalized medicine" could imply that treatments and preventions are being developed uniquely for each individual. After all, isn't medicine supposed to be personalized to begin with? On the other hand, the focus of "precision medicine" is to explore how treatment or prevention approaches can be developed based on the combination of genetic, environmental, and social factors which could be targeted to individuals or populations.[34]

Khoury's embrace of precision medicine and its focus on groups is certainly understandable given his public health orientation. But the "populations" to which he was referring, were being actively constructed by newly emerging federal initiatives that were foregrounding race and diversity.

Reflecting on the implications of this rhetorical shift for public health genomics, Juengst et al. went on to note that to operationalize group-based approaches to precision medicine, its advocates had to "equate genetic health risk groups" with "the kinds of human groups of concern to public health officials and policy-makers: visible groups with names, locations, and legitimate claims on public resources"; and these "visible groups" typically are organized around "socially discernible lines" such as "family, ancestry, community identity, ethnicity, and race." Finally, they argued that as these group concepts were extrapolated into public health initiatives, "the concepts of race and ethnicity and their links to health disparities remain badly tangled with the logic of genomic risk stratification in precision medicine's promotional discourse and public health initiatives."[35] The move from the individual to the group was not simply a matter of creating neutral, genetically characterized, diverse "populations"; it was and continues to be, almost inexorably, a matter of race.

Precision medicine then was *not* about the individual—at least not primarily. It was not, as a matter of research protocols and study design, about "the right drug, at the right dose, for the right person." It was about classifying people into "subgroups" or "subpopulations." And therein lay the rub. Just what sort of "groups" or "populations" were they talking about here? Of course, ultimately, the goal was classification based on relevant genotype. But as with those who originally called the use of race in BiDil a "stepping stone" to move from phenotype to genotype, the foregrounding of group classification in precision paved the way for transforming such stepping stones into an institutionalized foundation for one of the largest biomedical research enterprises ever undertaken by the federal government—the Precision Medicine Initiative.

THE PRECISION MEDICINE INITIATIVE

In his 2015 State of the Union address, President Obama announced the launch of a new "Precision Medicine Initiative" (PMI) that aimed "to bring us closer to curing diseases like cancer and diabetes, and to give all of us access to the personalized information we need to keep ourselves and our families healthier." Ten days later, the White House committed $215 million to the PMI in its 2016 budget.[36] As the National Institutes of Health PMI working group later reported, the president's vision was, "to enhance innovation in biomedical research with the ultimate goal of moving the U.S. into an era where medical treatment can be tailored to each patient."[37] Co-chairing the working group was Kathy Hudson who, as leader of Johns Hopkins's Genetics and Public Policy Center, had previously played a prominent role in the efforts to survey and enroll participants in earlier large-scale population studies. By the time of the launch of the PMI she had moved on to become deputy director for science, outreach, and policy and the NIH.

At the heart of the PMI was a drive to create a massive database of volunteers who would supply their genetic and other health information. Initially dubbed the "PMI Cohort,"[38] this initiative has since been rechristened as the *All of Us* program. It initially described itself as, "creating a research community of one million people who will share their unique health data. This will include answering survey questions and sharing electronic health records (EHR). Some participants may also be asked to provide blood or urine samples. We'll ask you to answer more questions from time to time. It's up to you to decide how much information you want to share."[39]

Creating such a cohort involved many complex issues including privacy, consent, community participation and control, return of results to participants, commercialization of participant data, and so forth.[40] Of particular interest here is how the program aimed to construct and structure the million-person cohort itself. In contrast to earlier major initiatives to develop genetic databases, from its inception those overseeing the *All of Us* program emphasized the central importance of creating a "diverse" cohort.

While much of the public-facing publicity for the *All of Us* recruitment program emphasized the "individualized" aspect of precision medicine, it also stressed the need "to engage a community of participants that reflected the diversity of America."[41] What did diversity mean in this context? According to the *All of Us* materials, it meant people "from all ages, ethnicities, and walks of life" and "from all regions of the country." This approach made good sense for many reasons, not the least of which was that capturing such significant social and cultural variables as race, class, and environmental exposures can help us better understand persistent health disparities in America. The basic idea was that with the rich array of data, researchers would be better able to understand the complex interactions between genes and social/environmental factors that would provide deeper insight into the causes and progression of a wide range of diseases.

It is one thing to declare an interest in looking at interactions among genes, race, and environment, it is quite another to follow through in practice. Even as, over the past decade, biomedical researchers have declared an increasing interest in exploring the complexities of gene environment (GxE) interactions in relation to racial groups, in practice, the environment often falls out of the picture such that we are left with findings built around relations between genes and race that, once again, raise the dangerous specter of reifying race as a genetic category.

From the outset, we see in the *All of Us* recruitment materials a foregrounding of racial diversity over environment, class, or region. Thus, for example, the program produced ethnic-specific brochures targeted at "Asian American," "African American," and "Hispanic

American" communities.[42] Notably, the only thing differing across these brochures were the pictures of people serving as exemplars of potential volunteers. Perhaps even more notable is the absence of any brochures targeted specifically at "White" or "European" Americans, while there is one for the "General Community"—yet another example of the category "White" being used as an unmarked norm in biomedical research and practice. Nor were there brochures specific to any other identified "diverse" categories, such as region or age.

"DIVERSITY" AND THE REENTANGLEMENT OF RACE AND GENETICS

This brings us back to the concept of diversity and its deployment in the construction of precision medicine generally and the *All of Us* program in particular. What exactly were advocates of precision medicine talking about when they invoked diversity, and what were their rationales for structuring genomic enterprises around this concept? Broadly speaking there were four basic types of rationales for diversity: technical, medical, political, and commercial.

We see three of the rationales expressed concisely in the single assertion from leading NIH researchers that "Increased attention to diversity will increase the accuracy, utility and acceptability of using genomic information for clinical care."[43] The focus on accuracy goes to the idea of improving the technical aspects of genomic discovery; utility goes to serving the medical ends of improving health outcomes; and acceptability addresses political concerns that "underrepresented" groups have a seat at the table of genomic advancement, particularly as it relates to addressing persistent race-based health disparities.[44] The commercial rational for diversity is evident in the idea that one core goal of the PMI "is to reinvigorate stagnating therapeutic R&D," in part by using subgroup analysis to rescue otherwise failed drugs.[45] We also see this in statements from the FDA itself, which noted that the use of subgroup analysis and related clinical trial "enrichment" strategies "promises to enhance medical product development by improving the probability of success."[46]

Speaking to the question of diversity's meaning in 2014, sociologist Troy Duster noted:

There is yet a prior question of what is meant by "diversity," and on that matter, it is vital to be really clear on the substantive meaning of genetic diversity. If the goal is to capture genetic diversity, the strategy might aim to obtain samples from people who are presumably as genetically different from each other as possible. If that is the goal, then the researcher is simply trying to capture a wide range of specifiable variation. Here is where we must get to substance, because there are numerous dimensions and levels on which people can vary from each other genetically. Thus, the idea of a high degree of variation may not be meaningful because it is not likely to capture the type of genetic variation in which the researcher is most interested. Or conversely, such a strategy might capture a lot of variation in which the researcher has little interest, for example, variation of little apparent relevance to health outcomes.[47]

One group of leading researchers from the NIH defined "genomic diversity" simply as "variation in alleles and allele frequency" that was "inherent within and across human populations."[48] But clearly, when the *All of Us* materials referenced diversity they were speaking of something very different from simple genetic diversity. They did not have brochures directed at people defined by their genetic makeup but rather by their socially defined racial or ethnic groups. This, for the simple reason, so clearly identified earlier by Juengst et al., that state-sponsored public health initiatives, as a practical matter, are invariably going to be organized around socially identifiable "visible groups" such as race.[49]

The same group of NIH researchers who provided the above referenced definition of genomic diversity went on to explain that in the context of critiques of the "representativeness" of genomic databases and related research initiatives the term "diverse" could "refer to the characteristics related to peoples' ancestry and to the physical and social environments in which they live and receive health care."[50] Similarly, as one proposed "research roadmap" for the PMI put it, "A critical challenge . . . is to accurately identify medically relevant variation in the context of an ethnically and geographically diverse and admixed target population." Here we have the additions of physical and social environment and geography to the concept of diversity.[51]

With specific reference to building the cohort of volunteers to populate what would soon become known as the *All of Us* program, in 2015 the NIH's Precision Medicine working group argued that "The PMI cohort should broadly reflect the diversity of the U.S." and went on to define this diversity as follows:

> The U.S. is a diverse society. Americans comprise diverse social, racial/ethnic, and ancestral populations living in a variety of geographies, social environments, and economic circumstances. The U.S. population includes people with extreme wealth and people living in abject poverty and with varying access to social, educational, and other resources. . . .
>
> Racial and ethnic diversity do not just reflect genetic ancestry; they are social constructs rooted in cultural identity and shaped by historic and current events, which influence an individual's behavior, place of residence, and life opportunities.[52]

The working group provided an explicit caveat cautioning that racial and ethnic diversity were "social constructs" yet, nonetheless, it deemed them to "reflect genetic ancestry."

If we now connect discussions of the meaning of diversity back to the rationales for diversity, we see some potentially problematic ambiguity in terminology that presents a heightened danger of revitalizing misbegotten understandings of race as genetic. One of the critical rationales for this drive to diversify the cohort was not only to drive medical progress in general but to address social justice concerns of race-based health disparities in particular.[53] In this regard, the intermixing of social and genetic constructs might seem warranted insofar as health disparities are products of complex social and biological factors. On the one hand, this all seems a fairly straightforward approach to the wide array of variables that need to be taken into account to address the complex dynamic of gene x environment interactions (GxE). On the other hand, when dealing with the interplay of social and genetic factors in relation to race it is imperative to rigorously maintain clear boundaries to avoid the dangers of geneticizing race in general, and health disparities in particular. There is already some problematic category slippage going on here.

Note that genomic diversity was earlier defined by leading NIH researchers as "variation in alleles and allele frequency." This is not quite the same thing as variation of "genetic ancestry." Variation due to genetic ancestry is only one subset of the wider variety of genetic variation that can occur through such mechanisms as random mutation, genetic drift, and basic clinal variation. This is not the end of such slippage. The underlying logic of the idea of combined technical, medical, political, and commercial rationales ran something like this: in order to find efficient and cost-effective medical interventions to address the genomic basis of disease, particularly those diseases disproportionately affecting underrepresented minority populations, we need to find those genetic variants that play a causal role in such diseases.

The goal was to find relevant causal genetic variants. To do this, researchers needed to construct databases that contained a sufficient range of genomic variation—i.e., diversity—to identify causal variants. Genomic diversity then became constructed in terms of "genetic ancestry," which in turn became "continental ancestry," which became African, Asian, European, and Indigenous American, which in turn became race. Even those researchers sensitive to using race in such contexts nonetheless seemed to find no other way to accomplish their goals. And so, for example, we had Ramos, Callier, and Rotimi arguing in 2012 that, "Although race and ethnicity are inadequate proxies for individual genetic variation, the success of personalized medicine will depend on how well we understand genetic variation among ethnically diverse populations."[54] Race, it seems, was the proxy we could not do without.

But what exactly was it a proxy for? Continental ancestry? Perhaps, but continental ancestry itself was really a proxy for genetic ancestry; and genetic ancestry itself did not get at what they really wanted but was only a means to the end of identifying patterns of differing allele frequencies. And yet, it was not simply differing allele frequencies that they wanted; they wanted those differences that correlated with particular health conditions; and then they did not want simple correlation, they wanted causative genetic variants. Next, even if they identified causative variants they needed to understand about their penetrance and expressivity; and finally, once they figured all that out, they needed to then begin the entirely distinct process of trying to come up with some sort of interventions that would actually be of clinical utility—which of course, was the ultimate promise behind the billions of dollars being committed to the PMI.

Further complicating matters was that fact that advocates of diversity in biomedical research were using the social category of race not only as a proxy for genetic factors but also for other nonbiological factors such as environment, economic status, or social experiences of racism, all of which can influence disease etiology. This piling of proxy upon proxy facilitated the reification of race not only by blurring critical distinctions between social, geographical, and biological constructs, but also by facilitating the elision of the very social and geographical factors that diversity was supposed to capture in the search for understanding the complexities of GxE interactions in shaping human health. This accords well with the observation by Panofsky and Bliss of a "persistent and indiscriminate blending of classification schemes" by geneticists "that has led the practical definition of 'population' to become more ambiguous rather than standardized over time."[55]

Advocates of diversity in precision medicine research (PMR) have acknowledged this danger. For example, Sabetello et al. noted that "Although PMR is expected to extend beyond race and genomics to include environmental and lifestyle factors on health interventions and outcomes, the latter have historically been neglected."[56] But in a 2016 review of research purporting to explore gene-environment interactions (GEI), Darling et al. tellingly showed how, "GEI researchers' expansive conceptualizations of the environment ultimately yield to the imperative to molecularize and personalize the environment." This molecularization of the environment occurs as researchers "seek to 'go into the body' and re-work the boundaries between bodies and environments," which, in turn, leads to a shift from "efforts to understand social and environmental exposures outside the body, to quantifying their effects inside the body."[57]

In programs such as *All of Us*, race served not only as a proxy for genetic ancestry (or continental ancestry, etc.), but also for social circumstances such as environment and class. Using race as a social proxy has been central to much of the enterprise of addressing health disparities. Consider, for example, Khoury et al.'s discussion of how diversity in research cohorts promotes health justice:

> To learn what interventions work for whom, data on each individual needs to be compared with data from large, diverse numbers of people to identify population subgroups likely to respond differently to interventions.... In addition, new biomarkers promise to improve the understanding of disease natural history. For example, epigenetics is providing insights into the impact of the environment on gene expression throughout life with the possibility of targeted interventions. There is also strong suspicion that cumulative epigenetic changes due to environmental stressors may explain population health disparities in the burden of various diseases among disadvantaged populations.[58]

Here, Khoury et al. connected diversity to different environments, but then reduced it to its epigenetic impact within the bodies of diverse populations—the sort of molecularization discussed by Darling et al. That is, race was used as a category to construct a diverse cohort that captured variable environmental and social conditions, but then those conditions were reduced to molecular epigenetic impacts within the bodies of cohort participants—bodies that had been racially marked. And so, any resulting genetic differences found became racially coded. The environment faded out of the picture and molecularized race rose to the fore.

The unfortunate dynamic we are left with then can be summarized roughly as follows:

1. In the 1990s, even as the Human Genome Project (HGP) was leading toward President Clinton's confident declaration that "the only race we are talking about is the human race,"[59] the federal government was developing standards and guidelines that encouraged, and in some cases mandated the use of racial categories in biomedical research, largely in a laudable effort to address persistent health disparities.
2. After 2000, as the HGP largely failed to deliver on the grand promises of truly personalized medicine, race emerged as a stepping stone, an interim proxy for refining genomic research, again in service to the promise of ultimately achieving truly personalized genomic medicine.

3. In the mid-2000s, diversity momentarily receded from research agendas as the CD/CV hypothesis drove campaigns to develop massive genetic data banks.
4. After 2011, with the erosion of the CD/CV hypothesis, and in part fueled by the rationale of addressing the same health disparities that were the target of various nongenomic federal initiatives of the 1990s, the stepping stone of race became a foundation of all future research as the concept of diversity became a central feature of constructing massive new federal databases seeking to uncover differing allele frequencies among certain human "subpopulations." The focus here was less on uncovering the genetic basis for common diseases and more on diversity as a means to further "health equity" and realize a new promise of genomics.
5. In practice, the sorts of GEI studies that provided a central rationale for racially diverse research cohorts molecularized the environment by directing their gaze inward to the racialized bodies of affected individuals.

What resulted was a sort of double-pronged re-reification of race as genetic: First, through the conflation of socially defined racial groups with distinctive and discrete genetic clusters; and second, through the molecularization of social phenomena that disparately impact racial groups—such as environmental racism. The former threatened to reinvigorate dangerous, reductive, and racist constructions of race as genetic; the latter threatened to geneticize health disparities themselves, both blaming the victims and directing attention away from the sociopolitical initiatives that needed to be undertaken in order to address such problems.

These were not new problems, since the completion of the HGP many others, including myself, have drawn attention to these dangers.[60] Each new iteration of genomic advancement seemed to present new challenges and opportunities for resurrecting these old problems. In this latest version, reification was being driven by the concept of diversity. It was a concept ready-made for conflating racial and genetic categories because diversity itself had roots in both worlds; a dual pedigree of biodiversity—or as E. O. Wilson's popular 1992 book characterized it, "The Diversity of Life"—and social diversity as a concept driving racial justice, ensconced into jurisprudence, law, and policy through Justice Powell's opinion in *Bakke v. Regents of the University of California*, wherein he cited diversity as a compelling state interest that might justify affirmative action programs.[61] Diversity is a useful concept, but like race itself, it can be hard to define and even harder to use in a productive way that avoids the dangers of genetic essentialism and racial reification. Hence the need to identify and address these challenges in each new manifestation.

CHAPTER EIGHT

Genetic Entanglements of Sociolegal Diversity in the 2010s

While genomic researchers were making the shift from personal to precision medicine, conservative critics of affirmative action were continuing to develop new arguments based in ideas of genetic diversity and the social construction of race. These arguments proceeded along two lines: first, trying to deploy knowledge from genetic ancestry testing to challenge affirmative action and minority set-aside programs; and second, trying to appropriate liberal ideas about the socially constructed nature of race to argue that because race was not genetic, it could not be used as a legitimate basis for any legal classifications. Entanglements of biological and sociolegal concepts of diversity were not the exclusive province of the Right. Well-meaning liberals also deployed ancestry tracing and related concepts of population-based genetic variation to produce understandings of diversity that tread dangerously close to biologizing race.

RALPH TAYLOR'S STORY: USING DNA TO CLAIM MINORITY STATUS

As Lynwood, Washington, insurance contractor Ralph Taylor tells it, sometime around 2009 or so: "I was in a bar... and some visually Caucasian guy... was talking about how he got money for being a minority, which piqued my interest." This guy "had gotten a big chunk of money," so Ralph "looked into the different avenues and . . . found OMWBE," Washington State's Office of Minority and Women's Business Enterprises. Ralph himself "looked Caucasian" but he had recently taken a DNA ancestry test that said he had some African ancestry, and this gave him an idea. Until that time, the forty-seven-year-old Taylor had always thought of himself as White, but he now began to look into using his DNA test as a basis for claiming that his business, the Orion Insurance Group, was a minority-owned business enterprise and hence eligible for funding via minority set-aside contracts.[1]

At first, there was some back-and-forth with the Washington State office about the sufficiency of his first DNA test, so in 2010 he took a test offered by AncestrybyDNA™, part of Genelex corporation. The test estimated that he was, "90% European, 6% Indigenous American, and 4% Sub-Saharan African."[2] Despite the test showing he had more "Indigenous American" ancestry, he chose only to claim he was "Black" in his application for Minority Business Enterprise (MBE) status under Washington State law. In 2013, after an initial rejection, Washington granted his application for MBE certification.[3]

The next step was to obtain federal certification as a Disadvantaged Business Enterprise (DBE). Under the DBE program the U.S. Department of Transportation (USDOT) set aside a certain amount of funding from federal contracts to go to small businesses owned by "socially and economically disadvantaged individuals."[4] According to the Code of Federal Regulations (CFR), "socially disadvantaged individuals" are those "who have been subjected to racial or ethnic prejudice or cultural bias within American society because of their identities as members of groups and without regard to their individual qualities. Social disadvantage must stem from circumstances beyond their control."[5]

The federal DBE certification was particularly attractive to Taylor because it would allow him to gain more deals providing liability insurance to contractors under multimillion-dollar federal contracts. He noted that, under the federal certification, he could get access to insuring contractors on programs "like the Seattle Tunnel project . . . which would have been a substantial amount of income to [his] agency."[6] His application for DBE certification under the federal program was to be evaluated by the same state office that granted him MBE status, so he understandably assumed approval would be forthcoming. Taylor was, therefore, surprised and more than a bit put out when Edwina Martin-Arnold, the OMWBE certification analyst assigned to his case, denied his initial application in 2014 and requested that he submit additional evidence that he was a member of a disadvantaged group.[7] The problem, according to Taylor, was that Martin-Arnold "did not think [he] looked black enough on [his] driver's license" that was submitted as part of his application.[8] In due course, Taylor submitted his DNA test results along with some other genealogical records, but Martin-Arnold deemed them insufficient. And so, in 2016 Taylor sued the OMWBE and the USDOT in federal court.[9]

On its face, this might seem like an example of some officious bureaucrat denying the claimed racial identity of an applicant based on a crude stereotyped assessment of his phenotype. It turns out, however, that the federal guidelines direct an examiner to "require the individual to present additional evidence that he or she is a member of the group" if the examiner has a "well-founded reason to question the individual's membership in that group."[10] Washington State had no such directive, so the examiner would not have had similar grounds for questioning Taylor's MBE application (which, after all, had also been initially denied). In evaluating his application for federal recognition, Martin-Arnold, in looking at the totality of the materials (including the DNA test), concluded that she had a "well-founded reason to question" Taylor's claim of group membership.[11]

The federal guidelines address the understandable concern that when millions of dollars are at stake some applicants may falsely claim membership in a disadvantaged group. Nonetheless, such situations also involve the uncomfortable reality that, under any racial preference program, some person or office must have the authority to review and adjudicate claims of racial membership.[12]

THE NEW ROLE OF DNA TESTING IN DETERMINING LEGAL RACIAL CATEGORIES

Such controversies long predate Taylor's application. In 1975, not long after the first affirmative action hiring programs were implemented, Philip and Paul Malone, twin brothers from Boston, applied to be firefighters but were not hired because of their low civil service test scores. They reapplied in 1977, this time changing their self-identified racial classification from "White" to "Black" and were hired the following year. This particular gaming of the system did not come to light until ten years later when the brothers applied for promotion and the commissioner, who knew the twins personally, saw that they listed their race as "Black." A hearing ensured and the twins were dismissed for committing "racial fraud."[13] After the Malones' case, investigations within the fire and police departments of Boston and other cities uncovered similar cases of questionable claims about racial identity.[14] As recently as 2019, the *Wall Street Journal* reported the story of a college counselor who, as part of a scheme to get wealthy students into elite colleges, urged them to falsely identify as racial minorities in their application materials.[15]

Then there is the case of onetime U.S. Republican Speaker of the House Kevin McCarthy's brother-in-law, William Wages, who claimed Native American ancestry to win more than $7 million in no-bid federal contracts.[16] Wages asserted in 1998 that he was Cherokee Indian, but an investigation in 2018 by the *Los Angeles Times* found no Cherokees among his ancestors in birth and census records examined going back to the 1850s. It turned out that Wages made his claim based on certification from the so-called "Northern Cherokee Nation," which is not a federally recognized tribe. As the *Los Angeles Times* noted, "All three Cherokee tribes with federal recognition consider the Northern Cherokee group illegitimate." "It's very much a con," said David Cornsilk, a Cherokee genealogist and citizen of the Cherokee Nation, the largest of the recognized Cherokee tribes. After the exposure by the *Los Angeles Times*, Wages stopped identifying his company as Native American–owned in government records.[17]

What set Ralph Taylor's case apart from these cases of racial fraud is that he submitted the results of his 2010 AncestryByDNA™ test to support his claim of racial group membership (notably, Wages said he considered getting a DNA test to bolster his claim but said he opted not to because the tests were unreliable for Native Americans.[18] Given the absence of any documented Cherokee ancestry in his genealogical records going back to 1850, there may have been other reasons for avoiding such a test). Taylor was not simply checking a box; he was augmenting his claim with an appeal to genetic science. The issue, therefore, was not one of racial fraud, but of the sufficiency of the genetic evidence he presented to establish his claim.

The OMWBE found the DNA test results insufficient for two primary reasons. First, Genelex specified that the test results held a margin of error of 3.3 percent, meaning that Taylor's ancestry could be as little as 2.7 percent Indigenous American and 0.7 percent sub-Saharan African. The OMWBE further noted that from reviewing the information on the AncestrybyDNA™ website, it was unclear if the website's use of the term sub-Saharan African corresponded to the definition of Black American in the CFR, which referred to "persons having origins in the Black racial groups of Africa." Second, the sufficiency of the Genelex test was further undermined by Taylor's decision to submit *additional* DNA ancestry evidence in the form of a test his father took that estimated he was 44 percent European, 44 percent sub-Saharan African, and 12 percent East Asian.[19] In considering the implications of the significant divergence between the father's and son's test results, the OMWBE noted:

> Mr. Taylor submitted a DNA test to prove he is 4% Sub-Saharan African and 6% Native American. The test results for Mr. Taylor and his father are highly inconsistent and incomplete. Half of a son's DNA comes from his father and half comes from his mother. OMWBE acknowledges that the pieces of DNA from each parent are random and will not equal exactly half from each parent. The two DNA tests between father and son should, however, be related. Without a complete picture of Mr. Taylor's mother's DNA, OMWBE contends that the tests are not reliable to determine ethnicity. This information fails to prove that Mr. Taylor is a member of a minority group or regarded as a member of a minority group.[20]

The OMWBE focused primarily on issues of statistical error and divergence between the father's and son's test results. Ironically, Taylor may have hurt his case by submitting the additional test results from his father because it called into question the reliability of the technology overall. Yet the OMWBE left open the possibility that *more* DNA information—i.e., from the mother—might cure the reliability problem.

The company that provided Taylor's test, Genelex, also warrants a bit more scrutiny. As Kim TallBear has noted, Genelex was advertising its services as early as 2004 to "confirm that you are of Native American descent... if your goal is to assist in validating your eligibility for government entitlements." Such claims are particularly problematic in relation to Native American identity, where, as TallBear also observed, tribal membership is not determined with reference to genetics but involves varying and complex political and social histories that are grounded in tribal sovereignty claims and practices.[21] Yet it is evident from advertisements such as this, that Taylor was hardly the first to think about using DNA ancestry testing to make a claim for government benefits. He was, however, apparently the first to bring a federal lawsuit to challenge the denial of claims based, in part, on such testing.

In appealing the OMWBE decision to the USDOT, Taylor argued that the regulations defined "Black Americans" to include "persons with 'origins' in the Black racial groups of Africa" (this terminology being derived from the U.S. Census Bureau classifications), and that his DNA test showed such "origins." The USDOT, however, noted that the broader definition of a "socially and economically disadvantaged individual" also required that the person "have been subjected to racial or ethnic prejudice or cultural bias within American society because of his or her identity as a members [sic] of groups and without regard to his or her individual qualities."[22] Here the USDOT was, in effect, noting the importance of social context and lived experience for the construction of racial membership sufficient to meet the purposes of the DBE program, which are "to level the playing field by providing small businesses owned and controlled by socially and economically disadvantaged individuals a fair opportunity to compete for federally funded transportation contracts."[23]

Elaborating upon Taylor's claim about having genetic "origins" in Africa, the USDOT further noted that:

Construing the narrower definition as broadly as Orion advocates would strip the provision of all exclusionary meaning. It is commonly acknowledged that all of mankind "originated" in Africa. Therefore, if any (Black) African ancestry; no matter how attenuated, sufficed for DBE purposes, then this particular definition would be devoid of any distinction-which was clearly not the Department's intent in promulgating it. There is little to no evidence that Mr. Taylor ever suffered any adverse consequences in business because of his genetic makeup.[24]

Here the USDOT brought in the temporal aspect of the genetic construction of ancestry. The racial estimates provided by genetic ancestry tests are premised not only on "where" your ancestors might have been from but also "when"—that is, at what point in time we fix their purported place of origin. All human populations have their origins in Africa many hundreds of thousands of years ago, but there are also estimates placing the most recent common ancestors of all current humans on Earth around 1400 BCE, fewer than 3,500 years ago.[25]

Taylor conceded that he had neither held himself out nor regarded himself as a member of a covered group before 2010 but rather "grew up thinking of himself as Caucasian." In response to the OMWBE's request for additional information to support his claim of being Black, Taylor stated that he had subsequently joined the NAACP, subscribed to *Ebony* magazine, and had "taken a great interest in Black social causes." The OMWBE found such assertions to be insufficient to meet the regulatory standard for inclusion as a member of a recognized racial group under the regulation and so it denied his application.[26] The tenuousness of such claimed personal identification with the Black community was underscored by Taylor's own admissions outside of court that his initial interest in claiming a Black identity was to gain access to federal funds.[27] The federal district court ultimately agreed with the OMWBE's reasoning and upheld its actions, dismissing all of Taylor's claims in 2017.[28]

Undeterred, Taylor appealed to the U.S. Court of Appeals for the Ninth Circuit and in the interim had his California birth certificate amended to change his father's race from "Caucasian" to "Black, Native American, Caucasian."[29] At oral argument, Judge Fletcher found this additional bit of evidence less than dispositive, noting that California did not require any evidence to support such a request and in any event finding it *post hoc* to the case at hand. Fletcher also noted in passing that while DNA testing might have evidentiary value in certain forensic contexts where DNA from a crime scene was being matched with a suspect, the value of DNA ancestry testing in affirmative action contexts was highly questionable.[30] In some respects, this echoed the court in Deadria Farmer-Paellmann's earlier reparations case commenting on the limitations of DNA ancestry testing in legal proceedings. In any event, the court of appeals had little trouble affirming the district court's dismissal of Taylor's claims, noting that neither the OMWBE nor the USDOT acted in an arbitrary or capricious manner in rejecting Taylor's claims.[31]

The reference to an absence of arbitrary or capricious action is common in appellate review. It is, nonetheless, significant here because a secondary aspect of Taylor's argument involved his using the DNA evidence to challenge the very legitimacy of the racial classifications being employed by the OMWBE and USDOT. While he may have originally sued in order to get access to federal funds, his claim evolved into something of a crusade against affirmative action programs.[32] This ties Taylor's story to other conservative attempts to undermine affirmative action programs that contrast the purported arbitrary nature of socially based racial classifications to assertedly more robust or "real" classifications based on genetic science.

After the OMWBE rejected his initial application for federal certification of Orion Insurance as a minority-owned business enterprise, Taylor shifted tack and started arguing that the rejection of the evidence he presented, particularly his DNA ancestry evidence, showed that the use of racial categories in affirmative action programs was itself arbitrary and capricious, or alternatively that the federal regulations specifying the categories were unconstitutionally vague. Taylor's complaint was grounded in the assertion that "the OMWBE decision was arbitrary and capricious because it did not find him to be 'Black enough' based on his appearance on his driver's license." He went on to argue that his DNA test showed that this initial phenotype-based determination was erroneous. The district court found, however, that in developing this claim, Taylor's "reliance on [his] genetic makeup, without regard to his appearance, is misplaced and does not demonstrate that OMWBE acted arbitrarily or capriciously in finding that there was insufficient evidence that Mr. Taylor was a member of either the Black or Native American groups."[33]

Related to his claim of arbitrary and capricious action, Taylor asserted a potentially more powerful and far-reaching claim that the definitions of "Black American" and "Native American" as used in the DBE program and the CFR were themselves void for vagueness, presumably contrary to the Fifth and Fourteenth Amendments' due process clause.[34] This claim, if upheld, would not only have had implications for Taylor, but could have undermined the viability of the entire DBE program and, indeed, potentially *any* program using racial classifications.

Taylor's argument was premised, in part, on the idea that the genetic information from his ancestry test provided a clear and objective measure of race, whereas the criteria that led to the denial of his application were based on vague and amorphous standards of social understandings of race. He argued, for example, that he, "was denied inclusion in the Black and Native American minority categories because OMWBE believed he had insufficient minority DNA, though they had no written guideline or policy delineating a DNA limit under which a person would not be considered a minority."[35] He also argued that "genotype is a more stable indicator of race than phenotype," the latter being susceptible to change (e.g., "a Caucasian person can sit in the sun for a few years and look more Black, or someone like Michael Jackson can get skin treatments to look more Caucasian"), while the genotype is unchanging. His brief also drew upon an antisemitic canard that Jews are a genetic race, asserting that "people considering themselves German were likewise swept into the holocaust because they had Jewish blood (genotype), not because of how they looked (phenotype)."[36] In all of this, there is an eerie echo of racist laws of hypodescent from the Jim Crow era, that declared a person Black for purposes of the law if they had "one drop" of Black blood.[37] In any event, it reified race as genetic, while dismissing social understandings of race as untenable and epiphenomenal.

Unconvinced, the court dismissed Taylor's claims, noting that "considering the purpose of the law, the regulations clearly explain to a person of ordinary intelligence what is required to qualify for this governmental benefit."[38] The court here was willing to accept contextual and common sense understanding of both the purpose of the law and the socially understood boundaries of the group categories to which it applied. The court's recognition that racial categories can constitutionally be employed in a commonsense manner grounded in contextual social understandings of race may seem obvious, but it is critical. It goes to the heart of broader conservative efforts to use advances in genetic science and testing to undermine precisely such commonsense administration of affirmative action programs.

A NEW GENETIC POLITICS OF AFFIRMATIVE ACTION?

Legal scholar Mary Ziegler noted this trend in 2018, arguing that since the early 2010s: "antiaffirmative-action amici and activists have developed a new argument: a claim that if race is a social construct, race-conscious remedies are arbitrary, unfair, and likely to reinforce existing stereotypes."[39] Ziegler focused in particular on the cases of *Schuette v. Coalition to Defend Affirmative Action* and *Fisher v. University of Texas*.[40] In *Schuette*, the Supreme Court upheld an amendment to the Michigan state constitution that forbade the use of racial preferences by any university system or school district.[41] Ziegler drew attention to the fact that in the plurality opinion in *Schuette*, "the Court ... questioned whether it was possible any longer for the racial categories used in affirmative-action programs to have any value. '[I]n a society in which [racial] lines are becoming more blurred,' Schuette explains, 'the attempt to define race-based categories ... raises serious questions of its own.'"[42] Similarly, Ziegler noted that the conservative dissenters in Fisher, a case challenging the use of racial categories as part of the admissions process at the University of Texas, built upon the argument from Schuette "insisting that racial categories are 'ill suited for the more integrated country that we are rapidly becoming.'"[43] Ziegler astutely observed that, "Far from denying claims that race is a construct, opponents of affirmative action now use those claims to their advantage. For anti-affirmative-action amici and activists, the idea that race is a social construct now militates in favor of color blindness. Since race is a social construct, it is argued to be devoid of meaning. Any use of race, in this account, becomes an unfair and incoherent allocation of government benefits."[44] As understandings of race as a social construction gained ascendance, conservatives adapted their arguments against using racial categories in law or policy accordingly.

The idea of race as a social versus genetic construct has a complex and contested history. The ascendance of the social constructionist view of race was grounded in the work of anti-racists on the Left who have been working for decades to discredit the biological understandings of race used to rationalize racial hierarchy.[45] This view seemed at its apogee in 2000 at the White House ceremony celebrating the completion of the first draft of the human genome.[46] The following decade, however, saw the steady rise of race-based studies in genetics, biomedicine, and pharmaceutical development promoting the idea that because statistical correlations of different frequencies of certain genetic variations seemed to cluster by racial groups, that genetic constructions of race have merit.[47] As Dorothy Roberts noted in this period, "the liberal faith in scientific objectivity has generated an approach to the genetic definition of race that sounds remarkably similar to the conservative one. Like conservatives, liberals separate racial science from racial politics to retain a supposedly scientific concept of race as a genetic category."[48] In biomedicine this was perhaps most clearly evident in the case of BiDil, which in 2005 became the first drug ever approved by the FDA with a race-specific indication—to treat heart failure in a "Black" patient.

BiDil had the support of the Congressional Black Caucus, the NAACP, and the Association of Black Cardiologists as a means to address purported health disparities in mortality from heart failure. Notably, however, it was also embraced by Newt Gingrich and Sally Satel, a doctor affiliated with the conservative American Enterprise Institute. For the likes of Satel, who wrote a high-profile piece in the *New York Times Magazine* titled, "I am A Racially Profiling Doctor,"[49] BiDil was used as evidence that race-based health disparities had more to do with biology than with racism and social injustice.[50] The more liberal proponents of BiDil saw it both as a way to help a community long-victimized by the biomedical establishment and as a stepping stone to personalized genomic medicine. For them, the idea was that even if they did not accept the idea of race as genetic, they nonetheless saw it as a useful proxy for certain important genetic variations. Here, a well-meaning liberal attempt to address racial health disparities could also be flipped by conservatives to undermine the premise that such disparities were caused by social forces rather than biology.[51] Even while being promoted to address issues of racial justice, the case of BiDil is also an example of conflating race and genetics to gain tactical advantage in the market.

By juxtaposing Ziegler's article with the case of BiDil, we can see that even as some conservatives have come to embrace the idea of race as a social construct in order to challenge affirmative action, some liberals have come to accept the utility of using social race as a proxy for genetics as a tactic to realize the promise of personalized genomic medicine. This strange political inversion has no specific cause but is testament to how both race and genetics operate in fluid, contestable domains that can be variously deployed to serve diverse and sometimes conflicting ends.

For most political purposes, it seemed that the language of race as a social construct had triumphed; yet in the realm of scientific inquiry and biomedical practice race has continued to be used as a proxy for genetic populations. Roberts has further observed about this phenomenon in the context of biomedical practice:

> With the new distinction between biological and social race... conservatives now have a way to speak about racial difference while maintaining a color blind approach to social policy. They find it acceptable to refer to race explicitly as long as it has a biological meaning because that use of race is purportedly scientific and unbiased.... Genomic science, conservatives argue, frees us from political correctness so we can act on racial differences in genetics that determine our health. In this ingenious twist of political logic, those who criticize racial biomedicine because of its impact are seen as interfering with health out of loyalty to racial ideology.[52]

Where Roberts showed how some conservatives invoked the idea of genomic race to undermine progressive efforts to address health disparities, Ziegler provides a critical perspective vantage on other conservatives who invoked the idea of social race to undermine any progressive attempts to use racial categories in law and policy. Ziegler, though, did not fully appreciate how conservatives were distinctively employing genetic concepts to buttress these claims.

Extending Roberts's idea of conservatives' "ingenious twist of political logic" beyond biomedicine to affirmative action more broadly, I would characterize the conservative framing as being something along the lines of: Race as a social construct is not "real" and hence cannot be the basis of legitimate legal policies; conversely, race as a genetic construct is real, and further shows the arbitrary and unconstitutionally vague nature of using race as social construct in law and policy. Such conservative framings have been enabled and reinforced by the dynamic Roberts identified (evident in the story of BiDil) whereby "liberals separate racial science from racial politics to retain a supposedly scientific concept of race as a genetic category."[53]

This framing also informed Taylor's claims. After he was denied access to benefits as a disadvantaged person operating a business, he essentially attempted to burn down the entire edifice of the DBE program, arguing that the racial categories it employed were unconstitutionally vague. He did this not simply by asserting that "race is a social construct... devoid of meaning,"[54] but also by taking the additional step of juxtaposing the supposedly arbitrary social categories of race against the purportedly more real, objective, and scientific construction of racial membership afforded by his genetic ancestry test. Such use of genetics to undermine affirmative action received scant attention in Ziegler's article but adds an important dimension to the broader phenomenon she identifies. Beyond Taylor's case, we see this most prominently in the genealogy of the *Fisher v. University of Texas* affirmative action case.

JUDGE GARZA AND *FISHER V. UNIVERSITY OF TEXAS*

In 2008, Abigail Fisher and Rachel Michalewicz each filed suit against the University of Texas at Austin after they were denied admission to the undergraduate program there.[55] The plaintiffs, both characterized by the court as "Caucasian female[s]," challenged the university's affirmative action program, contending that the "admissions policies and procedures currently applied by Defendants discriminate against Plaintiffs on the basis of their race in violation of their right to equal protection of the laws under the Fourteenth Amendment of the United States Constitution, and federal civil rights statutes."[56] The Federal District Court for the Western District of Texas, applying the standard set forth in Grutter v. Bollinger for evaluating affirmative action programs in higher education, found no liability and granted summary judgement to the university.[57]

The plaintiffs would go on to pursue the case up through the Fifth Circuit Court of Appeals and finally to the Supreme Court, where, ultimately, they would fail to prevail.[58] Along the way, Judge Emilio Garza, a conservative jurist appointed to the Fifth Circuit by George H. W. Bush in 1991, would write a concurring opinion with a curious allusion to biological race and a remarkable footnote citing liberal anti-racist scholars of race and genetics.[59] Garza was no fan of affirmative action programs and made it clear that he disagreed with the holding in *Grutter* even as he felt bound to apply its principles. Nonetheless, he wrote separately in hopes that his reasoning might provide a basis for the Supreme Court to reconsider its logic and "rectify the error" of *Grutter*.[60] A key component of his argument was his assertion that:

> The idea of dividing people along racial lines is artificial and antiquated. Human beings are not divisible biologically into any set number of races. A world war was fought over such principles. Each individual is unique. And yet, in 2010, governmental decisionmakers are still fixated on dividing people into white, black, Hispanic, and other arbitrary subdivisions. The University of Texas, for instance, segregates student admissions data along five racial classes. *See, e.g., 2008 Top Ten Percent Report* at 6 (reporting admissions data for White, Native–American, African–American, Asian–American, and Hispanic students). That is not how society looks any more, if it ever did.[61]

Garza upended the liberal constructivist critique of race essentialism that had challenged White supremacy. He appealed to the authority of "science" not to challenge racial hierarchy but sustain it. He denied that racial categories themselves could be used to challenge racial hierarchy.

This brings us back to Ziegler's observation about how conservatives have begun to flip the liberal idea that race is a social construction against liberal affirmative action policies by arguing that social construction is somehow equivalent to or synonymous with incoherence and arbitrariness.[62] Ziegler, however, overlooked Garza's telling footnote,

Genetic Entanglements of Sociolegal Diversity

quoting a 2004 article by prominent law professor, Larry Alexander who, with his colleague Maimon Schwarzschild, cited philosophers of science to support their contention that "racial (and ethnic) classifications are unscientific, arbitrary, and often nearly meaningless."[63] Perhaps most telling is the footnote's additional citation to Joseph Graves Jr.'s 2001 book, *The Emperor's New Clothes: Biological Theories of Race and the Millennium.* Graves is not a philosopher of science. He is a professor of biological science at North Carolina Agricultural and Technical State University and was the first African American to receive a PhD in evolutionary biology in this country. In the book, Graves made it clear that his goal "is to show the reader that there is no biological basis for the separation of human beings into races and that the idea of race is a relatively recent social and political construction." His project was unabashedly anti-racist. He was confident that "demolishing the idea of biological race lays bare the fallacies of racism."[64] Graves was a signatory to two subsequent amicus briefs in the *Fisher* case filed in support of the University of Texas's affirmative action program.[65]

Graves himself later wrote a rejoinder to Garza and other similarly minded conservatives who would seek to use his scholarship to undermine affirmative action. Noting that people he characterized as "color-blind racists" had "co-opted" scientific findings to assert that "racism can no longer exist, since we have no biological races," Graves argued that the mere scientific fact that "the human species does not really contain biological races ... has absolutely nothing to do with the ongoing racial discrimination faced by persons with dark skins in the United States."[66] Directly responding to Garza's citation of his work, Graves asserted:

> The problem with Garza's reasoning is precisely that it confuses biological and socially defined racial categories (and their impacts). Garza is correct in pointing out the nonexistence of biological races. Indeed, he cited important scholarly literature supporting that fact (including my own work). However, the past-discrimination that the University of Texas (and other affirmative action) plans attempts to redress are based on how socially defined races suffered past and are suffering ongoing discrimination in American society. It is also not just the government that divides Americans

into socially defined racial groups, it is virtually all lay Americans who continue this practice.[67]

Garza tried to leverage the scholarship of the anti-racist, pro-affirmative action evolutionary biologist Graves, to argue that if racial categories have no firm biological basis, then they cannot be used as a basis for coherent state policy. For Garza, it was not simply that race as a social construct was arbitrary; it was that it was arbitrary *in comparison to* and because it had *no grounding in* biology. The reference to genetic science gave this argument its distinctive bite.

The logic is clear: if race *were* genetic, then perhaps there would be a way to use racial categories in a manner that was not arbitrary; but *because* race is a social construct, the use of racial categories in law and policy is inherently arbitrary. The appeal to scientific authority was foundational to this new challenge to affirmative action. Ultimately, Garza was no more successful in convincing his fellow judges that the absence of a genetic basis for race made it an untenable legal category than was Taylor in his case. In both instances the decision makers, in effect, echoed Graves's (and Judge Fletcher's) pragmatic understanding of the historic use and impact of racial categories in social and legal domains.

GARZA'S APPROACH TO RACE PROLIFERATES

While repudiated by Graves (and largely ignored by the other Fifth Circuit judges and the Supreme Court), Garza's arguments found a sympathetic ear among other conservative legal scholars. Most notably, as the appeals process in *Fisher* worked its way through the courts subsequent to Garza's special concurrence in the 2011 case (the case went up to the Supreme Court in 2013, back down to the Fifth Circuit in 2014, and back up again to the Supreme Court in 2016), amicus briefs filed by Judicial Watch and by the American Center for Law and Justice took up Garza's genetic argument and expanded upon it.[68]

Garza's argument first reappears in the 2012 amicus brief of Judicial Watch, which describes itself as "a conservative, non-partisan educational foundation, which promotes transparency, accountability and integrity in government, politics and the law."[69] Judicial Watch has a long history, going back to its founding in 1994, of pursuing investigations and litigation to attack liberals and promote conservative policies. A key early supporter was Richard Mellon Scaife, a deeply conservative Pittsburgh billionaire who bankrolled various anti-Clinton crusades, including investigations of "Filegate" during President Bill Clinton's administration, and seeking emails related to Secretary of State Hillary Clinton's role during the killings at the U.S. Embassy in Benghazi.[70]

In its brief, Judicial Watch began by referring to race as an "intellectually impoverished concept" that when introduced into law foments "racial and ethnic resentment and intolerance." It went on to explain that race was an impoverished concept because of "inherently ambiguous social constructs that have no validity in science."[71] Or, as it put in their later brief with direct reference to the legal standard of strict scrutiny requiring state use of racial categories to be narrowly tailored to serve a compelling state interest: "Government policies such as the policy enacted by the University, which seeks to classify applicants by crude, inherently ambiguous, and arbitrary racial and ethnic categories to promote diversity can never be narrowly tailored to further a compelling government interest. Attempts to categorize individuals by racial and ethnic groups necessarily lead to absurd results."[72] This broad argument has implications not only for affirmative action in higher education, but for the use of racial categories in any law or state policy.

Judicial Watch appealed to genetics not only to make its specific case for Abigail Fisher but also to attack the controlling Supreme Court precedent of *Grutter v. Bollinger,* which allowed for the use of racial categories as a "plus" factor in a holistic admissions process to help achieve a diverse student body.[73] Pointedly, Judicial Watch noted:

> In the same year that marked the completion of the Human Genome Project, the Court [in *Grutter*] upheld the Law School's use of race—a concept that has been rejected by science and for centuries has been used to divide, impoverish, oppress, and enslave people—as a "plus" factor weighing in favor of admission. *Id.* at 335–43. In its ruling, the Court assumed that race was a meaningful proxy for diversity without addressing the issue in any direct way. The Court also assumed that race presented a fixed, natural, and unambiguous means of distinguishing between groups of

people such that individual Law School applicants could be assigned a particular racial classification and awarded—or not awarded—a "plus" factor based on race.[74]

The explicit invocation of the Human Genome Project is telling. The HGP stood as a symbol of scientific and technological prowess; a transformative scientific advancement akin to the Copernican or Darwinian revolutions.[75] Beyond conventional legal arguments, Judicial Watch was trying to leverage the authority of science to challenge the empirical underpinnings of the holding in *Grutter*. Perhaps developing a diverse student body was a compelling state interest sufficient to justify the use of race in university admissions, but the lack of a scientific foundation to the racial categories out of which such diversity was to be constructed meant that the use of racial categories could never be narrowly tailored to serve that end, and so all race-based programs failed the test of strict scrutiny.

To support such sweeping contentions, Judicial Watch moved beyond Garza's initial references to scholars such as Graves to cite public statements on the nonbiological nature of race from the American Anthropological Association (AAA).[76] It also quoted an extensive discussion of "'Race' as a Biological Fiction," from an opinion by liberal judge and civil rights pioneer Jack Weinstein, which noted:

> DNA technology finds little variation among "races" (humans are genetically 99.9% identical), and it is difficult to pinpoint any "racial identity" of an individual through his or her genes. International gene mapping projects have only "revealed variations in strings of DNA that correlate with geographic differences in phenotypes among humans around the world," the reality being that the diversity of human biology has little in common with socially constructed "racial" categories.[77]

Clearly, Judicial Watch was taking the genetic ball from Judge Garza and running with it.

The appeal to science was foundational to its assault on affirmative action. This is also evident in the brief's assertion, following its discussion of the AAA statements that "although science may have rejected race long ago, law and public policy, and in particular the University's admission policy, have yet to catch up. It is time that they did so. Race has no place in either."[78] Judicial Watch thus presented the use of racial categories in law and policy as untenable not simply because they were vague but because they were unscientific. It grounded their purported vagueness or arbitrariness precisely in the fact that "science" had rejected them.

THE LIBERAL RESPONSE

Perhaps simply to taunt pro-affirmative action liberals, in its 2015 brief Judicial Watch also highlighted the controversies around Senator Elizabeth Warren's claim of Native American ancestry as further illustrating the point, "that racial categories are generally too crude to convey accurate and useful information about individuals and groups." Judicial Watch feigned sympathy for Warren noting that "based on nothing more than 'family lore' and 'high cheek bones,' Ms. Warren claimed, perhaps quite sincerely, that she was 1/32nd Cherokee and therefore a Native American and a minority." The brief went on to note that "many people predictably expressed doubt that classifying Senator Warren as a 'Native American' based on a system of racial self-identification made any sense, much less served a legitimate purpose." It then considered what might happen if someone with Warren's claims applied for admission to the University of Texas and asked: "How much additional 'holistic diversity' would UT have achieved by deciding to admit these hypothetical Elizabeth Warrens based at least in part on their self-identification with a particular race or ethnic group?"[79]

Ironically, two years later, as she was preparing to run for the Democratic nomination for president, Warren herself had a DNA ancestry test performed for her by MacArthur prize-winning population geneticist Carlos Bustamante. Warren was clearly trying to make use of the authority of genetic science to quell the controversy surrounding her claims of Native American ancestry. The results, while not exactly robust, did indicate that Warren likely had a Native American ancestor in the range of six to ten generations ago.[80] Both the conservative Judicial Watch and the liberal Warren were looking to leverage the authority of genetic science to make claims about racial identity. While

194   *Genetic Entanglements of Sociolegal Diversity*

Warren was careful to maintain that her assertion of genetic ancestry was in no way related to any claims of legal tribal membership, there was nonetheless an implicit understanding that genetic constructions of racial membership somehow validated her social claims. Judicial Watch was using the same logic to make a counterargument—namely, absent such genetic constructions racial categories had no coherent meaning. Both privileged genetic constructions of racial identity.

Echoing Clinton's and Venter's statements from the ceremony marking the completion of the first draft of the human genome, the briefs from the American Center for Law and Justice (ACLJ) highlighted the idea that "there is likewise no difference in kind between black, white, Asian, or other ethnic groups of human beings. There is one race—the human race."[81] In 1990, the televangelist Pat Robertson founded the ACLJ as a conservative counterbalance to the American Civil Liberties Union. The most notable thing about the ACLJ briefs, perhaps, is that they were submitted by Jay Sekulow, who would soon gain prominence as one of President Donald Trump's lead personal attorneys during his first impeachment trial.[82] Sekulow was so enamored by the brief's arguments that he later published an article in *University of Miami Business Law Review* largely recapitulating their main points.[83]

Asserting that "racial categories are both arbitrary and porous," one ACLJ brief contended that the University of Texas "cannot use race to attain a 'critical mass' of minority students if it cannot even intelligibly define what a minority student is. Hence, the deliberate use of race-conscious admissions cannot be a narrowly tailored means to achieve diversity."[84] To further buttress Judge Garza's opinion, the ACLJ quoted extensively from the 1987 Supreme Court civil rights case of *St. Francis College v. Al-Khazraji*, where Justice White observed: "Many modern biologists and anthropologists, however, criticize racial classifications as arbitrary and of little use in understanding the variability of human beings. It is said that genetically homogeneous populations do not exist and traits are not discontinuous between populations; therefore, a population can only be described in terms of relative frequencies of various traits."[85] Here, in contrast to Kiara Bridge's concerns that Justice White's opinion left the door open to legal recognition of race as genetic, the ACLJ flipped the logic of the anti-racist critique of race essentialism to undermine racial justice. The conservative characterization of racial categories as arbitrary depended, critically, not simply on the fact that they were socially constructed, but that they were socially constructed *in contrast* to purportedly "real" genetic population groupings.

Ironically, these conservatives invoked genetic science to undermine the legal foundations of affirmative action in much the same manner that liberal, self-styled "behavioral realists" invoke the cognitive sciences behind implicit bias to try to reinforce and extend affirmative action. Behavioral realists are an interdisciplinary group of scholars who have tried to reinvigorate antidiscrimination law by drawing upon research in the social and natural sciences about the cognitive foundations on individual attitudes and biases. Acknowledging the difficulty of overturning existing legal precedent that focuses on discriminatory intent rather than impact, behavioral realists instead turn to empirical scientific measurements of "implicit bias" to argue that "intent" (rather than legal doctrine) must be reconfigured to incorporate unconscious intent as a basis for assessing the legality of particular actions or systems.[86]

As conservatives such as Garza or Judicial Watch compare genetically grounded population groupings to purportedly incoherent social constructions of race, so too do liberal behavioral realists contrast the science of implicit social cognition to what they see as relatively uninformed people or commonsense understandings of how humans make judgments or form biased intent. As Jerry Kang and Mahzarin Banaji put it, "Behavioral realism identifies naïve theories of human behavior latent in the law and legal institutions. It then juxtaposes these theories against the best scientific knowledge available, to expose gaps between assumptions embedded in law and reality described by science. When behavioral realism identifies a substantial gap, the law should be changed to comport with science."[87] While behavioral realists are far more rigorous in their exploration, development, and understanding of the relevant science, they nonetheless share a common tactic of presenting apparent rigor and robustness of scientific findings as superior to, or more "real" than, "naïve" common sense understandings of social experience—whether it be in the form of racial bias (for the liberal behavioral realists) or of race itself (for the conservatives). Each denigrates or marginalizes the qualitative, interpretative, and narrative bases of interpretation necessary to understand and address racial categories in social and historical contexts, while elevating the scientific findings of "true" empirical reality as a means to challenge existing Supreme Court precedent. They share the idea that if a legal standard is controlled by precedent you do not like, you challenge the empirical assumptions underlying the logic of the holding. It is a high-tech version of the old saw, "if the law is against you, argue the facts."[88]

Behavioral realists have a sincere belief in the power of cognitive science and have taken the time and trouble to actually master it. The fact that conservatives may be more cynically exploiting the liberal embrace of science to serve their ends points up the fact that elevating biology as somehow more "real" than society is a double-edged sword. Conservatives can exploit the tactic of casting commonsense social constructions as "naïve" or less real than scientific claims in a manner that hearkens back to the classic 1987 declaration of Tory British Prime Minister Margaret Thatcher that "there is no such thing as society."[89]

BEYOND AFFIRMATIVE ACTION

Beyond affirmative action in education, Judicial Watch and other prominent conservatives, such as Ward Connerly, have raised similar arguments about race and genetics to challenge the continued use of racial and ethnic categories in the U.S. census. Connerly is founder and president of the American Civil Rights Institute, "a national, not-for-profit organization aimed at educating the public about the need to move beyond race and, specifically, racial and gender preferences."[90] In the 1990s, Connerly played a major role in the passage of California's Proposition 209, which effectively eliminated affirmative action programs from all state institutions. In 2004, he led the campaign for Proposition 54, or the "Racial Privacy Initiative." This proposition would have prevented the State of California from classifying individuals by race, ethnicity, color, or national origin.[91] In these campaigns Connerly did not make much reference to biology or genetics. Echoing the color-blind ideology of many anti-affirmative action advocates, he focused primarily on the idea that racial classifications themselves were antithetical to ideals of American equality and individualism.[92]

By 2017, Connerly had added another arrow to the quiver of his color-blind ideology— genetics. In a *Washington Post* opinion piece reviving the ideas of his "Racial Privacy Initiative," Connerly (together with Mike Gonzalez of the conservative Heritage Foundation) attacked the idea of using racial categories in the U.S. census. Calling for President Trump to make changes to the 2020 census, Connerly and Gonzalez argued that the current system of racial and ethnic classification, "doesn't just ignore science. It also completely overlooks a burgeoning 'mixed-race' population that resents arbitrary racial straitjackets."[93]

Like Taylor and Garza before them, they appealed to science to assert the arbitrariness of social categories. They advocated, "getting rid of the official categories and asking simple national-origin questions ('are your ancestors from Ecuador, Germany, Japan? Check as many boxes as apply') and, perhaps, questions on races identified by anthropologists instead of bureaucrats." There is an echo here of Judicial Watch's appeal to the AAA statements on race in its amicus brief in *Fisher*. They then invoked genetic science explicitly, noting that "Today, you can spit into a vial, send it to genomics companies and discover that you are not 'Irish,' as you thought, but instead 60 percent English. Or you could be roughly 30 percent German, 45 percent Slavic, 15 percent Native American and 10 percent Bantu. The census's official categories ignore this rich diversity."[94] The precision of these genetic estimates of ancestry contrasts strongly to highlight the asserted arbitrariness of the broad racial and ethnic categories of current census. Note the irony of Connerly and Gonzalez embracing "diversity," a concept that lies at the heart of modern affirmative action jurisprudence and practice. Again, there appears to be an attempt to turn the liberals' own terms against them.

Connerly and Gonzalez seemed to think that national or ethnic markers, such as German or Slavic, were somehow less socially constructed than the racial categories of the census. Test results from ancestry tracing companies in the early 2000s were often presented in terms of the census categories but were later refined, largely in response to critics concerned that such framings might reinforce the idea that race was genetic. However genetic ancestry tracing companies characterize their results, such determinations have no objective claim to authority over other characterizations of socially constructed groups used for the purposes of governance.

Disconcertingly, similar initiatives have also been promoted by actors identifying more with liberal causes. This includes activists such as educational consultant Carlos Hoyt, who was profiled in a 2023 *Washington Post* article. As the article noted, Hoyt and his compatriots "recoil at the idea of being confused with people who call themselves 'colorblind.'" To the contrary, Hoyt distanced himself from those "who are trying to deny that there is racism in the world." Very reasonably he argued that the government must account for harms caused by "race," but without resorting to debunked categories that suggest it is biological. For Hoyt, the revelations of modern genomics made it clear that race is not a biological category. In the article, Hoyt acknowledged that the U.S. Census Bureau itself recognizes that "the racial categories included in the Census Questionnaire generally reflect a social definition of race recognized in this country, and not an attempt to define race biologically, anthropologically, or genetically." He went on to lament that "To recognize that race ... is a false concept but to keep doing it anyway, there's something intellectually problematic about it." But to recognize that a category is not biological does not mean it is a false concept. Such framings reinforce conservative ideas that being "merely" social, racial categories cannot provide a legitimate basis for legal redress of other state action.[95]

There were, then, contending conceptualizations of diversity vying for legitimacy here. The old diversity of the world of *Bakke*, grounded in basic census categories of race, and a newer, genetically inflected and fractured construction of diversity that had the potential to undermine all legal and policy rationales for redressing past and continuing inequities based on those earlier census categories.

In 2017, Judicial Watch submitted comments to the Office of Management and Budget (OMB) opposing a proposal to add a new category of "Middle Eastern and North African" to its "Standards for Maintaining, Collecting, and Presenting Federal Data on Race and Ethnicity" (which includes the census). Marshalling arguments that echoed both Connerly and its own briefs in *Fisher*, Judicial Watch again invoked statements from the American Anthropological Association in arguing that the proposal would lead to: "less precise and more arbitrary data" because "human race and ethnicity are inherently ambiguous social constructs that have no scientific validity." Judicial Watch then asserted that "science [has shown] the concept of race to be hollow."[96] At first blush this may sound like a simple restatement of the idea that race had no "scientific validity" but in taking the additional step of asserting that the concept of race itself was "hollow" it reinforced the idea that if a category was merely social (i.e., not scientific) it had no legitimacy whatsoever. Legitimacy, in this schema, could only come from "science." This played into the larger conservative program of delegitimizing the use of racial categories in any and all contexts.

Robert Popper, the director of Judicial Watch's "Election Integrity Project," filed the comments on its behalf. Popper also submitted an amicus brief to the Supreme Court in the case of *United States Department of Commerce v. New York*, arguing for the inclusion of a citizenship question on the census.[97] Consider the relation between voting rights and the use of racial categories. Challenging the collection of racial data would undermine the ability of the federal government to track race-based discrimination in voting rights (among many other areas of civil rights), while adding a citizenship question to the census was widely understood as likely leading to undercounting undocumented, largely non-White, immigrants and hence diluting the representation of (and allocation of federal benefits to) the states wherein they reside.[98] Without the ability to collect data by race, there would be no way to identify racial discrimination in voting, housing, employment, education, or policing; no way to identify health disparities or develop policies and practice to address them.[99] All this, it seems, would suit conservatives just fine.

LEGALLY RACIALIZING INDIGENEITY VIA GENETICS

The recent scientific consensus that race is not viable as a genetic construct, has not stopped diverse actors—from both the Left and the Right—from using concepts of genetic ancestry to make legal claims regarding the status and rights attendant upon certain claims of Indigenous identity. In the years since the completion of the Human Genome Project, a complex and fraught series of legal contestations have played out in the United States around issues of race, genetics, and indigeneity. As anthropologist Jennifer Hamilton has observed:

The legal history of Indian identity in the United States is both complex and often contradictory.... At various times throughout U.S. legal history, establishing who is and is not Indian has been central to determining collective and individual identities. These identities are, in turn, tied to questions of land and resource distribution, property, inheritance, treaty payments, state and federal benefits, civil and criminal jurisdiction, tribal membership, and certain political rights.[100]

Tribal membership determination is a critical attribute of sovereignty.[101] Scholars of genetics and Indigenous identity, such as Kim TallBear and Krystal Tsosie, and Native American tribal leaders themselves have made clear that tribal identity is not determined genetically.[102] As Tall-Bear puts it, "The tribe is not strictly speaking a genetic population. It is at once a social, legal, and biological formation, with those respective parameters shifting in relation to one another."[103]

In 2004, around the time that Alexander and Schwarzschild were publishing their article on how modern genetics undermined the logic of race-based affirmative action, biologist Rick Kittles (who was also at that time working with Farmer-Paellmann on the reparations case), approached several leaders of the Cherokee Freedmen to offer his services in the quest for recognition and inclusion as members of the federally recognized tribes of the Cherokee Nation. The Cherokee Freedmen were descendants of black slaves kept by the Cherokee until 1866, when they gained their freedom and were granted tribal citizenship. In 1983, the Cherokee tribe adopted a rule effectively limiting citizenship to individuals who could trace direct descent from an ancestor listed on the Dawes Roll, a 1906 federal census of the United States. Indians that largely excluded freedmen. As a result, thousands of black members were expelled from the tribe and denied access to benefits and reparations money. In the following years various freedmen brought a series of lawsuits trying to obtain tribal membership and the attendant benefits, all to no avail as U.S. courts have generally avoided meddling in such affairs.[104]

Kittles's services took the form of free DNA ancestry tests, which many Cherokee Freedmen embraced "in hopes that science would succeed where rhetoric, litigation, and historical documents have failed." Ultimately, Kittles's tests proved less than definitive, finding that the degree of Indian ancestry among the freedmen was on average around 6 percent, about the same as an East Coast African American population.[105] And so, in the end, little was changed by the appeal to genetics. Nonetheless, in the years since Kittles approached the Cherokee Freedmen, some twenty-five different companies emerged offering DNA ancestry tests to help consumers discover their ancestral origins. Of these, five offered a test solely for Native American heritage that they claimed might be used for asserting legitimacy in a tribal enrollment process.[106]

The Cherokee Freedmen's attempt to use DNA tests to gain access to the material benefits of tribal membership may sound similar to what Taylor would try to do several years later in Washington State. But in this case, there was a clear and long history of the freedmen holding themselves out as being tribal members. Complicating matters further, TallBear notes that the entire concept of blood quantum as a basis for determining tribal membership has a complex and contested history. It was U.S. federal agents in the nineteenth century who "settled upon blood as a mechanism to break up collectively held Native American land bases." The Cherokee policy of 1983 was not grounded in blood quantum per se, or genetics, but in lineal descent from someone listed on the Dawes Roll; yet the Dawes Roll, in turn, was constructed around ideas of blood quantum.[107] Other scholars note that the very concept of "Native American DNA" is scientifically misleading because "the genetic markers commonly used to identify this type of ancestry are also found in other populations at lower frequencies."[108]

The Indian Child Welfare Act

One gets a fuller sense of the double edge of invoking genetics in such contexts by considering that a decade or so after the Cherokee Freedmen tried to use genetics to gain access to membership in particular sovereign Native American tribes, conservative activists would apply a similar genetic logic in an attempt to undermine the sovereign power of Native American tribes more generally. In 2017, a non-Native American couple filed suit in federal court to have the Indian Child Welfare Act

(ICWA) declared unconstitutional. The couple, Chad and Jennifer Brackeen, sought to adopt "A.L.M.," an "Indian child" under the terms ICWA. The child's biological mother was an enrolled member of the Navajo Nation and his biological father was an enrolled member of the Cherokee Nation.[109] ICWA was passed in 1978, "to address rising concerns over "abusive child welfare practices that resulted in the separation of large numbers of Indian children from their families and tribes through adoption or foster care placement, usually in non-Indian homes."[110] ICWA does not bar non-Native American families from adopting or fostering Native American children outright but requires them to show "good cause" that the child cannot or should not be adopted by other Native Americans before gaining custody.[111] More specifically, the ICWA requires that "a preference shall be given, in the absence of good cause to the contrary, to a placement with: (1) a member of the child's extended family; (2) other members of the Indian child's tribe; or (3) other Indian families."[112] Under the law, an "Indian Child" is defined as, "any unmarried person who is under age eighteen and is either (a) a member of an Indian tribe or (b) is eligible for membership in an Indian tribe and is the biological child of a member of an Indian tribe."[113]

In 2018, Judge Reed O'Connor of the Federal District Court for the Northern District of Texas, found in favor of the plaintiffs and struck down ICWA as unconstitutional. At the core of his ruling was his assertion that the classifications used in ICWA were "racial" in nature and therefore had to satisfy strict scrutiny to satisfy under constitutional equal protection review.[114] To make this finding, O'Connor had to distinguish this case from the 1974 case of *Morton v. Mancari*, where the Supreme Court had found that ICWA involved *political* not *racial* classifications and hence upheld the ICWA under the less stringent "rational basis" standard as "rationally related to the federal government's desire to protect the integrity of Indian families and tribes."[115] O'Connor focused on the term "biological child" to make this distinction, finding that the definition of "Indian Child" used in the ICWA was not based on the political identity of any tribe but on whether "the child is related to a tribal ancestor by blood."[116] Citing *Adarand Constructors v. Pena*, he concluded that "the ICWA's jurisdictional definition of 'Indian children' uses ancestry as a proxy for race and therefore 'must be analyzed by a reviewing court under strict scrutiny.'"[117] He essentially found that being based in biology the preference amounted to a racial preference—both casting race as a biological construct and using the reference to biology as grounds for recasting the preference from a political one (based on tribes being semi-sovereign entities under U.S. law) to a racial one.

In reaching this conclusion, O'Conner echoed arguments made the previous year by the conservative Pacific Legal Foundation (PLF) in filings made in support of writs of certiorari seeking Supreme Court review of two ICWA related cases.[118] In these briefs, the PLF argued that the strict scrutiny standard of *Adarand* should apply to ICWA because it, "regulates Indian children based solely on their genetic association and descendancy," and that "because ICWA equates 'Indian' with the tribe's blood quantum rules, it equates tribal interests with genetic interests, and therefore dictates that 'biology' and not 'social, legal, or political identification, makes a person Native American.'"[119]

Like Judge O'Connor in *Brackeen*, the PLF was arguing that being based in genetics, the category of "Indian Child" had to be construed as racial in character. Yet, less than a decade prior to this, in 2008, the PLF was arguing in the case of *Hawaii v. Office of Hawaiian Affairs*, that genetic science actually showed racial classifications themselves to be arbitrary.[120] *Hawaii v. Office of Hawaiian Affairs* involved the question of whether Congress stripped the State of Hawaii of its authority to alienate its sovereign territory by passing a joint resolution in 1993 to apologize for the role that the United States played in overthrowing the Hawaiian monarchy in the late nineteenth century.[121] In its amicus brief, the PLF (together with the Cato Institute and the Center for Equal Opportunity) focused on whether this resolution required the State of Hawaii to reach a political settlement with Native Hawai'ians regarding the status of contested lands. Central to its argument that it did not, was its contention that the category of "Native Hawai'ian" was an incoherent and divisive racial classification and hence could not be constitutionally employed without meeting strict scrutiny. Anticipating some of the arguments later used by Judge Garza in *Fisher*, the PLF argued that "considerable doubt exists whether race can even be quantified scientifically," and that "there is no taxonomic basis in biology or physiology to support racial distinctions used by the U.S. Census." It concluded that "because there is no single Hawaiian tribe or nation that can make this determination, the state and federal governments have answered this question with arbitrary distinctions."[122] The genetically based arguments here were less fully developed than they would become in hands of Garza and Judicial Watch (and in the PLF's own later briefs), but the seed was planted.

Returning to *Brackeen*, O'Connor's opinion with respect to the racial nature of the classifications in ICWA was overturned at the Fifth Circuit Court of Appeals by a panel of three judges in 2019.[123] It was then reheard en banc by the full Fifth Circuit Court of Appeals which upheld some of the panel's findings but was evenly divided as to whether ICWA's other preferences—those prioritizing "other Indian families" and "Indian foster home[s]" over non-Indian families—unconstitutionally discriminated on the basis of race. It thus effectively affirmed the district court's ruling that these preferences were unconstitutional.[124]

The case then made its way up to the Supreme Court under the name of President Biden's Secretary of the Interior Deb Haaland, as *Haaland v. Brackeen*.[125] In June 2022, a trio of conservative legal organizations, the Goldwater Institute, the Cato Institute, and the Texas Public Policy Foundation, filed an amicus brief in support of the plaintiffs that foregrounded genetics, stating: "ICWA categorizes based on genetics alone-not culture, political affiliation, or treaty rights. It therefore applies to children who may never become tribal members. That means it creates not a political, but a racial classification."[126] Similarly, one of the petitioners, the State of Texas, led by Republican Attorney General Ken Paxton, filed a brief that opened with a declaration that that the ICWA "creates a child-custody regime for 'Indian child[ren],' a status defined by genetics and ancestry." The brief went on to argue that the "ICWA violates the Constitution's equal protection guarantee by categorizing children based on genetics and ancestry and potential adoptive parents based on their race."[127]

The transmogrification of Indian identity from political to genetic lay at the core of the effort to abolish the ICWA. These arguments were developed by focusing on the reference to "biological child" in the ICWA's definition of "Indian Child" but ignored the preceding section stating that an Indian child may also be defined simply by membership in a tribe.[128] As Abi Fain and Mary Kathryn Nagle made clear in their analysis of similar arguments made by the Goldwater Institute in the related 2015 case of *A.D. v. Washburn*: "ICWA's 'Indian child' definition renders the identity of an 'Indian child' contingent upon the political citizenship of one of the child's biological parents—not ancestor—or the biological parent's decision to enroll his or her child, if the child is already a tribal citizen at the time of the adoption proceedings."[129] This distinction was central to the Fifth Circuit's overturning of O'Connor's opinion, wherein it concluded that "contrary to the district court's determination, that ICWA's definition of 'Indian child' is a political classification subject to rational basis review."[130]

Justice Alito's opinion in the 2013 ICWA case of *Adoptive Couple v. Baby Girl* gave advocates of protecting Native America rights under the ICWA concern. In that case, the reach rather than the constitutionality of the ICWA was at issue. In his opinion for the court, Alito held that the ICWA did not apply to a situation where the relevant parent never had custody of the child. Alito, however, opened his opinion by declaring that "This case is about a little girl (Baby Girl) who is classified as an Indian because she is 1.2% (3/256) Cherokee."[131] The degree of Cherokee ancestry was never at issue in the case, yet Alito set the entire frame for his opinion around the idea of biological descent as the basis for Indian identity and tribal membership. In highlighting the small percentage number, he implied there was something absurd about this result. And Alito was not alone in his concerns. Fain and Nagle observed that "several Supreme Court Justices [in *Adoptive Couple v. Baby Girl*] began to question whether a Tribal Nation could grant citizenship to a child of a tribal citizen if the child lacks sufficient blood quantum. As Chief Justice Roberts asked, 'is there at all a threshold' at which the child of a tribal citizen can no longer be considered eligible for citizenship in a Tribal Nation?"[132] In foregrounding what was, in effect, a blood quantum basis for characterizing tribal identity, Alito's opinion certainly could be read as laying the groundwork for adopting conservative arguments that the ICWA employed racial rather than political classifications.

This was not just about the ICWA. The genetic argument has potentially far-reaching implications for tribal sovereignty. Sarah Kastelic, the executive director of the National Indian Child Welfare Association, has noted that in challenging the ICWA, conservative think tanks such as Goldwater and Cato, "have a broader agenda about state rights and subverting or dismantling tribal sovereignty as part of their agenda."[133] Rebecca Nagle noted that the logic of the plaintiff's arguments in Brackeen "could gut native sovereignty," and that "the potential domino effect of the lawsuit rests in the way it seeks to reframe tribal membership as a racial rather than a political category."[134] This is one reason why some Native Americans reacted with such vehemence to Senator Elizabeth Warren's appeal to a DNA ancestry test to support her claims of Indian ancestry; connecting genetics to tribal identity could be (indeed was being) used to undermine tribal sovereignty.[135]

In the end, the Supreme Court, in a 7–2 decision written by Justice Amy Coney Barrett, affirmed the Fifth Circuit's determination that the ICWA was constitutional. In his concurrence, Justice Gorsuch emphasized the political nature of tribal entities, citing *Morton v. Mancari* for the proposition, "confirming... the bedrock principle that Indian status is a 'political rather than racial' classification."[136] Unsurprisingly, Justice Alito dissented, as did Justice Thomas. None of the opinions directly engaged with the issues of genetics or biology raised in some of the briefs, but in affirming the political nature of tribal status, Gorsuch's concurrence clearly indicated that, at least for the purposes of the ICWA, the identity of "Indian" was not genetic.

Judge O'Connor's district court opinion in *Brakeen v. Bernhard* sought to bring the ICWA under the same standard of strict scrutiny review as the racial preferences employed in affirmative action cases. These are the racial preferences other conservative think tanks such as Judicial Watch have been challenging as incoherent. The Supreme Court rebuffed this attempt to biologize the legal status of Native American tribes. But the point is not simply that some conservatives believe race is genetic and others believe it is social (or political). It is that conservative activists will tactically deploy either conception of race if it serves their interest of undermining what they perceive to be as any sort of racial preferences aimed at remedying past injustices.

THE TACTICAL DEPLOYMENT OF RACE AND GENETICS IN A POST-GENOMIC ERA

The conservative embrace of social constructionism in affirmative action has been tactical. Following the publication of Richard Herrnstein and Charles Murray's book, *The Bell Curve*, in 1994, the idea that race-based genetic differences accounted for differences in standardized test scores gained greater traction among opponents of affirmative action.[137] Of course, the idea of inherent differences among the races has been central to the construction of White supremacist racial order in the United States from its inception, but *The Bell Curve* brought the idea of *genetic* difference to a broad popular audience (especially after being featured in *The New Republic*) at a time of intense debate over affirmative action policies in the United States. Nonetheless, less than a decade after the publication of *The Bell Curve*, the Supreme Court had reaffirmed the validity of affirmative action in *Grutter v. Bollinger*.[138] The Supreme Court upheld the use of racial categories in admissions programs in *Grutter* less than two years after Craig Venter, standing beside President Clinton at the White House, declared that the completion of the first draft of the human genome made it clear "that the concept of race has no genetic or scientific basis."[139] The Supreme Court itself made no reference to *The Bell Curve* or genetics in its opinion. Less than a year later, Alexander and Schwarzschild wrote the law article that would provide the basis for Garza's genetically informed arguments in *Fisher*.[140] Before the corpse of Herrnstein and Murray's genetically based assault on "racial preferences" had even cooled, the conservative pivot to leverage findings from genetic science and embrace social constructionism as a means to undermine affirmative action had begun.

On a more mundane level we see a similar dynamic at work in Ralph Taylor's case. He began by trying to use the results of his DNA ancestry test to claim membership in a group that would qualify Orion Insurance as a Disadvantaged Business Enterprise. When the state failed to find his genetic evidence sufficient and required additional evidence relating to his social standing and practices, he pivoted to attack the entire premise of racial preferences as arbitrary.[141] From someone who began his quest inspired by a "Caucasian looking guy" who had gotten "a big chunk of money" by claiming to be a minority, Taylor evolved into a self-styled crusader "fighting for a greater good, exposing flaws with affirmative action programs," embraced by conservative talk show hosts and highlighted on White supremacist websites such as *VDARE* and *American Renaissance*.[142]

The key theme connecting the conservatives' tactical embrace of race as social in opposition to affirmative action and embracing it as biological in characterizing health disparities was the basic idea that race should not be invoked when addressing social inequalities. It was not simply about law or policy but about the ends to which those laws and policies were applied. This connects the racial essentialist arguments of *The Bell Curve* to the social constructionist arguments of Garza and Judicial Watch. The former look to the "reality" of race to make *racism* seem irrelevant to policy; the latter look to social constructionist statements to make the case to make *race* seem irrelevant. Thus, in *The Bell Curve*, racial disparities were cast as due to biology not racism: there was no racism. For Garza and Judicial Watch, affirmative action was invalid because there was no coherent category of *race*, again, validating the racial status quo. The apparent inconsistency of the conservative embrace of race as genetic in some contexts and as social in others is made coherent when we understand that the underlying purpose it to maintain a preexisting racial order.

These arguments are grounded in false premises. The assertions in the realm of biomedicine concerning purported genetic bases for health disparities ignore not only the scientific consensus that race is not genetic, they also fly in the face of voluminous epidemiological evidence that such disparities are products not of genetics but of social, historical, and environmental forces. The alternative claim that because race is not genetic, because it is "merely" social, it therefore is not "real," suffers from the delusion that social constructions somehow do not exist or have an impact. Certainly, this would come as a surprise to anyone who uses money, which, after all, is a social construct valuable only because we, as a society, have deemed it so.[143] There is no inherent value to a dollar bill, yet it exists, it has a worth, and people know how to exchange and use it. That its value may be fluid, may change across time and space, does not make it any less real. Similarly, the idea that a social construction such as race might be fluid, change over time and space, or be subject to contestation does not make it arbitrary or incoherent. Over our history, these characteristics have given race a distinctive power and adaptability in both law and society. Nonetheless, while these spurious arguments have as yet achieved limited success, they are out there like a loaded gun, waiting to be picked up and fired by a sympathetic court or policymaker. Eternal vigilance is the price of adapting to new challenges to racial justice.

CHAPTER NINE

Diversity and the Frames of Representation

CONSTRUCTING ANCESTRAL DIVERSITY

In September 2015, the National Human Genome Research Institute (NHGRI) convened a roundtable to discuss the opportunities and challenges associated with the inclusion and engagement of underrepresented populations in genomics research. The roundtable focused on "the lack of ancestrally diverse populations represented in genomic studies and how that results in limitations for genomic science and medicine." Particular attention was devoted to research being done "to increase the inclusion of underrepresented ancestral populations in genomics research and address health inequities in genomic medicine." Foremost among its recommendations was that the NHGRI "provide[] leadership in promoting genomics research that addresses minority health and health disparities."[1]

The primary focus of the roundtable, however, was on "the lack of ancestral diversity in genomic studies."[2] "Ancestral diversity" is a very particular type of diversity. It hearkens back to the sort of categorization provided by the STRUCTURE program and to the idea of AIMs—ancestry informative genetic markers. Ancestry here was used as a distinctively genetic concept reflecting the transmission of genetic variations through generations over time. It is, technically speaking, very much not a racial concept. But as with the rise of the commercial, direct-to-consumer ancestry tracing business, genetic ancestry in practice becomes all too readily conflated with race. And such was the case here, because the roundtable directly connected the lack of "ancestral diversity" in genomic studies to the idea of "minority [i.e., race-based] health and health disparities."

The roundtable constructed and operationalized diversity in two ways: first, as a function of underrepresentation which necessarily involved concepts of social groups and their presence or absence from existing databases; second, as a function of genetic ancestry. But what did this really mean? Ancestry itself became relevant as a function of what was absent from databases. One might think of ancestral groups as preexisting in the natural world prior to social groups, but there is a difference between "ancestry" and "ancestral groups." Ancestry may be constructed as a function of lineal descent, but "ancestral diversity" as used here necessarily implies the idea of "populations" or "groups" that were underrepresented in existing databases. Such groups (necessarily) are constructed, they are a matter of whose ancestry gets grouped with others, over both time and space. Ancestral groups may sound as if they are less socially constructed than races, but they are only constructed differently.

In terms of representation, the idea of ancestral genetic diversity is grounded in the concept of homology—derivation from a common biological source. In this sense, an "African" allele does not represent all those living today with that same allele, it re-presents (in the sense of making present again) some original ancestral allele from the past which is literally present in others of the same descent. The tricky thing is that in a database or clinical trial the actually human sources of the allele are not just representing that allele or even all those who currently have the allele—they are representing full and unique genomes as well, not to mention the racial group to which that allele has been assigned, not by reason of the molecules, but of the social, racial designation. That is, any one genome "re-presents" many different ancestral alleles while being in its totality wholly unique and unrepresentative with respect to the entire complement of genomic DNA.

Moreover, given the complex dynamic nature of how DNA actually functions and is expressed in environmental and epigenetic context, even if you have a particular allele—particular alignment of certain base pairs in a particular location—it does not mean that they will function in the same manner as the ancestral allele—i.e., representing isolated molecules is not necessarily the same as re-presenting their function in vivo.

THE LIMITS OF DIVERSITY MANDATES

In 2012, Congress passed the Food and Drug Administration Safety and Innovation Act (FDASIA) directing the FDA to produce a report that took a closer look at the inclusion and analysis of demographic subgroups in drug trials.[3] Reiterating a theme that had been articulated in various forms at least since the National Institutes of Health's (NIH) Revitalization Act of 1994, the report noted that "it remains important that clinical trials include diverse populations, whenever possible and appropriate," and that the "FDA's next step will be to use the information gathered for this report as a starting point for developing an Action Plan" to address these concerns.[4] The action plan, sadly, highlighted the case of BiDil as a positive example of how to use subgroup analysis in drug development.[5]

Significantly, however, the plan took pains to address the dangers of conflating social categories of race with genetic categories of ancestry, noting: "Stakeholders commented to the public docket that there are issues concerning the validity, both scientifically and clinically, of patient identification by race (phenotype) or ethnicity (a socio-cultural quality), rather than ancestry or genetic make-up, where appropriate."[6] Stakeholder comments on the proposed action plan were notable in their different approaches as to how best to diversify clinical trials. Numerous nongovernmental organizations wanted the FDA to set specific guidelines requiring certain levels of racial and ethnic diversity in clinical trials. For example, The American Heart Association together with representatives from the Society for Women's Health Research, WomenHeart: The National Coalition for Women with Heart Disease, and the National Women's Health Network, urged that the FDA "require that representative proportions of women, minorities, and older adults are included in clinical trials at all phases, consistent with the disease's prevalence in the underlying population."[7] Stunningly, these comments directly geneticized race, expressing that diversity was important because, "it is critical that patients and clinicians have information that describes the impact of genetic factors such as sex and race on the efficacy of new devices and treatments."[8]

In contrast, many pharmaceutical corporations expressed concern over affirmatively mandating such diversity. For example, the Biotechnology Industry Organization (BIO), while supporting the goal of diversifying enrollment, opposed any direct mandate. It strongly urged that FDA approval for any drug, "should not be delayed if the Sponsor has made a good faith attempt to include . . . relevant but historically under-represented subpopulations in clinical trials of investigational medicines."[9] Far better, it asserted, for the FDA to provide "incentives" for enrollment, akin to the use of "pediatric exclusivity" provisions that gave added market protection to companies agreeing to conduct trials for safety and efficacy in children. Similarly, Boehringer Ingelheim Pharmaceuticals Inc., while pledging to continue "our goal toward increase diversity in our clinical trials," nonetheless urged that "consideration of genotypic profiling versus phenotypic categorizations may begin to take precedence over current guidelines regarding race and ethnicity."[10]

Multinational pharmaceutical corporations were particularly concerned that any mandated inclusion criteria would be based on the FDA's adoption of the U.S. census categories, which do not translate to an international context. Similarly, ideas of proportional representation for enrollment would change dramatically depending on the local population where a particular drug trial might be conducted. For example, in posing the example of calculating representational enrollment for a type 2 diabetes drug trial, Boehringer noted: "If the denominator is shifted from a U.S. population to an international population, then the proportions will change accordingly"; therefore it urged that "flexible guidelines would be useful during project planning stages and throughout trial conduct."[11]

Here multinational pharmaceutical corporations were recapitulating many of the arguments they made about diversifying clinical trials over a decade earlier in response to the proposed 1998 International Conference on Harmonization E5 guidelines on "Ethnic Factors in the Acceptability of Foreign Clinical Data."[12] For them, it was imperative that diversity be discretionary. All the talk was on incentives but not mandates. In this, Big Pharma's approach echoed arguments in favor of discretionary affirmative action made broadly by major corporations. In the end, the FDA went with the idea of discretionary diversity. For all the talk of how essential it was to diversify clinical trials in order to serve concerns of equity and address health disparities, when it came to corporate recruitment, organizations such as BIO were clear that justice had no role to play. Diversity was a nice idea but must be subordinated to the discretion, judgement, and "good faith" efforts of corporations.

THE REACH OF TECHNICAL RATIONALES FOR DIVERSITY

In January, as the *All of Us* program was rolling out its recruitment program, the FDA declared 2016 to be, "The Year of Diversity in Clinical Trials." The declaration listed several of the FDA's leading activities to realize such a plan, such as having its Office of Minority Health develop tools to support clinical trial participation including a multimedia campaign highlighting the importance of clinical trial participation. Again, the concept of "underrepresentation" undergirding such initiatives. The primary rationale being that "certain groups of patients may respond differently to therapies."[13]

Later that year, Alice Popejoy and Stephanie Malia Fullerton garnered a great deal of attention with an article in *Nature News* titled, "Genomics is Failing on Diversity." They noted that a 2009 analysis had found that only 4 percent of participants in GWAS (genome-wide association studies) were of not of "European descent." By 2016, the proportion had risen to 20 percent but much of that was attributable to studies being conducted in Asia on populations of Asian ancestry. The proportion of subjects in GWAS trials identified as African American, Hispanic, or Indigenous had "barely shifted."[14]

The authors were concerned that such limited databases were missing much of the world's genetic variation and with it opportunities to uncover "important information about disease biology." Such concerns were not grounded in the idea that the same genes might behave differently in different populations (although that might be possible given different environmental exposures) but that, given the new focus on the importance of rare variants on significant impact, some actionable variants might be missed while others thought to be actionable might actually be false positives. Diverse databases were essential to capture such variation and help distinguish between true and false positive associations between certain genetic variations and particular health conditions.[15]

POLITICAL RATIONALES FOR DIVERSITY

Beyond this technical scientific reason for diversifying databases, the authors also asserted a political or social justice rationale for diversity in genomics:

> All genomics researchers need to recognize the importance of studying under-represented populations to ensure that the benefits of research are distributed fairly and to maximize the potential for discovery. On a practical level, training programmes and new infrastructure, such as good health-care clinics that provide genetic testing in predominantly black or Hispanic neighborhoods, could enhance trust and allow people to engage in projects as stakeholders rather than as study participants.[16]

While certainly a laudable concern, there is also a complex of hope and assumptions bundled into this statement. First, the term "underrepresented" here stands in for diversity—or rather it implicitly recharacterizes diversity as the function of levels of representation in genetic databases—as opposed to actual diversity of genetic variants. Second, it assumes that increasing social representation in databases was necessarily connected to ensuring the equitable distribution of the benefits of such research to those same populations. That may or may not be the case. Certainly, merely increasing representation in databases in no way ensures such outcomes; more than this though, it is not even clear that increasing such representation is a necessary precondition to obtaining those benefits. As in the earlier efforts to build generic databases in the 2000s, the main appeal here was to "maximize the potential for discovery."

Again, potential was being invoked to make demands to get more bodies. Only here the potential was tied to making demands on (or for) racially or ethnically marked bodies. Finally, there was the explicitly social focus on training programs and infrastructure to serve underrepresented communities. Again, laudable, but in this case providing such resources was not an end in itself but a means to increase "trust" and thereby facilitate the enrollment of "diverse" bodies into genomic studies. This, too, hearkens back to the recruitment efforts explored by the Genetics and Public Policy Center (GPPC) nearly a decade earlier but here the impetus for gaining trust was centrally tied to the perceived need for diversity.

DIVERSITY AND HEALTH DISPARITIES

The following year, Director of the Center for Research on Genomics and Global Health at the NHDRI Charles Rotimi, alongside his colleagues Amy Bentley and Shawneequa Callier, published an article titled, "Diversity and Inclusion in Genomic Research: Why the Uneven Progress?" Expanding on Popejoy and Fullerton's call for diversifying genetic research, the article emphasized that "conducting genomic research in diverse populations is... important in ensuring that the genomic revolution does not exacerbate health disparities by facilitating discoveries that will disproportionately benefit well-represented populations."[17] The article described two main goals motivating the drive to promote genomic research in diverse populations:

> First, as a matter of justice, individuals are expected to benefit most from genomic research conducted in individuals with a similar ancestral background to them. Failure to fully engage diverse populations at all levels of genomic research perpetuates already considerable health disparities. Second, including diverse populations in genomic research is not just the right thing to do for reasons of equity, it is a scientific imperative. The genomes of diverse individuals harbor a treasure trove of humanity's responses to challenges experienced by some or all populations: changes in climate, infectious diseases, diet, etc.[18]

Again, these are worthy goals. Nonetheless there is a significant blurring of the relationship between genetic and social diversity here. The justice concerns focusing on health disparities are clearly oriented around socially constructed racial categories, groups of people who have suffered disproportionate burdens of morbidity and mortality primarily due to social and historical factors. These disparities are not due to differences in genetics. Yet, the article confidently asserts without support that "individuals are expected to benefit most from genomic research conducted in individuals with a similar ancestral background to them." Here, diversity in a genetic database is purported to represent "ancestral background." But disparities are neither tracked nor are they caused as a function of "ancestral background." This concept emerged out technologies such as the STRUCTURE program and the analysis of Ancestrally Informative Markers (AIMs).

One might more readily assert that for any given medical condition in the United States with significant health disparities, say diabetes or hypertension, "individuals are expected to benefit most from genomic research conducted in individuals *with similar medical conditions.*" Now, it might be that a genetic database in the United States made up of individuals predominantly of European ancestry might represent a group of individuals with a lower incidence of those conditions than one including more individuals of African ancestry. But there is no evidence that that would be because the *genetics* of those individuals was any different. Rather, it would be because the historical, social, and environmental experiences of those individuals led to a higher incidence of disease. Having more people with diabetes in a database might help uncover genetic contributions to diabetes. Having a more "diverse" database might result in having more people with diabetes providing samples. But those samples would be valuable not as samples from a "diverse" population but as samples from individuals with diabetes. Justice is not served by merely diversifying genetic databases along racial or ethnic lines. Justice is served and health disparities are addressed by ensuring that the databases, however constructed, provide a basis for sustained and productive analysis of those conditions where health disparities are most prevalent. More to the point, justice is ultimately only served if those "diverse" populations suffering from disparities have practical and effective access to any therapies derived from such research. More radically, justice is also served by addressing the underlying structural and historical conditions driving disparities, wholly independent of any downstream biomedical interventions. None of that is contingent upon increasing the representation of African Americans *qua* African Americans (or other minoritized groups) in genetic databases.

The Rotimi et al. article then turned to the issue of recruiting a more diverse genomics workforce as a necessary antecedent to achieving more diverse databases. For the authors, "an overarching concern" was that "more diverse representation at all levels of research is sorely needed—from the participants included to the reviewers of proposals to the scientists conducting the research. Scientists from low-resource settings, minority backgrounds, and diverse ethnic groups are grossly underrepresented in all of these areas." In the end, this all comes down to using diversity as representation of marginalized groups in order to cultivate trust, for as the authors noted, "empirical evidence has shown, for instance, that trust is a critical component in the participants' views regarding important ethics and legal challenges in genomic research."[19]

BIOMEDICAL DIVERSITY BLURS INTO DEI

In 2019, the American Society of Human Genetics (ASHG) announced a, "Program to Increase, Network, Mentor, and Retain Diverse Early-Career Researchers." This initiative aimed at increasing and supporting "workforce diversity in the human genetics and genomics research community" by providing a two-year mentoring and skill building experience for scientists from "underrepresented backgrounds."[20] The ASHG mentioned neither race nor ethnicity, only "underrepresentation." This initiative was clearly informed by the goals and rhetoric of affirmative action and its more recent iteration as a movement for "diversity, equity, and inclusion" (DEI). But it was DEI with a twist. Where the sort of affirmative action endorsed by diversity trainers and the Supreme Court in cases such as *Grutter* focused on the commercial and efficiency benefits of a diverse workforce in a global economy, the benefits of diversity and more specifically "representativeness," were here linked (as they were by Popejoy and Fullerton and then Rotimi et al.) to enhancing recruitment of similarly diverse subjects into genetic research.

As the ASHG press release noted:

> Greater diversity is a recognized need in the human genetics community. Findings show that historically underserved populations

are under-studied and underrepresented in nationwide and global genetic research databases, highlighting the de facto exclusion of diverse populations from many health studies and the resulting benefits. Workforce studies demonstrate the need to adopt more effective strategies to recruit, retain, and promote diverse researchers in research.[21]

In other words, recruiting, retaining, and promoting researchers from underrepresented populations would presumably facilitate the recruitment of subjects from those same populations. In this, the press release was merely representative of many other similar calls in a range of articles, many spurred by the PopeJoy and Fullerton piece, similarly calling for the diversification of the genetics workforce in order to facilitate the diversification of genetic databases.[22]

Ultimately, this focus on redressing disparities was cast as contributing "to realizing the benefits of human genetics for all people."[23] Again, there is an echo of the language of corporate diversity management's argument that diversity improves the bottom line for all; but this rationale also leaves the unsettling impression that merely redressing the injustice of racial health disparities was not enough. To truly merit full support, such initiatives had to benefit "all people," which in this context can only mean including those people who do not count as "diverse," i.e., White people.

On the one hand, this might seem like a canny invocation of Derrick Bell's idea of "interest-convergence": that racial progress is unlikely to be attained except in those areas where the interests of the majority and minority group are understood to converge.[24] On the other hand, because genomic research deals with people not only as social groups but as embodied beings, the convergence inevitably implied some sort of essential biological difference underpinning the groups. Remember that when the FDA approved BiDil as a drug for "Black" people, it did so on the grounds that the clinical trials were conducted only in Black people. The implicit idea being that Black people, biologically speaking, were somehow less than fully representative of humanity. Here, the idea that research on "diverse" genomes would ultimately benefit "all people" necessarily implied that genomes from a particular "diverse group," e.g., Black people, were not coextensive with or representative of the genomes of "all people." Such a framing while not exactly the same as that involved in the approval of the race-specific heart failure drug, BiDil (a drug tested only in Black people could only be approved for Black people, while a drug tested in White people could be approved for everybody), nonetheless invokes a similar troubling logic of racialized subgroups as being somehow "less than" the full category of humanity.

SLIPPAGE BETWEEN SOCIAL AND GENOMIC DIVERSITY

When it came to diversifying the workforce, genomic research here crossed over into the arena of affirmative action. Unlike affirmative action in education or employment, however, the rationale for diversity in the genomic workforce was not limited to increasing productivity, creativity, or other goals largely internal to the enterprise. Here diversity served the additional, and in many ways primary, goal of facilitating the recruitment of a secondary set of diverse individuals—research subjects. Diverse workers in a multinational corporation might facilitate access to broader global markets, but those markets are valued *as* markets—their diversity is secondary, even irrelevant. But in the genomic enterprise, diverse workers are presumed to facilitate access to research subjects who are valued precisely because of their purported diversity—insofar as they represent diverse ancestral genetic backgrounds.

The focus on diversifying the genomic workforce to help diversify databases is notably different in this regard from earlier calls to diversify the healthcare workforce more generally in order to improve quality of care and address disparities. For example, the Democratic Party platform of 2008, in a section on its "Commitment to the Elimination of Disparities in Health Care," declared that "we will support programs that diversify the health care workforce to ensure culturally effective care."[25] Here diversity had nothing to do with ancestral genes, or genetic populations in any way whatsoever. It had a predominantly social meaning and was being employed to address a problem of health disparities that was similarly understood to be social in character not biological.

Consider further how the ideas of voice and representation were embedded in traditional affirmative action programs. Under the *Bakke* and *Grutter* visions of affirmative action, diverse student bodies were valued because of the distinctive viewpoints and experiences the subjects brought to bear. The subject's voice was valued and provided a basis for legitimizing the program. Precisely to avoid the idea of stereotyping, any given individual was understood *not* to represent the whole of their identity group when it came to exercising their voice. In these genomic initiatives, a central goal of affirmative action in the workforce was to diversify databases, where individual subjects were valued precisely because they are understood as representing something essential about their "ancestral background." Moreover, the research subjects themselves largely had no voice at all. Rather it is their DNA that speaks—or is made to speak through the intervention of the diverse workers (and others) who facilitated their recruitment. In politics, representation means delegating authority to your representative. In genomic research, representation means giving your racialized body, or at least your DNA, to your scientists. The scientist does not represent you and is not accountable to you.

Around this same time, the NHGRI was touting a major new publication from the Population Architecture Using Genomics and Epidemiology (PAGE) Consortium with a press release titled, "Putting Diversity Front and Center."[26] The gist of the study involved identifying new genomic variants associated with health conditions such as high blood pressure, diabetes, and chronic kidney disease. Based on a GWAS of "26 clinical and behavioural phenotypes in 49,839 non-European individuals" (specifically Hispanic/Latino, African American, Asian, Native Hawaiian, Native American, or Other) the study found that the frequency of genomic variants associated with certain diseases could vary from one group to another.[27] GWAS, of course, had been coming under increasing criticism over the prior decade for showing lack of actionable results.

In this study, PAGE foregrounded diversity as distinguishing this new GWAS and arguing for its relevance in addressing the politically prominent issue of health disparities. The authors framed the study in the abstract by asserting that GWAS, "have laid the foundation for investigations into the biology of complex traits, drug development and clinical guidelines"; they immediately went on to note that "the majority of discovery efforts are based on data from populations of European ancestry." Already, ancestry was being identified in continental terms, which are all too readily conflated with racial categories. This conflation allowed for the connection of genetic diversity to race-based health disparities, as the framing abstract for the article went on to assert that "in light of the differential genetic architecture that is known to exist between populations, bias in representation can exacerbate existing disease and healthcare disparities," and that "in the United States—where minority populations have a disproportionately higher burden of chronic conditions—the lack of representation of diverse populations in genetic research will result in inequitable access to precision medicine for those with the highest burden of disease." The PAGE study itself used "strategies tailored for analysis of multi-ethnic and admixed populations, [to] describe a framework for analysing diverse populations."[28]

The slippage here is palpable. In the one paragraph statement of the abstract, the study characterized diversity in terms continental ancestry, genetic architecture, minority populations, ethnicity, and genetic admixture. Perhaps most striking is the blurring of concepts of statistical bias with bias as a form of racial discrimination. When the article spoke of "bias in representation," it was describing a fundamentally statistical phenomenon whereby one population of interest was overrepresented with respect to another. The study then connected this statistical bias to the social bias that manifested as health disparities in "minority populations." It asserted that addressing the former would directly ameliorate the latter.

In all of this, the idea of racial animus or racism was completely left out of the picture. This was a purely technical fix for a fraught and complex social and historical problem. But, of course, the study merely asserted rather than established that diversifying genetic databases would ultimately address disparities. Finding rare variants of significant impact might possibly help address certain diseases that disproportionately impact minority or underrepresented populations. This is all to the good. But this framing implied that the basis for such disparities was primarily molecular rather than social. It located the problem of disparities in the bodies of those suffering from them and moreover imposed the burden of addressing those disparities on those same bodies—demanding that they be enrolled in genetic research studies at higher rates. Perhaps, as the ASHG initiative indicates, it was a responsibility of the genetic research community to do a better job of recruiting such bodies—but this approach located the problem of addressing the burden of disease not in society, not in a history of racism, but in the underrepresentation of those bodies in genetic studies.

There is much resonance here with the diversity of corporate recruiting and affirmative action. In these social contexts, the animating idea of diversity management is that the mere presence of more minorities in education or employment itself addresses certain concerns of justice and equity. Similarly, here, the mere presence of "diverse bodies" in research studies is presumed to alleviate disparities. But "representation" functions differently in legal and social contexts where the absence of minoritized people is itself part of the injury. Representation in social contexts involves recognition and voice; recognition as a coequal member of the community and an opportunity to be heard. It involves potential access to power, if not always power itself. Affirmative action in hiring or education provides real goods to underrepresented groups. In biomedical research, it is the underrepresented groups, their bodies, that are themselves the goods being sought after. These studies may make it sound as if underrepresented groups are being provided access to valuable insights about their genetic makeup that might help ameliorate disparities, but in actuality they are the objects not the subjects of access here.

It is additionally notable that "bias in representation" was derived from the idea that most GWAS had been conducted in "European populations." This appears as a bias (statistically speaking) primarily if one is concerned to find rare variants that may have different frequencies across different "populations." But there may be different frequencies of rare alleles *within* groups as well—particularly if they are defined as broadly as "European populations," even more so if we are talking about genetic variation within a continentally defined "African" population, which would contain more variation than the rest of the world's populations combined.[29]

Running in parallel to PAGE at this time was the African Genome Variation Project (AGVP). In a major article published in *Nature* in 2015, the AGVP framed its findings at the outset by declaring that "Globally, human populations show structured genetic diversity as a result of geographical dispersion, selection and drift." Here the focus was on "structured genetic diversity," that was a function of geography and the genetic dynamics of selection and drift, not race. The AGVP article noted how previous efforts, "examining African genetic diversity have been limited by variant density and sample sizes in individual populations, or have focused on isolated groups, such as hunter gatherers limiting relevance to more widespread populations across Africa."[30] This framing of diversity hearkens back to Allan Wilson's original call for the Human Genome Diversity Project to conduct its sampling by a geographic grid method rather than focus on population "isolates." In contrast to the PAGE study's idea that genetic databases are "biased" because they lack representation of broadly grouped non-European populations, the AGVP study had a much more fine-grained, nonracialized understanding of diversity that was less easily entangled with social ideas of racial bias. Here we see how the uses and conceptualizations of diversity are contingent and capable of appearing and reappearing in different forms, at different times, conveying different messages. In this framing, genetic diversity avoids conflation with race; indeed, it implicitly breaks down the identification of genetic diversity with racial or continental groupings by showing the great range of diversity *within* sub-Saharan Africa.

THE PERSISTENCE OF REGRESSION

Ignoring or glossing over the genetic diversity within Africa can have significant implications for broader understandings of, or arguments about, the relation between genetic variation and the sociohistorical conditions of different racial groups. This was painfully evident in a book published by *New York Times* science reporter Nicholas Wade in 2014 titled, *A Troublesome Inheritance: Genes, Race and Human History*.[31] At the core of this book was an argument that there are discrete and definable genetic differences across groups that we can label as biological races. As biological anthropologist Agustin Fuentes noted in his critical review of the book: "Wade suggests that believing in biological races (especially African, Caucasian, and East Asian) is both common sense and solid science. He asserts that evolved differences in these races are the key explanation for social differences in histories, economies, and trajectories in societies; why 'Chinese society differs profoundly from European society, and both are entirely unlike a tribal

African society.'"[32] As sociologist Philip Cohen noted, Wade's argument was largely contingent on his treating all of sub-Saharan Africa as one racial group. Citing earlier work by geneticist Sarah Tishkoff that presaged the findings of the AGVP, he noted that "The vast diversity within Africa should have warned Wade away from lumping Africans into one racial category, but he ignores this fact."[33]

More damningly, a letter to the *New York Times* signed by over 140 geneticists and evolutionary biologists condemned the book, stating:

> Wade juxtaposes an incomplete and inaccurate account of our research on human genetic differences with speculation that recent natural selection has led to worldwide differences in I.Q. test results, political institutions and economic development. We reject Wade's implication that our findings substantiate his guesswork. They do not. We are in full agreement that there is no support from the field of population genetics for Wade's conjectures.[34]

Nonetheless, Charles Murray, coauthor of *The Bell Curve*, wrote a glowing review of the book as did notorious White supremacist Jared Taylor.[35] As biological anthropologist Jennifer Raff noted, much of Wade's argument (and Murray's embrace of that argument) was grounded in a misreading of how the STRUCTURE program worked to create genetic clusters via a predesigned sorting algorithm. To support his idea that genetic diversity reflected racial groupings, Wade uncritically focused on how the STRUCTURE program could be used to cluster genetic samples into five continental groupings. Wade liked continental groupings because they supported his ideas about the "reality" of racial groups. But such groupings were not "natural'"; rather, they reflected parameters set by researchers. As Raff noted, "rather than being defined by empirical criteria, as Wade had asserted so confidently earlier in the book, it really is just a subjective judgment call. The differences between groups are so subtle and gradual that no objective lines can be drawn, so Wade draws his own on the basis of his own preconceptions. In other words, he can't define distinct races. He just knows them when he sees them."[36]

Arguments by the likes of Wade explicitly miscast racial diversity in genetic terms and were therefore susceptible to these clear and direct critiques. More complicated and difficult to untangle is the sort of work that ideas of diversity and representation were doing in the PAGE program and in related initiatives such as *All of Us*. While driven by underlying conceptions of race, race itself was rarely mentioned but was subsumed under the rubric of "underrepresented populations." This in itself was not necessarily a bad thing but when connected to the idea of addressing "deep health disparities" it became far more problematic because it elided the historical injustices and racism underlying those disparities. Instead, it transmuted disparities into a function of lack of diversity in genetic databases. It implicitly constructed lack of representation in research as itself a form of injustice and thereby obscured the presence and dynamics of racism.

BUILDING A GLOBAL PANGENOME

Building upon initiatives such as the HapMap Project, the 1000 Genomes Project, PAGE, and the AGVP, in May, 2023, the international Human Pangenome Reference Consortium, a group funded by the NHGRI announced the release of a "new high-quality collection of reference human genome sequences that captures substantially more diversity from different human populations than what was previously available." The original reference genome coming out of the HGP was at this point over twenty years old. Researchers were concerned that it was "fundamentally limited in its representation of the diversity of the human species, as it consists of genomes from only about 20 people, and most of the reference sequence is from only one person.... [U]sing a single reference genome sequence for every person can lead to inequities in genomic analyses." This new "pangenome" reference included genome sequences of forty-seven people with plans to increase the number over time to 350.[37] Each of the studies was framed in terms of showing how researching more "diverse" genomes could yield useful insights into disease.

Directly drawing the link between representing "diverse" genomes and addressing health disparities, NHGRI Director Eric Green, declared that "basic researchers and clinicians who use genomics need access to a reference sequence that reflects the remarkable diversity of the human population. This will help make the reference useful for all people, thereby helping to reduce the chances of propagating health disparities." Green highlighted the importance of "striving for global diversity in all aspects of genomics research" as being "crucial to advance genomic knowledge and implement genomic medicine in an equitable way."[38]

The Human Pangenome Reference Consortium published an earlier article on the project in *Nature* in April, 2022, where they characterized the pangenome as, "a global resource to map genomic diversity." It characterized its goal as "identify[ing] of individuals from diverse genomic and biogeographical backgrounds to include in the pangenome reference."[39]

The article was quite sensitive to ethical issues involved in characterizing diversity, noting: "Researchers have demonstrated a lack of diversity in genomics research using biogeographical ancestry groupings at the continental level, as well as sociocultural categories such as racial and ethnic identities. It is important to distinguish biological and sociocultural diversity, as sociocultural labels are not derived from genotype data, and vice versa."[40] Such cautions are important, thoughtful, and necessary. But, alas, they echoed those articulated by the researchers behind the International HapMap Project nearly twenty years earlier that were soon forgotten or glossed over by researchers. The use of the idea of diversity is understandably driven by concerns to include important data that could help further biomedical advances. Unfortunately, the term diversity itself is the problem here. Precisely because it has uses in both sociocultural and biological frames, in practice it will almost always provide an avenue for confusion and conflation of what is being represented by these categories—sometimes through sloppiness, sometimes by accident, sometimes willfully.

Thus, for example, the subheadline to the *New York Times* article covering the announcement of the Pangenome recharacterized the stated goal seeking out individuals from "diverse genomic and biogeographical backgrounds" as "47 people of diverse ethnic backgrounds." As has been all too common in *New York Times* coverage of genomics over the decades, researchers' concerns to distinguish sociocultural from biological framings of diversity get lost in translation.[41] Indeed, neither did the NHGI press release announcing the study, nor the study itself, even use the words race or ethnic. The study spoke instead of "genetic and biogeographic diversity."[42] Yet the *New York Times* article immediately went to race and ethnicity, stressing that this "more accurate and inclusive edition of our genetic code" promised "to deliver treatments that can benefit all people, regardless of their race, ethnicity or ancestry."[43]

A STEP IN THE RIGHT DIRECTION

Sensitive to many of these issues, in 2023 the National Academies of Science, Engineering, and Medicine (NASEM) issued a report titled, "Using Population Descriptors in Genetics and Genomics Research: A New Framework for an Evolving Field." The NIH charged the interdisciplinary committee, "to conduct a study to review and assess existing methodologies, benefits, and challenges in using race, ethnicity, ancestry, and other population descriptors in genomics research." The report explicitly cautioned "against the use of typological categories, such as the racial and ethnic categories established by the U.S. Office of Management and Budget in Statistical Directive 15, for most purposes in human genomics research"; noting that "the fundamentally sociopolitical origins of these categories make them a poor fit for capturing human biological diversity and as analytical tools in human genomics research." More specifically, the report's very first recommendation states: "Researchers should not use race as a proxy for human genetic variation. In particular, researchers should not assign genetic ancestry group labels to individuals or sets of individuals based on their race, whether self-identified or not." The report was also very attentive to the importance of the assurance that environmental factors do not fade from consideration as studies progress. It's fourth recommendation states, "Researchers conducting human genetics studies should directly evaluate the environmental factors or exposures that are of potential relevance to their studies, rather than rely on population descriptors as proxies."[44] The concern in both recommendations to avoid the pitfalls of casually using proxies that might, over time, come to be reified as the thing they represent is well-taken. The report represents precisely the sort of deliberate engagement with the issues that is necessary to keep social and biological conceptualizations of diversity in proper juxtaposition so as to allow their mutual use in genomic and biomedical research without them becoming entangled.

Nonetheless, in February 2024, a set of papers from the *All of Us* research program published together in the journals *Nature, Communications Biology,* and *Nature Medicine,* used explicitly racialized language in purporting to identify "gaps in genetic research in non-white populations."[45] One figure in particular was published in the highest-profile journal, *Nature,* and drew attention from geneticists who expressed concern that "it could be misinterpreted as reinforcing racist beliefs."[46] The figure itself was intended "to showcase the diversity of the first 250,000 genomes included in the All of Us database" by plotting out groupings of sampled genomes along an x-y axis.[47] The researchers used an algorithmic program called UMAP to visualize genetic relationships among the samples. UMAP, in some respects, might be understood as an updated version of the STRUCTURE program first developed by Jonathan Pritchard over twenty years earlier. The researchers explicitly associated the samples with "Race" and employed the terms "White," "Black," "Asian," "Middle Eastern or North African," and "Native Hawaiian or Other Pacific Islander" in plotting out the relatedness of the samples.[48] The resulting figure made it look as if racialized samples clustered neatly into discrete genetic groups.

Pritchard, who had since come to have deep concerns about how STRUCTURE had been used to reinforce mistaken ideas that race was genetic, was among the first to call out the problems, posting on the social media platform X (formerly Twitter): "It's a pity that All of Us used UMAP to visualize ancestry variation in their new marker paper, out today in Nature. The UMAP algorithm, by design, exaggerates the distinctiveness of the most frequent ancestries, a message that can be misinterpreted by the public."[49] Further commenting on the figure, statistical geneticist Roshni Patel, a colleague of Pritchard's, reiterated the well-known argument that "genetic variation is a continuum, and thus genetic ancestry cannot be objectively carved out into discrete groups." As a subsequent news article in *Nature* reported, "To a layperson, the chart shows several distinct colourful blobs that could be misinterpreted as supporting genetic essentialism." In response, one of the coauthors of the paper asserted that this was the opposite of what the data showed, and that "Our analysis reaffirms that race and ethnicity are social constructs that do not have a basis in genetics." This is precisely the sort of lack of attention, lack of "care of the data" where race is concerned, that sustains and even exacerbates the reification of race as genetic. As Roshni Patel commented to *Nature*, the researchers should have followed more closely the recommendations of the 2023 NASEM report.[50]

UNPACKING "REPRESENTATION"

Sociologist Steven Epstein has convincingly shown how the trope of "underrepresentation" has informed calls for inclusion in clinical trials at least since debates over the NIH Revitalization Act in 1993. He observed how this trope has been a site of overlap between reform efforts in the biomedical and social domains where, "by claiming numerical underrepresentation, reformers have been able to construct and analogy between their cause and the successful (if contested) struggles to institute affirmative action measures in other social domains."[51] The idea of representation embedded in the NIH Revitalization Act is nicely encapsulated by political scientist David Plotke, who wrote in 1997 that "the opposite of representation is not participation. The opposite of representation is exclusion."[52]

Epstein goes on to discuss four conceptualizations of representation embedded in call for diversifying clinical trials:

1. Representation in the statistical sense: Different groups should be included in clinical research populations relative to their proportion of the overall U.S. population or incidence or prevalence of the disease.
2. Representation in the sense of social visibility: Different groups should be represented within subject populations (or, for that matter, in the research community), so that medical research ... "looks like America."
3. Representation in the sense of political voice: Different communities of stakeholders have a right to demand that researchers ... study their particular needs and condition and not just those of more privileged social groups.

4. Representation in the symbolic sense . . .: Biomedicine must make a positive contribution to the shaping of ideas and imagery that surround our cultural understandings of what different social groups are like.[53]

In more recent genetic studies these dynamics of representation take on somewhat different valences. As Lee et al. observed in their recent analysis of precision medicine research aimed at addressing a so-called "diversity gap" in genetic samples:

> The language of representation is used flexibly to refer to two objectives: achieving sufficient genetic variation across populations and including historically disenfranchised groups in research. We argue that these dual understandings of representation are more than rhetorical slippage, but rather allow for the contemporary collection of samples and data from marginalized populations to stand in as correcting historical exclusion of social groups towards addressing health inequity.[54]

Related to this idea of "standing in" for correcting historical exclusion, the concept of representation here also obscured the dynamics of power embedded in the concept of representation.

In her foundational work, *The Concept of Representation*, Hannah Pitkin argues there are two basic categories or typologies of representation. The first involves the formal arrangements which precede and initiate it: authorization and accountability. It assumes that such representation must be done by human beings. The second involves representation as "standing for" rather than "acting for." Pitkin notes that this may be accomplished either by humans or inanimate objects. This latter form of representation can be "descriptive" or "symbolic."[55] In the case of the various initiatives to increase diversity both among researchers and in databases, these concepts overlap and blur into one another in ways that not only divert attention from issues of historical exclusion but undermine those aspects of representation grounded in ideas of authorization and accountability, thereby obscuring the power dynamics embedded in and enabled by such initiatives.

When it comes to genetic databases, the concept of representation initially deployed is largely descriptive, a situation in which, a Pitkin notes, "a person or thing stands for others 'by being sufficiently like them.'" In this sense, she continues, "what matters most is not their actions... but what they are, or are like."[56] The problem is, genomic representation does not give subjects any say or active role in the research enterprise. They are valued primarily as representative objects—in the sense that their genes "stand in" for the larger group to which they have been assigned.

In terms of diversifying the corps of genetic researchers, such researchers are valued insofar as they derive from underrepresented groups, but the relationship is largely symbolic and in no way involves direct authorization or accountability. These initiatives are certainly laudable and deserving of support, but they are perilous insofar as they imply a more substantive form of political representation based on ideas of power. Instead of substantively empowering the represented group, diversity initiatives are promoted as a means to facilitate the extracting of DNA from such groups so as to diversify gene databases. Authorization exists only insofar as the sources of genetic samples sign consent forms. The actual authorization comes from the researchers who are seeking out subjects for the databases. The sources of the DNA do not actually convey the genetic information. It is the researchers who extract the information from them and make it representative through various technical interventions that isolate, analyze, and categorize it. Accountability is largely nonexistent. The databases thereby become more descriptively representative of diversity without altering power dynamics.

In these more sophisticated and well-meaning initiatives, the concept of "representation" in relation to diversity thus does several things. First, there is the idea that "underserved" populations are "underrepresented" in research. Lack of representation is presumed to exclude affected populations from the benefits of such studies (also a concern noted by Popejoy and Fullerton); but there is really very little actual research to support this contention. Additionally, there are no studies showing that a specific genetic variant functions differently in one "population" (generally understood as a racial or ethnic group) than another. Certainly, particular alleles may have different frequencies across different populations groups, and in this regard certain kinds of diversity in genetic databases can indeed be useful in deepening our understanding and, as Popejoy and Fullerton noted, avoiding false positive associations. But none of this has anything to do with access to benefits. Similarly, some drugs may be shown to have different levels of efficacy or rates of metabolism on average across certain populations, but there is no drug that is race or ethnic specific—the FDA's approval of BiDil to the contrary notwithstanding.

Second, the researchers targeted by programs such as the ASHG initiative are defined in terms of their identity as representatives of groups that are underrepresented in genetic research. There is thus a dual valence to representation. There is the representation (or lack thereof) of certain groups in genetic databases; and then there is the representation of researchers who are from those underrepresented groups.

Third, the representative researchers are valued insofar as their status as "representative" of underrepresented groups will foster trust in those groups and facilitate their recruitment to genetic studies. This is an instrumental use of representation in one social arena (the workplace) to foster representation in a distinctly biological arena (a genetic database). This echoes the idea in corporate diversity management that increasing diversity in the workforce will facilitate access to globally diverse markets, but in this context it entangles the social with the biological.

Fourth, there are additional levels of displaced representation embedded here as well. The groups who are underrepresented in databases (generally characterized in terms of race and/or ethnicity) are valued because they, in turn, are presumed to "represent" or embody different frequencies of particular rare alleles that might be of interest to researchers. It is identifying those alleles that was the true goal of the research. In the sorts of clinical trials examined by Epstein, where there were generally direct links being made between interventions and medical conditions (as in a drug trial), whole embodied individuals were standing in for other people in their socially identified (usually racial) group. In the case of genetic databases, the bodies are really standing in for genes that are the true object of interest.

More generally, unlike calls for inclusion in biomedical research such as drug trials, there is very little evidence of concerns for political voice involved in claims for inclusion in genetic databases. Many earlier calls for inclusion in biomedical research involved initiatives from affected groups themselves seeking entry into research as a means to gain access to benefits and have their conditions recognized and acknowledged as worthy of consideration. Much of this can been seen in the model of AIDS activists seeking access to both early drug trials and related early therapies. Representation in massive genetic databases (or even in related genetic studies), operates at a much farther remove from potential benefits and there are no equivalent interest groups advocating for inclusion as was the case in the agitation behind the NIH Revitalization Act. Instead, calls for inclusion largely emanate from the researchers themselves.

There is here a sort of inversion of the traditional concept of representation grounded in the idea of a principle and an agent—a represented and a representer. In most situations where representation is involved, one typically thinks of the principal, the represented, choosing their agent, the representative. This is what Pitkin was referring to as the act of authorization.[57] Authorization provides the basis for subsequent accountability. What was given can be taken away. But with genetic research, the researchers, who are in effect agents, are choosing the principals—the research subjects. At the level of descriptive representation, these subjects (and their ascribed groups) are, strictly speaking, being represented by their DNA. But there is also a political valence to this act of representation. Subjects are also being sought insofar as the diversity they embody is also social and political. These initiatives all argue that they address inequities in the social arena that manifest as health disparities. These inequities were the result, in large part, in prior (and ongoing) deficiencies in the system of *political* representation where minoritized groups lacked adequate power to authorize and hold accountable state representatives.

This is the dynamic that is embedded in the observation made by Lee et al. that allows: "samples and data from marginalized populations to stand in as correcting historical exclusion of social groups towards addressing health inequity."[58] But in the structure of contemporary genomic initiatives, it is unauthorized and unaccountable researchers who have the power to define, engage, and provide representation for "underrepresented" groups. In this, these more recent initiatives share a fundamental problem with the original and similarly well-meaning researchers behind the Human Genome Diversity Project decades earlier: they are interested in defining and exploring genetic diversity for the betterment of humanity, but they cannot seem to do so in a manner that cedes any real power to the groups they are seeking to study. Instead, they rely on increasingly sophisticated rhetorical flourishes that conflate concepts of representation so as to give the appearance of voice and accountability without the substance.

CHAPTER TEN

Political Valences of Contemporary Genetic Diversity

Concerns over representation, equity, and power might give one pause before invoking "diversity" too casually or too broadly; but apparently not prominent Stanford geneticist and MacArthur fellow, Carlos Bustamante. In 2018 (around the same time he was conducting the analysis of Senator Elizabeth Warren's Native American genetic ancestry), the *MIT Technology Review* ran a profile of Bustamante titled, "DNA Databases are too White. This Man Aims to Fix That." When posed with the question, "You walk a tricky line, though, don't you? You're pointing out the importance of variance between different populations, but you don't want to reinforce old categories of race," Bustamante responded, "I'm actually an optimist. I think the world is becoming a less racist place. If you talk to the next generation of people, millennials on down, those abhorrent ideologies are thrown away."[1] Notably, Alice Popejoy who had coauthored the influential 2016 article on the need for greater diversity in genetic data banks was at this time a post-doctoral researcher in Bustamante's lab.

CONTEMPORARY CARE OF THE DATA

Earlier that year, Harvard geneticist David Reich had published *Who We Are and How We Got Here: Ancient DNA and the New Science of the Human Past*, a mostly thoughtful, insightful, and popularly accessible book on recent revelations on the history of human migrations and development derived from advances in the study of ancient DNA. Reich discussed the diversity of the samples from which he drew his analysis, noting, for example, that using a microchip designed for the purposes of studying the human past he "initiated a project to survey the diversity of the world's present day populations" that sampled over 1,000 population groups—truly an impressive range of diversity. In discussing his 2012 study of Native American population history he noted that his lab "published data on fifty-two diverse populations."[2]

In the penultimate chapter, however, Reich departed from his area of expertise to address, "The Genomics of Race and Identity." He also published the gist of this chapter in a widely read op-ed piece in the *New York Times* titled, "How Genetics Is Changing Our Understanding of 'Race.'" Reich framed his article with a nod to Lewontin's work and an acknowledgment that "race is a social construct." To his credit, Reich tried to distance his work from Nicholas Wade's 2014 book, *A Troublesome Inheritance*, and noted that he was among those geneticists signing a letter to the *New York Times* protesting that Wade had distorted their work to reach dangerous conclusions about purported genetic bases for cross-racial behavioral differences. Yet Reich nonetheless asserted that while, "I have deep sympathy for the concern that genetic discoveries could be misused to justify racism ... it is simply no longer possible to ignore average genetic differences among 'races.'"[3]

In response to Reich's book and op-ed, a group of sixty-seven scholars from disciplines ranging across the natural sciences, medical and population health sciences, social sciences, law, and humanities (including myself) published an article arguing that Reich's approach to race and genetics was seriously flawed. The core of this critique centered on the casual and problematic way Reich (like many other geneticists and biomedical researchers) understood, constructed, and used the categories of "race" and "population." The article noted that among the examples Reich cited to reinvigorate the idea of a genetic basis for race, was one that included average genetic differences between Northern and Southern Europeans, while another involved higher frequencies of a particular gene that conveyed a higher risk of prostate cancer among certain West African populations. Such differences certainly were "real," but they in no way supported the idea that such variation was somehow racial in character. To the contrary, the article argued, "we need to recognize that meaningful patterns of genetic and biological variation exist in our species *that are not racial.*" The problem was not with finding genetic difference, it was with how that difference was organized and classified. As the article continued, "This is not to say that geneticists such as Reich should never use categories in their research; indeed, their work would be largely impossible without them. However, they must be careful to understand the social and historical legacies that shape the formation of these categories and constrain their utility."[4] This was especially striking in light of Reich's acknowledged characterization of sampling individuals "from more than a thousand populations worldwide."[5] Clearly under certain circumstances Reich was capable of understanding the difference between carefully characterized genetic population groups and broad, heterogenous, and often amorphous socially defined racial groups. One can only wonder how Reich could possibly go from the idea of characterizing diversity in terms of a thousand populations to facilely declaring that traditional racial groups themselves constitute populations that can be characterized by genetics.

This brings us back to the issue of "Care of the data," discussed earlier in relation to race and forensic DNA. Precisely because race is a commonsense, culturally accessible construct that we all use daily, often without thought, scientists such as Reich think it is relatively obvious and transparent. They do not think of the historical contingency of race, how its meaning and boundaries change over time and space, from culture to culture. They simply think that the way they use race is clear and does not need the same sort of scientific care or examination that they apply to the tools of their own trade, genetics. This results in the inconsistent and sometimes incoherent deployment of race in relation to genetic diversity that leaves the door open to dangerous reifications of race as genetic and the misguided conflation of social diversity with biological diversity.

Bustamante's optimism and Reich's deep concern are all well and good, but just a year before this, violent White Nationalists held the Unite the Right rally in Charlottesville, Virginia, after which Donald Trump declared there were "very fine people on both sides."[6] By the following year, the Southern Poverty Law Center (SPLC) would identify "a surging white nationalist movement that has been linked to a series of racist and antisemitic terror attacks and has coincided with an increase in hate crime," concluding that " the number of white nationalist groups identified by the SPLC rose for the second straight year, a 55 percent increase since 2017, when Donald Trump's campaign energized white nationalists who saw him as an avatar of their grievances and their anxiety over the country's demographic changes."[7]

GENETICS, DIVERSITY, AND WHITE SUPREMACY

In Bustamante and Reich's own chosen realm of genetics, this same period saw a striking rise of interest in DNA ancestry tracing among White supremacists. A 2016 article in *Vice* observed that soon after 23andMe® dropped the price for its services to $99 in 2012, "screenshots of DNA testing results began appearing on white nationalist message boards—first on Stormfront, then occasionally on subreddits related to white nationalism, and most frequently on 4chan's 'politically incorrect' board/pol/, from which many alt-right memes originate." Posters used these screenshots to prove their "Whiteness" genetically, or, as the article noted, "to serve as an invitation to trash talk others."[8] That year, Jedidiah Carlson, a bioinformatics graduate student at the University of Michigan, stumbled upon a link to a forum post on the White supremacist website *Stormfront*.[9] He found users there involved in detailed discussions of various scientific articles about genetic diversity and race that distorted the findings produced by programs such as STRUCTURE to argue that race was genetically real and further their arguments about racial hierarchy.[10]

By 2018, the *New York Times* was writing about, "The Racial Spectacle of DNA Test Result Videos," and noting that "White supremacists are drawn to DNA testing to prove their racial purity."[11] In the aftermath of his election in 2016, President Trump increasingly gave permission to broad swaths of American society to be more open about their racial animus and their antipathy to liberal ideas about the value of diversity (not least in his response to the Charlottesville rally or later in his clarion call to the White Nationalist Proud Boys to "stand back and stand by" during his 2020 presidential debate with Joe Biden).[12] The

New York Times article noted this context, stating that "It's probably not a coincidence that these mechanisms of biological self-discovery are on the rise now, amid seething tensions over racism, immigration and what constitutes a 'real American.'" Where liberals seemed to be using the tests as methods for, "performing racial harmony and assuaging white guilt," White supremacists were clearly using the tests to establish their "racial purity."[13]

Not all DNA tests came back with the pure results that White supremacists might hope for. Sociologists Aaron Panofsky and Joan Donovan explored how White supremacists managed such confounding results in discussion posts on websites such as *Stormfront*. They found that "posters exert much more energy repairing individuals' bad news than using it to exclude or attack them." In these cases, the information provided by direct-to-consumer (DTC) ancestry tracing companies has provided White supremacists with unsettling information about the character of genetic diversity within their own bodies. They must adopt distinctive strategies, both individually and collectively, to characterize and address this diversity. Sometimes this simply takes the form of dismissing the science, but in other, more complex encounters, the posters engage with the science but also go through conceptual contortions to "reinterpret racial boundaries and hierarchies" so as to make the results more manageable and palatable within the community.[14] Panofsky and Donovan refer to this as a form of "racial repair." It might also be understood as a type of "diversity management."

Along with a rise of interest in genetic ancestry tracing, White supremacists, seizing upon new developments in genetics to further their worldview, have embraced the term "human biodiversity." In 2016, journalist Ari Feldman characterized human biodiversity as, "the pseudoscientific racism of the Alt-Right," that appropriated scientific authority, "by posing as an empirical, rational discourse on the genetically proven physical and mental variation between humans."[15] He noted that Reddit hosted an invitation-only human biodiversity forum that had been flagged as racist. The term itself was appropriated by conservative blogger Steve Sailer in the 1990s from anthropologist Jon Marks, whose book *Human Biodiversity* actually provided an extensive analysis undermining claims of meaningful genetic diversity by racial group.[16] The original point of the term was to offer an alternative framework for conceptualizing patterns of human variation. The alt-right turned it into a euphemism for White supremacy and eugenics.[17]

The human biodiversity movement gained its most notable acolyte in the person of Steven Miller, a key architect of President Trump's harsh immigration policies; it gained one of its most brutal in the person of the perpetrator of the massacre of twenty-three Latinos at an El Paso Walmart in 2019. The shooter's manifesto posted on the online message board 8chan titled, "The Inconvenient Truth," stated, "I am against race mixing because it destroys genetic diversity and creates identity problems.... Cultural and racial diversity [from "Hispanic immigration"] are largely temporary. Cultural diversity diminishes as stronger and/or more appealing cultures overtake weaker and/or undesirable ones. Racial diversity will disappear as either race mixing or genocide will take place."[18] The manifesto juxtaposed cultural with racial diversity—constructing the latter as biological, hence susceptible to what he saw as dangerous dilution through "race mixing." Panofsky et al., noted that the shooter drew from the "Great Replacement" theory, "which holds that Western countries were facing 'White genocide' owing to a conspiracy to ensure permissive immigration policies, racial differences in birthrates, and the inherent violence of non-Whites." They went on to argue that "ideas from human biodiversity have served to inform this world view, but in particular it has offered a genetic rationale for White nationalist violence and a specific focus on the preservation of White biodiversity as a goal."[19]

In 2020, Carlson, together with geneticist Kelley Harris, published a study of engagement with scientific manuscripts on social media. Most discussions on a variety of topics were conducted primarily among scientists themselves. However, they found that 10 percent of the manuscripts analyzed had, "sizeable audience sectors that are associated with right-wing white nationalist communities," even though none of the manuscripts "appeared to intentionally espouse any right-wing extremist messages."[20]

2020 also saw Charles Murray adding to this toxic stew with a sort of sequel to *The Bell Curve*, titled *Human Diversity: The Biology of Gender, Race, and Class*.[21] In many respects the book was a rehash of *The*

Bell Curve's arguments about the genetic basis for social stratification. As science writer Philip Ball noted in his review of the book:

> There is no sign in *Human Diversity* that the ensuing controversy [following the publication of *The Bell* Curve] and criticisms have swayed Murray an inch, although here he is so prolix, and his facts and figures are so undigested, that racists scouring his book for arguments to support their prejudices might get bored or confused before they find what they want. Even the title dog-whistles the term "human biodiversity," widely used among the far right today as code for the alleged genetic distinctions (meaning hierarchy) of races.[22]

Murray updated his old arguments with reference to advances such as STRUCTURE and GWAS. "The sequencing of the human genome changed everything," he declared.[23] But psychologist and behavior geneticist K. Paige Harden referenced the book as an example where the, "results on the heritability of human individual differences were misused to advance the idea that economic and racial inequalities were immutable."[24] The book thus recapitulated many of the distortions for which the 165 geneticists had criticized Nicholas Wade's *A Troublesome Inheritance* just a few years earlier. For example, ignoring the critiques of the limitations of STRUCTURE put forth by the likes of Bolnick et al., Murray seems to have fallen in love with how algorithms can sort single nucleotide polymorphisms SNPs in a manner that looks like it comports with race. But as Bolnick had shown over a decade earlier, that all depended on how you set the parameters for such algorithms.[25] Ball additionally noted:

> The idea that a person can be identified as "black" or "east Asian" from their biology alone needn't surprise us: we only need to look at them. The fact that SNPs cluster this way is unremarkable too. There are gradations of SNP profiles at all levels: in places where populations have remained relatively isolated culturally or geographically, you could equally distinguish populations of cities or traditional tribal groupings this way. There is nothing fundamental

about such a definition of race—it's just one data point on a continuum of degrees of relatedness among populations at all scales.[26]

Murray, nonetheless, seized on new data produced by massive databases of the sort being developed by the *All of Us* and Million Veterans programs. For him, the diversity of these databases provided the key for studying, among other things, "the genetics of mental disorders in other populations as well, which means studying the ways in which they differ from Europeans." Focusing in particular on AIMs technology to blur distinctions between diversity in genetic ancestry and racial diversity, he argued that "virtually all traits, whether physiological, related to disease, or related to cognitive repertoires, exhibit many large differences in target allele frequencies across continental populations."[27]

Not all reviews were negative. The *Mankind Quarterly*, a journal with deep ties to the eugenicist Pioneer Fund, and which has been called a "white supremacist journal," gave the book a glowing review. This should come as no surprise as Murray had relied on several articles from the journal in writing *The Bell Curve*. The only problem the reviewer had with the book was that "it d[id] not go far enough."[28]

Murray was not the only mainstream writer popular in these circles. Wade's and Reich's books were also seized upon by the alt-right. Jared Taylor, the founder of the White Nationalist group American Renaissance, in a video citing Reich's book, crowed that "Science is on our side."[29] Steve Sailer rushed to praise Reich's book on the White Nationalist website VDARE, and the Human Biodiversity webpage listed his book among its approved readings.[30] Reich himself did not wish to be bothered. When asked to lead a discussion on the topic at the annual meeting of the American Society of Human Genetics, he demurred, saying, "I really wanted to return to research."[31] Such a sentiment, while wholly understandable, brings to mind nothing so much as the refrain from political satirist (and mathematician) Tom Lehrer's song about ex-Nazi rocket scientist Wernher von Braun: "Once the rockets go up/who cares where they come down?/That's not my department/Says Wernher von Braun!"[32]

The misappropriation of scientific research that entangles race and biology can also literally have deadly consequences, as evidenced by manifestos issued by the White supremacists responsible for the mass murders in Christ Church, New Zealand, El Paso, Texas, and Buffalo, New York.[33] The Buffalo shooter, for example, cited, "recent developments in human molecular genetics to falsely assert that there are innate biological differences between races in an attempt to validate his hateful, white supremacist worldview."[34] Aaron Panofsky noted of these killers, that "of all the ancient racist tropes they could have drawn from, they picked this pseudoscientific package of ideas about genetics and replacement theory [a racist idea that White people are deliberately being replaced by people of color in places like America and Europe [35]]. . . . It's this language of science and genetics that they find empowering and convincing, and which they think legitimates their perspective."[36]

Geneticists engaging in the emerging field known as "sociogenomics" whose work was cited in these documents were understandably horrified, but they should not have been surprised. Sociogenomic research is a somewhat updated version of E. O. Wilson's 1980s idea of sociobiology and occupies a significant place in the area of behavior genetics research.[37] It often relies on so-called "polygenetic risk scores" derived from using large datasets to try to find correlations between the frequencies of certain genetic variations and a wide array of complex social behaviors, ranging from educational attainment to risk aversion and political inclinations.[38] The focus on things such as educational attainment, resonates strongly with the earlier work of Herrnstein and Murray. Historian Nathaniel Comfort has argued that "sociogenomics is opening a new door to eugenics."[39]

When asked in 2018 about the responsibility of scientists to address the appropriation of their work by White supremacists, Carlson stated:

> I don't think engaging them directly will work, [but nonetheless acknowledged that] there's growing recognition that we as scientists bear some responsibility for guiding the public interpretation of our work. In the broad scheme of things, people are excited about the work we're doing. The genomic revolution is still well underway. But I think because of precisely that, we need to think more carefully about not just our own interpretation of the work, but anticipating how our work might be misinterpreted and trying to preemptively patch up the holes in that logic."[40]

Nonetheless, as Wedow, Martschenko, and Trejo wrote in *Scientific American*, in the aftermath of the Buffalo shootings: "As hard as it might be, and it certainly will be challenging, scientists need to consider their moral responsibilities as producers of this research. Otherwise, we stay caught in the delusion that science can speak accurately for itself."[41]

Bustamante's optimism about racism echoed that of liberal implicit bias researchers, like Mahzarin Banaji and Anthony Greenwald, who in their 2013 book, *Blind Spot: Hidden Biases of Good People*, asserted that "it is a mistake to characterize modern America as racist."[42] In this respect, implicit bias research was a fitting heir to one of the founders of the modern diversity, equity, and inclusion (DEI) movement, R. Roosevelt Thomas, who argued in 1990, that "the realities facing us are no longer the realities affirmative action was designed to fix," because "prejudice... has suffered some wounds that may eventually prove fatal." Such sanguinity about race relations might have been reassuring but it was also dangerous. 2013 also saw Chief Justice John Roberts gut the 1965 Voting Rights Act (VRA) in *Shelby Country v. Holder*. He rationalized his holding in language that might easily have cited *Blind Spot* or Thomas as sources, by declaring that section 5 of the VRA requiring preclearance of voting procedure changes was no longer necessary because "Nearly 50 years later, things have changed dramatically."[43]

Denying the persisting reality of racism not only facilitated the sort of retreat from engagement displayed by Reich, it also paved the way for the likes of Murray, who would declare in *Human Diversity* that "My proposition is that racism and sexism are no longer decisively important in determining who rises to the top."[44] Removing racism from the equation was essential to his argument that social stratification reflected race- (and sex-) based genetic diversity, not social, historical, or environmental factors.

TRUMP AND THE END OF DIVERSITY?

Of course, with the rise of Donald Trump, the right-wing entanglement of social and biological conceptions of diversity reached new levels of prominence and power. Trump himself has repeatedly articulated bizarrely eugenic ideas about himself and other racial groups. In 2010,

Trump told CNN, "I'm a gene believer... hey when you connect two race horses you get usually end up with a fast horse.... I had a good gene pool from the standpoint of that so I was pretty much driven."[45] He repeated the "gene believer" comment in 2016, and the "race horse" theory while on the campaign trial in 2020 where he said, "You have good genes, you know that, right? You have good genes. A lot of it is about the genes, isn't it, don't you believe? The racehorse theory. You think we're so different? You have good genes in Minnesota."[46] As I wrote with Marcy Darnovsky and Jonathan Marks at the time, "Trump is clearly referring to the stereotypical Minnesotan of Scandinavian descent—the Norwegian bachelor farmers of 'A Prairie Home Companion' and their families. He is signaling to his supporters that they belong among the fit, the superior, the winners. He is most decidedly not speaking about—or to—Minnesotans like George Floyd," who had just been murdered a month earlier.[47]

Trump returned to similar themes in 2023 when he characterized undocumented immigrants as "poisoning the blood" of the country. Critics immediately noted how this echoed the genocidal declarations of Adolf Hitler. In response, Trump said he had never read *Mein Kampf*, but as an article in the *New York Times* noted: "a table by his bed once had a copy of Hitler speeches called 'My New Order,' a gift from a friend that Ivana Trump, his first wife, said she had seen him occasionally leafing through."[48] Trump's racist dog-whistling about the genetic superiority of some groups over others comports with the odious postings of Steve Sailer and Jared Taylor or the White supremacists who have milk-chugging parties to prove their genetic tolerance for lactose.[49] But it also finds support (even if only through distortion) in the work of those who are uncritically calling for diversifying genetic databases and those like Reich and Wade (not to mention Murray) who argue to some sort of genetic "reality" to racial groupings.

Even as Trump was buttressing fallacious ideas about the genetic basis of racial diversity, he was also decrying and directly attacking the ideas about the social nature and value of racial diversity. In 2017, the Trump administration prohibited officials at the Centers for Disease Control from using a list of seven words or phrases—among them was diversity.[50] On September 22, 2020 he issued an executive order effectively banning the use of diversity training, which it characterized as "divisive" and "un-American," by any company receiving a federal contract. Divisive concepts included, "Race or sex scapegoating," which it defined as, "assigning fault, blame, or bias to a race or sex, or to members of a race or sex because of their race or sex." The idea here was that DEI training (or, for that matter, critical race theory) was bad because it made White people feel guilty for being White. Trump's order also authorized the director of the Office of Personnel Management "to pursue a performance-based adverse action proceeding against" any supervisor or employee who authorized or approved such training.[51] This was part and parcel of a larger assault on "wokeism," and "critical race theory" that gained momentum in the backlash to the widespread protests in response to the killing of George Floyd earlier that year.[52]

Corporate America was not pleased. An article in *USA Today* noted that "Behind the scenes, individual companies and industry groups are supporting efforts to mount a legal challenge to the executive order."[53] And within months, a federal judge in the Northern District of California issued a nationwide preliminary injunction against the order.[54] Then, on January 20, 2021, President Biden, in one of his first official acts as president, issued an executive order revoking Trump's order. Biden's order opened by intoning the old mantra that "our diversity is one of our country's greatest strengths."[55] This used to be a fairly bipartisan position, but by the summer of 2023, Republican presidential candidate Vivek Ramaswamy was proudly declaring that "Diversity is not our strength. Our strength is what unites us across that diversity."[56]

Ironically, Ramaswamy might seem to be echoing Bill Clinton's sentiments from the 2000 White House ceremony celebrating the completion of the first draft of the human genome that "all human beings, regardless of race, are more than 99.9 percent the same," but in fact it was more akin to the briefs from the American Center for Law and Justice (ACLJ) opposing affirmative action in *Fisher v. University of Texas* which foregrounded the idea that "there is likewise no difference in kind between black, white, Asian, or other ethnic groups of human beings. There is one race—the human race."[57] In this latter framing, diversity is a challenge, even a threat to unity. We see this concern also expressed by the Supreme Court, not only in Justice Roberts's dismissive declaration in *Shelby County v. Holder* that "the way to stop discrimination on the basis of race is to stop discriminating on the basis of race" (repeated with slight variation in the 2023 affirmative action case, *SFFA v. Harvard University* as, "Eliminating racial discrimination means eliminating all of it"), but also, and in some ways more concerningly, in opinions such as Justice Kennedy's in the 2014 case of *Schuette v. BAMN* that the state's broad use of racial classifications, even if causing no stigma, "might become itself a source of the very hostilities and resentments based on race that this Nation seeks to put behind it."[58] I say more concerning, but the dynamic here is more subtle. As Reva Siegel has noted, this frame casts government classification by race—or more generally the legal consideration of racial diversity—as a threat "to social cohesion, threatening balkanization and racial conflict."[59] But in Kennedy's framing, the only "hostilities and resentments" deserving of legal consideration were those of the majority White population. Affirmative action or school desegregation programs had not been provoking anger or protests from America's African American community. Kennedy here was validating and empowering White resistance to racial justice initiatives, in language that readily translates into Ramaswamy's "our strength is what unites us across that diversity."

This focus, however, was not limited to avatars of the Right. The diversity management industry itself fostered this approach with "diversity training" seminars that, like Justice Kennedy, seemed concerned primarily about the individual attitudes of White employees rather than educating all employees about the larger structural and historical bases of injustice. Professor of African American Studies Eddie Glaude makes the very helpful distinction between seeing diversity "as a problem to be managed or as a value to be cherished":

> As a problem, diversity enters the picture when so-called "others" demand inclusion and "we" must decide how to deal with those demands. This "we" calls up a particular history grounded in the American dream, in the idea of hard work and merit, and self-reliance. But this "we" stands apart from the actual diversity of the country; it has, as James Baldwin noted, "almost nothing to do with what or who an American really is." Understood in this sense, diversity is something to be managed, because the very idea of "we" narrows our view of who we are.

> Seeing diversity as a value to cherish, however, orients us differently. We begin with a recognition that diversity is constitutive of who we are, and our aim is to reflect it in our institutions and in our civic arrangements. The irony is that we often think we are treating diversity as a cherished value, but we are really trying to manage it. We end up checking boxes, more concerned about compliance, and less interested in the value itself. An add-on. Not something fundamental to who we are that becomes a critical part of how we assess whether we are fulfilling the overall mission of the institution.[60]

Ramaswamy's framing reflects the culmination of an approach to diversity that sees it as a problem to be managed or overcome, rather than a value to be cherished. Of course, most DEI programs do some of both. But being grounded in an enterprise that promoted diversity "management" as a means to promote corporate productivity and profitability, the DEI industry laid the foundations for the ascendant Right's assaults on diversity because it values diversity instrumentally rather than as "constitutive of who we are."

THE SUPREME COURT WEIGHS IN

By the time Trump was issuing his executive order, an affirmative action case originally filed by an organization called Students for Fair Admissions, Inc. in 2014 was working its way toward an initial decision in the United States First Circuit Court of Appeals. In the case, *Students for Fair Admissions, Inc. v. President and Fellows of Harvard University*, the conservative Pacific Legal Foundation (PLF) took up the argument that racial categories were inherently arbitrary and therefore could not be narrowly tailored to serve a compelling government interest sufficient to survive strict judicial scrutiny. The PLF did not make as explicit reference to genetic science but framed their amicus brief to the court with the opening assertion that "racial classifications are *inherently* arbitrary."[61] While focusing more on the social diversity within racial groups, the PLF made it clear that the fundamental problem of affirmative action was that "there is no sound system for classifying on the basis of race."⁶² In November, 2020, the First Circuit Court of Appeals, largely following the Supreme Court precedent in *Fisher*, affirmed the lower court's ruling upholding the admissions plan and did not address the issue of the arbitrariness of racial categories put forth by the PLF.

In 2022, *SFFA v. Harvard* made its way to the Supreme Court, which, in June 2023, struck down the affirmative action plans of both Harvard and the University of North Carolina (from a companion case), effectively (though not explicitly) overturning *Grutter*.⁶³ Much can (and has and will) be said about how Justice Roberts's majority opinion and the concurring opinions use (and abuse) legal precedent, history, and broader understandings about the nature of stigma and the conservative ideal of color blindness. For the purposes of understanding the intersections of legal understandings of social and biological diversity, the key aspect of the case was Roberts's embrace of the idea that "racial diversity," whether or not it was an interest sufficiently compelling to justify affirmative action, was too amorphous a concept to withstand strict scrutiny by the Supreme Court. In this, Roberts in effect adopted the arguments proffered below by the Pacific Legal Foundation. After noting that both Harvard and the University of North Carolina used the racial and ethnic categories from the census, Roberts stated:

> For starters, the categories are themselves imprecise in many ways. Some of them are plainly overbroad: by grouping together all Asian students, for instance, respondents are apparently uninterested in whether South Asian or East Asian students are adequately represented, so long as there is enough of one to compensate for a lack of the other. Meanwhile other racial categories, such as "Hispanic," are arbitrary or undefined. . . . And still other categories are under-inclusive. When asked at oral argument "how are applicants from Middle Eastern countries classified, [such as] Jordan, Iraq, Iran, [and] Egypt," UNC's counsel responded, "I do not know the answer to that question.⁶⁴

Roberts here adopted the strategy, first articulated in the aftermath of *Grutter* by Alexander and Schwarzschild (and subsequently taken up by Judge Garza in his *Fisher* dissent) that inverted progressive arguments about the socially constructed nature of race to argue that precisely *because* race was social, not biological, it was somehow arbitrary and therefore insufficient to meet the demands of strict scrutiny.[65]

Roberts here implicitly relied upon the new genetic politics of affirmative action articulated earlier by anti-affirmative action crusaders such as Ward Connerly who invoked DTC ancestry tracing to challenge the basic legitimacy of the census categories.[66] Similarly, as *SFFA v. Harvard* was making its way up to the Supreme Court, conservative commentator Michael Barone invoked the work of Charles Murray and David Reich side by side to argue that "genetics is undercutting the case for racial quotas."[67] Here genetic diversity was to be celebrated—but only as a means to undermine the value and legitimacy of social diversity, and ultimately to undermine efforts to address racial injustice.

In his concurrence, Justice Gorsuch seized on the idea that census categories were incoherent even more aggressively. He dismissed the racial categories used in the Common Application for college admissions by asking, "Where do these boxes come from? Bureaucrats. A federal interagency commission devised this scheme of classifications in the 1970s to facilitate data collection." Gorsuch went on, drawing heavily on the work of conservative law professor David Bernstein, to argue that "These classifications rest on incoherent stereotypes. Take the 'Asian' category. It sweeps into one pile East Asians (e.g., Chinese, Korean, Japanese) and South Asians (e.g., Indian, Pakistani, Bangladeshi), even though together they constitute about 60% of the world's population."[68] On the one hand, the basic critique of the breadth of the category "Asian" has some merit. This lends the argument a patina of reasonableness. On the other hand, Gorsuch has framed the categories themselves as resting on "incoherent stereotypes." This is where Gorsuch employs a legalistic sleight of hand to blur the social and the biological. If we were talking about Asian (or "Hispanic" or any of the census categories) as *genetic or biological* groupings, the classifications would indeed be incoherent. But as *social* groupings they are self-evidently coherent, if not unproblematic. They are so coherent that they have been used for decades to amass and use data to inform and guide a range of policies at all levels of government. Before their formalization as census categories, they were, of course, coherent enough to form the basis for Jim Crow laws, and before that slavery.

For his part, Roberts did not deny the significant benefits of diversity, which he acknowledged include: "promoting the robust exchange of ideas; broadenings and refining understanding; fostering innovation and problem solving; preparing engaged and productive citizens and leaders; [and] enhancing appreciation, respect, and empathy, cross racial understanding, and breaking down stereotypes." Where Gorsuch focused on the incoherence of "stereotypes" related to race, Roberts focused on the idea that the myriad benefits of diversity were not "sufficiently coherent for purposes of strict scrutiny."[69]

Roberts employed a spurious concept of measurement to support this assertion. His idea was that you must be able to measure the benefits of diversity in order to have meaningful judicial review. His primary criticism of diversity was not that it was not a compelling state interest but that "it is unclear how courts are supposed to measure any of these goals." What he meant by "measure" and why it was the sine qua non of a forward-looking diversity program he did not really say, but we get a sense of it in his unusual inclusion in his opinion of a chart showing a breakdown of the percentage of students admitted to Harvard by race year over year. For him it was the numbers, these clear percentages, which seemed real. What mattered for him was that the percentages did not vary greatly over time. Roberts concluded that "Harvard's focus on numbers is obvious." But this was bad. For him, this implied Harvard employed "racial balancing" which was "patently unconstitutional."[70] Roberts used his fetishization of measurement to create a Catch-22: if you can't measure the benefits of diversity, it is too incoherent to survive strict scrutiny; if you can measure diversity, then it is "patently unconstitutional."

Yet in the arena of constitutionally approved race-specific remedies, measurement has not always been required. Look no further than *Brown v. Board of Education*, which Roberts sanctimoniously invoked to support his assertion that "The time for making distinctions based on race had passed."[71] In that case, the Supreme Court focused on the harm of stigma and noted that segregating students, "solely because of their race generates a feeling of inferiority as to their status in the community that may affect their hearts and minds in a way unlikely ever to be undone."[72] Is the goal of undoing stigma and feelings of inferiority "coherent"? Is it a goal that can be measured? Were Justice Roberts on the Court in 1954 and desegregation were called for as a remedy, "How," he might have asked, "is a court to know whether children have overcome their feelings of inferiority?"

In other words, the logic of this opinion could also put *Brown* on the chopping block. This would be much in line with Justice Thomas's more forthrightly color-blind concurrence and with long-standing campaigns by conservative activists such as Ward Connerly, who have worked to excise all mention of race or diversity from any and all state or federal programs. It bears remembering that Roberts clerked for Chief Justice Rehnquist, who himself, while a clerk to the Supreme Court in 1954, wrote a memo declaring: "I think *Plessy v Ferguson* was right and should be reaffirmed."[73] There is a legacy here—the implications of Roberts's opinion, therefore, reach back not only to *Bakke* but also to *Brown*.

This was of a piece with President Trump's banning of both the word and practice of diversity and broader Republican attacks on diversity equity and inclusion programs across many states where they held power. By the summer of 2023, some forty anti-DEI bills had been proposed in twenty-two states, seven of which had already become law, while seven more had obtained final legislative approval. These bills included provisions that would prohibit colleges from having DEI offices or staff, ban mandatory diversity training, and prohibit institutions from issuing diversity statements.[74] In Texas, Governor Greg Abbott issued a directive in February 2023 warning state universities that the use of DEI in hiring was illegal.[75]

A NEW DIVERSITY

But, of course, it was not really diversity per se that the Republicans were really railing about—it was what diversity had come to stand for in their minds. But just what this was is unclear. Like the term "woke," diversity developed into an ill-defined, symbolic placeholder for all sort of grievance, tied not only to racial backlash or a sense of zero-sum access to goods (such as college admissions) but also for a sort of populist resentment of the experts and elites who constructed and administered DEI program and their progeny.[76] Thus, the type of diversity that was truly amorphous or ill-defined was not that involved in college admissions but that invoked by the likes of Governor Ron DeSantis or anti-critical race theory crusader Christopher Rufo (who helped to draft Trump's anti-DEI executive order.)[77] Such criticisms were distinctly different from those offered by more liberal scholars such as law professor Lauren Edelman or sociologist Frank Dobbin, who had been challenging DEI-style programs for the previous twenty years, not because they were somehow discriminatory or fomented feelings of White guilt, but because they did not go far enough and because they represented a retreat from the more concrete and far-reaching accomplishment of explicit affirmative action programs, particularly in employment.[78] Nonetheless, Rufo and others seized on such critiques to undermine diversity, much as Judge Garza had seized on liberal arguments about race being a social construct to attack affirmative action.[79]

For these right-wing extremists, diversity was not explicitly about genetics or ancestry. It was about using whatever tools were available to maintain and reinforce racial hierarchy and political dominance. In their hands diversity became a tool to be used to stoke faux outrage and validate White grievance. Yet always not far from the surface, lurk old racist assumptions about inherent genetic differences across racial groups. Thus, for example, in January 2024, an article in *The Guardian* noted that Rufo, having just played a leading role in the ousting of Harvard's first Black woman president, Claudine Gay, was recommending to readers of his webpage a Substack newsletter called *Aporia* that trafficked in the work of scientific racists and eugenicists. These included many adherents of the distorted conception of "human biodiversity" first misappropriated from Jonathan Marks by Steven Sailer back in the 1990s to mask the idea of race science. The article went on to quote Andrew Winston, a professor emeritus of psychology at Canada's University of Guelph, who observed that "this kind of race science keeps coming back into the mainstream, gets criticized heavily, and then diminishes it for a bit, perhaps, and then returns in some new form, depending on the general social context."[80] The likes of Rufo are using diversity as a wedge to reintroduce and broaden the reach of scientific racism into mainstream consciousness.

To a certain degree, this can be understood as the diversity movement's focus on the idea that diversity programs were somehow more "natural" than affirmative action coming home to roost. Remember how at the outset of the diversity management movement, there were gurus such as R. Roosevelt Thomas characterizing affirmative action as having "an unnatural focus on one group" and an "artificial nature." In contrast, he argued that diversity was "the ability to manage your company without unnatural advantage" that allowed for a focus on "pure competence and character."[81] For contemporary right-wing critics of DEI initiatives, it is now diversity itself that is unnatural. Where, in the 1990s, Thomas cast diversity as moving from the backward-looking racial redress of affirmative action toward a forward-looking focus on "pure competence and character," by 2023, Christopher Rufo was noting that "the right sees the abolition of D.E.I. as a step toward meritocracy."[82] Both Rufo today, and R. Roosevelt Thomas thirty years ago, were elevating the "natural" over the "artificial." The "natural" all too easily becomes conflated with the biological, which is valorized as "real"; while the "artificial" becomes identified with racial justice policies, which are delegitimized.

Rufo's "DEI" was different from Powell's "diversity." For Powell in *Bakke*, diversity was primarily a rationale of sufficiently compelling state interest to justify the consideration of race in higher education affirmative action programs. In this context, racial diversity was merely one type of a vast array of diversities that might be evaluated during the admissions process. Powell largely deferred to how educational experts defined and explained the benefits of such diversity to their educational program. In the 1990s, "diversity management" emerged as a significant driver of corporate personnel practices throughout the United States. This model of diversity moved away from the ideas of affirmative action, particularly as related to the idea of government mandated hiring and set-aside programs of the sort struck down by the Supreme Court in *Richmond v. Croson* in 1989.[83] New corporate diversity managers treated affirmative action and diversity as totally separate things.[84] They saw diversity programs as being under their own control and serving the interests of corporate competitiveness in increasingly diversified global markets. In the educational realm, in 2003 the Supreme Court in *Grutter* both embraced and expanded upon Powell's original conceptualization of diversity, adopting the very pragmatic understandings of how diversity served larger societal goals of training students to work in an increasingly diverse workforce, and cultivating leaders with legitimacy in the eyes of the citizenry.[85]

But soon after *Grutter*, a new iteration of diversity emerged in the form of DEI programs. As DEI developed, it tended to focus less on justifying the consideration of race in admissions or set-aside programs and more on managing the social spaces and encounters within the institutions where such programs operated. By 2006, Professor of Human Resource Studies at Cornell University Quinetta M. Roberson was observing the, "emergence of a new rhetoric in the field of diversity, which replaces the term diversity with the term inclusion."[86] In 2009, the Society for Diversity was founded, "to provide education, training, and certification programs for the field of DEI."[87] The diversity of DEI adopted the old diversity management idea that "when people from different backgrounds are able to work together, they can bring new ideas and perspectives to the table. This can lead to better decision-making and innovation." But it also moved to emphasize that DEI, and in particular DEI *training*, "can help create a more welcoming and inclusive environment for everyone."[88] As such trainings proliferated during the 2010s, administrators became increasingly aware of a phenomenon of "DEI Backlash" leading to resentment and resistance among employees in a wide range of settings.[89] Ironically, as Dobbin and Kalev have pointed out, "hundreds of studies dating back to the 1930s suggest that antibias training does not reduce bias, alter behavior or change the workplace."[90] At best, as a comprehensive review article by Levy Paluck et al. suggests, we simply do not know whether diversity training works, or if it has negative backlash effects.[91]

While backlash may have been growing during the 2010s, one of the major precipitating factors behind the contemporary crusade against "wokeism" was not DEI per se but the publication of the *1619 Project* in a special issue of *New York Times Magazine* in August 2019.[92] The *1619 Project* was not about DEI and had no pretense to offer training of any sort in that regard. It was rather a collection of essays that, from a variety of disciplinary perspectives, reassessed America's racial history and identity. Its publication generated a firestorm of controversy. As the essays centered on the historical and continued centrality of racism to the American story, much of the criticism naturally came from the

Right. But there were also critiques from established academics (some self-identified as liberal) who challenged some of the assumptions and historical framings of the essays.[93] The *1619 Project* went on to develop a book and a curricular initiative to provide materials for teaching about race and American history in a wide range of educational settings.[94]

The *1619 Project* gained added attention the following summer in the aftermath of the killing of George Floyd in Minneapolis. It was around this time that Senator Tom Cotton of Arkansas introduced a bill called the Saving American History Act, which would, "prohibit federal funds from being made available to teach the 1619 Project curriculum in elementary schools and secondary schools, and for other purposes." While Cotton's bill did not get very far, it was followed by Donald Trump ordering the formation of an advisory committee called the "1776 Commission," to respond to the *1619 Project*. Linking the project to Christopher Rufo's great bugbear, Trump said, "Critical race theory, the 1619 Project and the crusade against American history is toxic propaganda, ideological poison that, if not removed, will dissolve the civic bonds that tie us together."[95] Then in June 2022, Florida Governor Ron DeSantis signed into law the Stop WOKE Act, prohibiting instruction that could make some parties feel that they bear "personal responsibility" for historic wrongdoings because of their race, sex, or national origin. Although soon enjoined by a federal judge, the Stop WOKE Act, like the 1776 Commission and other similar efforts in conservative jurisdictions across the country were far more emphatic and aggressive than any previous critiques of DEI programs generally.[96]

Why was it that the *1619 Project* elicited such a forceful response while DEI trainings had been generally accepted, if sometimes grudgingly? I have no definitive answer to this. Of course, much of the response can be explained by sheer political opportunism on the part of Rufo, Trump, DeSantis, and others. But they nonetheless saw in the *1619 Project* an opportunity that they did not see in DEI trainings. I would hazard the suggestion that the leaders of the MAGA Right saw the *1619 Project* as threatening precisely because it was about *history* not about individual personal attitudes or biases. DEI trainings might have annoyed conservatives, but they did not *threaten* them precisely because they were so generally inefficacious in bringing about substantive change. Changing people's understandings of history, however, was another matter. It might not alienate them or make them defensive in the same way that some ham-handed diversity training modules might. It might not change the way they thought about *themselves* so much as it might change the way they thought about their *country*—and that could be dangerous. When diversity was understood in relation structures and institutions, it presented much greater challenges to the existing order than when it was directed toward individual attitudes and behaviors. In this, the *1619 Project* was closer to the original affirmative action programs of 1960s and 1970s which actually started to make some substantive progress until dismantled by the Supreme Court in *Richmond v. Croson* in 1989, leaving higher education as one of the last arenas where true affirmative action persisted in any degree. And then, along with the attacks on the *1619 Project*, the Supreme Court dismantled even this last bastion in the 2022 case of *SFFA v. Harvard*.

The Unnaturalness of Diversity

For the Right, it is diversity that is now unnatural. Just as social race is not real. But of course, for many on the Right, genetic race, "natural" race, *is* real. Naturalizing (or denaturalizing) something is not just about semantics. It is about power.[97] The rise of the alt-right and the demise of *Bakke* mark the culmination of a two-pronged attack on the legacy of the Second Reconstruction and its attempts, however imperfect, to address problems of persistent racial inequality in the country. The denaturalization of social race, rendering it as "merely" social, provided the foundation for the Supreme Court's recharacterization of diversity as incoherent and hence insufficient to sustain the constitutionality of affirmative action programs in education. Instead of dealing with the historical and social realities of racial hierarchy, the Supreme Court, in cases such as *Shelby Co. v. Holder* piously intoned that "things had changed." To insure that people agreed, Republican governors started excising African American history from school curricula, while local activists sought to ban books about race, gender identity, and diversity from libraries.[98] Beyond education, other activists such as Ward Connerly have been continuously engaged in trying to excise consideration of race from all things public or political; and they have been invoking the new science of genetic ancestry tracing to buttress their efforts

Such extreme policy color blindness finds its champion in Justice Clarence Thomas, whose bitter concurrence in *SFFA v. Harvard* dripped with contempt for the educational experts behind the idea of diversity which he saw as a cancer metastasizing to all parts of academia. "It has become clear," Thomas declared, "that sorting by race does not stop at the admissions office." He went on to note that Justice Scalia, in his *Grutter* dissent, had "criticized universities for 'talk[ing] of multiculturalism and racial diversity,' but supporting 'tribalism and racial segregation on their campuses,' including through 'minority only student organizations, separate minority housing opportunities, separate minority student centers, even separate minority-only graduation ceremonies.'" This trend, he concluded, "hardly abated with time, and today, such programs are commonplace." He characterized the end point of these policies as the antithesis of racial harmony: "a world in which everyone is defined by their skin color, demanding ever-increasing entitlements and preferences on that basis."[99] Thomas argued that such pernicious factionalism was, "based on ever-shifting sands . . . because race is a social construct . . . university admissions policies ask individuals to identify themselves as belonging to one of only a few reductionist racial groups. . . . Whichever choice he makes (in the event he chooses to report a race at all), the form silos him into an artificial category. Worse, it sends a clear signal that the category matters."[100] Again, almost echoing R. Roosevelt Thomas, who so fervently argued for the shift from affirmative action to diversity thirty years before, Justice Thomas here focused on the artificiality, not only of affirmative action, but of race itself. The former Thomas hoped the concept of diversity would open more doors to people based on their race, allowing them to realize the full measure of their individual potential. The latter Thomas saw diversity as a subterfuge, a mask for unnatural categories that should be erased all together. For him, it was evil to make race matter in any way, shape, or form. The implication was that all consideration of race should be excised from the public square.

Just as some activists tried to appropriate and distort ideas about the socially constructed nature of race to undermine the concept of diversity as a basis for affirmative action, so too did other activists on the Right appropriate and distort scientific findings about the nature of human genetic diversity to attack other social programs aimed at redressing or ameliorating the ongoing effects of structural racism. As early as 1970, Richard Lewontin had seen the writing on the wall when he declared: "Professor Jensen has made it fairly clear to me what sort of society he wants. I oppose him."[101] From Herrnstein and Murray's incorporation of Jensen's work in *The Bell Curve* in the 1990s, up through Murray's *Human Diversity* in 2020, scholars on the Right have been trying to naturalize racial difference as genetically based, to make biological race "real" so as to challenge the efficacy of social programs aimed at ameliorating the effects of racism—because if racial hierarchy is due to "nature" not society, then merely social attempts to address it were doomed to failure. Shedding the patina of scholarly respectability, more recent incarnations of the Right, such as the Charlottesville Unite the Right marchers or White supremacist milk-chuggers, praise and appropriate the work of Reich or Wade (which itself was tied back to the work of researchers of the Human Genome Diversity Project and those who developed sorting programs such as STRUCTURE) to declare outright the genetic superiority of the White race. Both prongs are connected by their common aim of using the idea of diversity to undermine attempts to dismantle racial hierarchy.

These actors on the Right have been aided in this unwittingly by many of the scientists exploring the nature of human genetic variation through initiatives such as the Human Genome Diversity Project or programs such as STRUCTURE, as well as by biomedical researchers seeking to address health disparities through uncritical invocations of the need to "diversify" genetic databases and clinical trials—whether in the *All of Us* program or in new drug trials. All without paying sufficient, sustained attention to the need to take care in defining their terms of analysis, particularly with relation to the entanglements of social and biological conceptions of human diversity. Making use of genetic science, the new avatars of the alt-right have made it clear what sort of society they want. To paraphrase Richard Lewontin's response to Jensen, it is important that we oppose them.

Epilogue

Diversity's Pandemic Distractions: A Case Study of the Contemporary Uses of Diversity

Even as right-wing attacks on Diversity, Equity, and Inclusion (DEI) were gaining force in the late 2010s, issues of diversity came to the fore in new ways as COVID-19 emerged in 2020. The pandemic provides a powerful case study of some of the contemporary dynamics of managing racial diversity at the intersections of law, politics, and biology. Pandemic diseases have a nasty history of racialization,[1] and COVID-19 is no exception. In 2020, as SARS-CoV-2 took hold and spread through the United States, race quickly entered the narrative framing of those seeking to identify, understand, and respond to the virus. In some ways, the racialization of COVID-19 sadly and predictably echoed older racist tropes of some foreign "other" being responsible for the spread of disease.[2] Invocations of the "China virus" or the "Wuhan flu" were common, particularly among President Trump and other prominent Republicans, and thereby came to circulate widely among the public.[3] The tropes persisted, giving rise to increased anti-Asian violence throughout the United States.[4]

The racialization of COVID-19, however, has also proceeded along two other distinct and perhaps more subtly problematic trajectories. First, significant racial disparities in disease incidence and mortality became evident early on. This was not particularly surprising given long-standing disparities in underlying comorbidities and related socioeconomic factors affecting exposure and response to the disease.

Nonetheless, as these disparities became evident, a frame of race-based genetic differences emerged as a possible explanation. This is problematic for many reasons, not least of which being the simple fact that racial categories are variable and do not map onto any clearly identifiable genetic groupings.[5] To simply invoke the mantra "race is socially constructed" is not enough.[6] As anthropologist Alan Goodman recently put it, "Race is real, but it's not genetic."[7]

In the specific realm of health disparities, it is useful to recall epidemiologist Nancy Krieger's formulation that race-based health differences can be conceptualized as "biologic expressions of race relations."[8] The genetic frame diverts attention from the social, environmental, or historical conditions that account for COVID-19 disparities, focusing instead on the incorrectly racialized genetic makeup of the affected groups to locate responsibility at the molecular level. Such diversion is not merely incidental. It can have a critical and wide-ranging impact upon how we marshal and deploy resources in the short term; and in the long term it can impede efforts to address the deeper structural issues that COVID-19 disparities have so powerfully brought to light.

Second, following readily upon the heels of this bipolarization of racial disparities, as vaccine development ramped up, there came widespread calls for racially "diversifying" clinical trials for the vaccines being tested. The rationales for such diversification were varied but tended to reinforce genetic frames of racial difference. Most common was the assertion that vaccines might work differently in Black or Brown bodies, and so racial diversity in trials was imperative for reasons of safety and efficacy. The idea behind racial diversification was that results from trials enrolling predominantly White subjects might not be generalizable to other "populations." Related to this was a more politically inflected concern that equitable distribution of vaccines would more readily follow from diversified trials, especially because such diversity would encourage trust among vaccine-hesitant groups, particularly African Americans. Safety, efficacy, generalizability, equity, and trust thus rapidly came to characterize the drive toward racializing vaccine development.

It is appropriate, indeed imperative, to be concerned about the racially disparate impact of COVID-19 and also to work to make sure that vaccines are safe, effective, and accessible to all. The problem is not with these overarching concerns, but rather with their being framed by geneticized conceptions of racial difference. Such geneticization is enabled by the blurring, conflation, and confusion of biological and social conceptions of diversity. While many stories about race and COVID-19 emerged and circulated during the height of the pandemic, the dominant narrative located responsibility for racial disparities—both in disease impact and vaccine uptake—in the bodies and minds of those suffering disproportionately from COVID-19. This diverted attention from the historical and ongoing structural factors driving racial inequities in health and had profound implications both for biomedical understandings of race and for sociopolitical approaches to addressing issues of racial justice in health.

In the 2003 affirmative action case of *Grutter v. Bollinger*, the Supreme Court affirmed diversity as a constitutional rationale for considering race as a factor in higher education admissions.[9] As noted earlier in chapter 4, responding to the decision, Derrick Bell wrote a foundational article titled, "Diversity's Distractions." While acknowledging the importance of increasing minority enrollment colleges and graduate schools, Bell cautioned that "the concept of diversity . . . is a serious distraction in the ongoing efforts to achieve racial justice." For Bell, diversity itself was not a bad thing; however, as a rationale for affirmative action it "enable[d] courts and policymakers to avoid addressing directly the barriers of race and class that adversely affect so many applicants." Additionally, Bell argued, "the tremendous attention directed at diversity programs diverts concern and sources from the serious barriers of poverty that exclude far more students from entering college than are likely to gain admission under an affirmative action program."[10] Similar dynamics have been at play in the invocation of diversity to address disparities that have come to light during the COVID-19 pandemic. In the biomedical context of the pandemic response, however, there was the added problem, not confronted by Bell, that diversity not only distracted from important structural causes of health injustice, it also focused attention on genetics in a manner that had the potential to reinforce pernicious and false ideas of essential biological difference among racial groups.

This extended case study argues that an uncritical embrace of the idea of diversity in analyzing and responding to emergent public health crises has the potential to distract us from considering deeper historical and structural formations contributing to racial health disparities. First, I explore the dynamics through which initial responses to racial disparities in COVID-19 became geneticized. I then move on to unpack the rationales for such racialization, examine their merits (or lack thereof), and consider their implications for developing an equitable response to pandemic emergencies. Second, I examine the subsequent racialization of clinical trials for COVID-19 vaccines through the concept of diversity and will then move on to explore how the geneticization of COVID-19 racial disparities laid the foundations for a similar geneticization of race in vaccine development. Third, I argue that such framings are poorly supported and work in tandem with the geneticization of racial disparities in COVID-19 morbidity and mortality to locate the causes of disparities in the minds and bodies of minoritized populations. This distracts attention from the historical and structural forces contributing to such disparities. I conclude by recognizing a certain intractability to the problems of using race in biomedical research and practice, particularly in the context of public health emergencies. I offer some modest suggestions for improvement that could have significant practical effects if taken to heart by researchers, clinicians, and policymakers.

GENETICIZING RACIAL IMPACT

When COVID-19 gained a foothold in the United States in early 2020, President Trump and other prominent Republican lawmakers wasted little time in racializing the disease itself, repeatedly referring to it as the "Wuhan Virus," "Chinese Virus," and "Kung Flu." Similar recent examples of racialization of disease can be seen the responses to the 2013–2016 Ebola outbreak in Africa[11] and in invocations of "African AIDS" during the 1980s.[12] This is hardly a new phenomenon. Historian Keith Wailoo has explored how epidemics have long given rise to "distinctive, recurring racial scripts about bodies and identities." He argues that the opening act of such scripts, from yellow fever epidemics in the eighteenth century, to cholera in the nineteenth century, and influenza, tuberculosis, and AIDS in the twentieth, up to COVID-19 today, has created a moment of "racial revelation" where "health experts and authorities take note of Black people's experiences, illnesses, or mortality as a specific object of curiosity and social commentary." He asserts that "whether the moment of racial revelation focused on supposed Black immunity or Black susceptibility, the revelation became material for an ur-script...a plotline framed as a mystery of racial difference."[13]

President Trump declared COVID-19 to be a public health emergency on January 31, 2020 and issued two national emergency declarations on March 13.[14] Before long, the racially disparate impact of the virus became evident. On April 12, 2020, the COVID Tracking Project started collecting race and ethnicity data from every state that reported such data. On April 15, it launched the first iteration of the COVID Racial Data Tracker using that dataset.[15] Three days later, Fordham law professor Catherine Powell coined the term "Color of Covid" in an opinion piece for CNN, stating that people of color were being hit particularly hard by the pandemic.[16] The COVID Racial Data Tracker and other subsequent studies would go on to document massive racial disparities in morbidity and mortality from COVID-19 in the United States. For example, one study found that as of June 17, 2020, "among older adults, Blacks and Latinxs have death rates approximately three and two times higher than Whites, respectively."[17] By September, one report found that the infection rate for Hispanic patients was over three times higher than the rate in White patients (143 vs. 46 per 10,000), and the rate among Black patients was over two times as high (107 per 10,000).[18] Some targeted studies, such as one in New York City, found that Blacks were five times more likely to develop COVID-19 than Whites.[19]

As such racial differences became more evident, they also became geneticized. Anthropologist Lance Gravlee flagged some of the crudest forms of geneticizing COVID-19 racial disparities in a June 2020 blog post for *Scientific American*, where he noted:

> Louisiana Sen. Bill Cassidy, who was a medical doctor before entering politics, claimed, without providing evidence, that

"genetic reasons," among other factors, put African Americans at risk of diabetes and, therefore, of serious complications from COVID-19. Scientists writing in the Lancet, one of the world's leading medical journals, suggested—also without evidence—that ethnic disparities in COVID-19 mortality may be partly attributable to "genetic make-up" and speculated on a "genomically determined response to viral pathogens." Epidemiologists writing in Health Affairs noted that "there may be some unknown or unmeasured genetic or biological factors that increase the severity of this illness for African Americans."[20]

This last reference to "some unknown or unmeasured genetic or biological factors" is particularly notable as it is a rhetorical move repeatedly deployed to create a space for geneticizing disparities without evidence. It is always available because one can never know *all* of the causes of any given racial disparity. Much more reasonable and responsible would be to assume a null hypothesis: racial disparities are *not* due to genetics until proven otherwise.[21]

Ironically, some of the earliest discussions of racialized genetic difference involved spurious reports of Black immunity to the virus. On March 10, Reuters reported about false claims circulating on social media that "African Skin" resists the coronavirus.[22] Around the same time, a columnist with the *Chicago Tribune* noted similar rumors and referenced a Twitter Live video posted by actor Idris Elba (who had recently contracted the virus) pleading "with black people to stop spreading the 'scary' rumor that they are immune," calling it, "the quickest way to get more black people killed ... around the world."[23] These examples did not address disparities per se, but they nonetheless played into the moment of "racial revelation" where early on in the pandemic race was becoming geneticized in relation to the virus. While these early rumors of Black resistance to the virus were quickly challenged, the search for genetic bases of COVID-19 susceptibility by seemingly more reputable and prestigious biomedical researchers rapidly became racialized in new and problematic ways.

Social media was not the only platform that began looking to genetics to explain this observed phenomenon. In a September 2020 article published in the prestigious journal *Nature*, researchers claimed to have identified a genetic risk locus for respiratory failure after infection with SARS-CoV-2 that was inherited from Neanderthals and present at much higher frequencies in people from South Asia and Europe than Africa.[24] Similarly, a published report from the direct-to-consumer (DTC) genetics company, 23andMe®, found a "strong association between blood type and COVID-19 diagnosis." Specifically, it found, "that the O blood group was protective when compared to other blood groups."[25] Type O blood is found the world over but is more common in Africa than elsewhere.[26]

Subsequent research revealed a much more complex picture of possible genetic bases for variable COVID-19 susceptibility. For example, later studies cast into doubt the significance of ABO blood group variation,[27] and the same researchers who published the Neanderthal study in September 2020 published a second study in January 2021 identifying a different Neanderthal genetic risk locus, this time finding it to be protective against severe COVID-19.[28] Based on these studies, there were apparently some genetic variants inherited from Neanderthals that could make populations outside of Africa more susceptible to COVID-19, and others that could make them less so. In any event, given that only 2–3 percent of modern humans' DNA derives from Neanderthals, it seems likely that most genetic variations of relevance to COVID-19 susceptibility have little or nothing to do with our Neanderthal ancestry.[29]

One of the factors contributing to early rumors of Black immunity to contracting COVID-19 was the relatively low incidence rates in sub-Saharan Africans as COVID-19 spread in 2020. There were, however, very clear nongenetic reasons explaining the much lower case-fatality rate for Africa as compared to the rest of the world at that time. One BBC report from October 2020 noted five reasons in particular. First, "Quick Action": this included the swift introduction of public health measures such as avoiding handshakes, frequent hand washing, social distancing, and the wearing of face masks. Second, "Public Support": for example, in a survey conducted in eighteen African countries, 85 percent of respondents said they wore masks in the previous week. Third, a "Young Population"—and few old-age homes. Fourth, a "Favorable Climate." And fifth, "Good Community Health Systems": particularly those that were familiar with outbreaks of viruses such as Ebola and methods of containing them.[30]

Many of these measures were similar to those taken by countries such as New Zealand, which also had very low rates of COVID-19 early on.[31] But, of course, no one was conducting any genetic studies of predominantly European-descended New Zealanders to try to understand why they seemed so resistant to the ravages of the virus. When the unmarked racial category of "White" was involved, it was assumed that political and social interventions must have made the difference because other "Western" (read "predominantly White") countries without those interventions still suffered from COVID-19.

The search for genes to explain disparities did not end with Neanderthals and blood groups. As racial disparities became more evident, studies of possible candidate genes to explain the difference proliferated. Prominent among these were studies of genotypes encoding for the Human Leukocyte Antigen (HLA) system; the angiotensin-converting enzyme-2 (ACE2) receptor; transmembrane serine protease 2 (TMPRSS2) nasal gene expression; ion channel genetic variants; and prothrombin genetic mutations.[32]

HLA variants encode for the production of cell-surface proteins that are responsible for the regulation of the immune system and are known to vary significantly geographically across the globe. There are three major histocompatibility complex (MHC) class 1 genes: HLA-A, HLA-B, and HLA-C.[33] These genes do not directly correlate with any racial groups, but they (and their various alleles) do vary in their relative frequencies in different populations across geographic space.[34] Thus, for example, one 2020 study identified global allele frequency distributions by country for three representative alleles (HLA-A*02:02, HLA-B*15:03, and HLA-C*12:03) thought likely to be protective for SARS-CoV-2 and for three other three representative alleles (HLA-A*25:01, HLA-B*46:01, and HLA-C*01:02) with the lowest predicted levels of protection based on an *in silico* analysis. While all of the alleles were found across the globe (i.e., none were race-specific), the study did generally find a higher prevalence of the protective alleles in Africa and a higher prevalence of less protective alleles in Europe.[35] Whether or not this in silico analysis actually played out in vivo, it could not be used to explain racial disparities because, like the ABO blood group study and the first Neanderthal gene study, the results of this HLA study would indicate a protective advantage on average for African-descended populations. Additionally, there is tremendous allelic diversity in all major classes of the HLA system among populations *within* Africa, so variable immune response is not simply of matter of HLA diversity based on so-called continental populations.[36]

One obvious molecular candidate for possible race-specific variable response to COVID-19 was the ACE2 gene encoding the angiotensin-converting enzyme-2, which had been proved to be the receptor for the SARS-CoV-1 and so was naturally looked to as a possible host receptor for SARS-CoV-2. One study published in February 2020, just as COVID-19 was taking hold, observed variable frequencies of certain possible significant alleles across "East Asian," "European," "African," "South Asian," and "Ad-Mixed American," groups but also found "no direct evidence supporting the existence of coronavirus S-protein binding-resistant ACE2 mutants in different populations."[37] A later study, published in May 2020, looked at both ACE2 and HLA loci hypothesizing that "genetic variations within these gateways could be key in influencing geographical discrepancies of COVID-19." However, the study concluded that "currently, there are no genetic data to support ethnic/geographical variation of COVID-19 on global basis."[38] A study published in September continued to look at ACE2 and other genes as possible contributors to COVID-19 disparities and similarly acknowledged that "it is unknown how these genetic polymorphisms contribute to the disparate mortality rates." Nonetheless, the authors of this study hypothesized that "genetic and biological risk for highly relevant COVID-19 comorbid conditions may be critical to our ability to understand and therefore address the observed health disparities in the COVID-19 pandemic affecting US NHBs [non-Hispanic blacks]." In pursuing this geneticized approach to disparities, the authors tellingly wrote: "Population-specific risk in Black communities is clearly multifactorial; however, recent research on the prevalence and risk in the UK indicates that comorbidity and social determinants of health only tell part of the story when it comes to accounting for disease risk and mortality in vulnerable populations. Naturally, biological risk likely fills the void, at least in some part."³⁹ In noting that "population-specific risk" is "multifactorial," the authors acknowledged social determinants of health but went on to assert that those determinants tell "only part of the story." The other part of that story, it seemed, had to be biological; despite the fact that they had presented no hard evidence to support the inference.

Certainly, "biological risk" plays a role in COVID-19 prevalence, but the authors conveniently blurred the distinction between biological risk and *race-specific* biological risk and strangely reconfigured the absence of evidence of such racialized risk into evidence of presence. This is a common rhetorical move in biomedical studies seeking to establish race-specific biological difference as an explanation for disparities: authors often note that the social determinants they have identified as contributing to disparities fail to explain the entirety of the disparity, and they therefore assert that the residual disparities *must* be caused by underlying (yet unidentified) race-specific biological differences.⁴⁰

In this case, the authors posited "ancestral variations" in ACE2 as a source of race-specific "biological risk." They framed their discussion of ACE2 with a reference to research from the UK which purported to indicate that "social determinants only tell part of the story."⁴¹ The study was published in September 2020. Its reference for the UK claim came from a news story from May 2020, which offered this striking claim as its opening sentence: "People from Asian and black ethnic backgrounds are at increased risk of dying from covid-19 and, contrary to speculation, this can only be partly explained by comorbidity, deprivation, or other risk factors, according to data from the largest study to date."⁴² Yet, the barely one-page-long news story from May 2020 also stated in the fourth paragraph:

> More research was needed on whether some of the increased risk [of dying from COVID-19] is from greater occupational exposure with proportionally more people from black and minority ethnic (BAME) backgrounds working in sectors such as healthcare or transport. Research is also needed into a range of factors throughout the disease pathway including access to testing, treatment, and intensive care.⁴³

This assertion is hardly a convincing argument as to the likely genetic basis for such disparities.

In October 2020, just one month after this article referenced research in the UK to support the idea that race-specific biological risk likely played a role in disparities, the UK government issued its first quarterly report on COVID-19 health inequalities. The report found that "existing research suggests that biological factors such as genetics are unlikely to explain the inequalities in ethnic groups from COVID-19."[44] As Reuters described, the report found that "the increased risk to ethnic minorities from COVID-19 is largely driven by factors such as living circumstances and profession and not the genetics of different groups."[45]

Nonetheless, another study published in March 2021, well after the UK report, deployed now-common rhetorical frames to perpetuate the idea that genes, not social determinants, *might* be responsible for COVID-19 disparities. To support its observation that "both biological and nonbiological factors *seem to* contribute to racial disparities in COVID-19," this article referenced a study purportedly finding that African Americans "are *likely* more susceptible" due to a polymorphism in the androgen receptor gene. It also went on to reference yet another study finding that "African populations are genetically predisposed to lower expression of ACE2 and TMPRSS2 and *may* therefore be less susceptible to the coronavirus."[46]

Putting aside the fact that these two studies (like the Neanderthal gene studies) would seem to cancel each other out in terms of explaining disparities, the authors' strategic use of words such as "might," "likely," and "seem to" is notable. On the one hand, such terminology is common in scientific papers seeking to explore possible hypotheses. Here, however, they are doing a different sort of work. For example, the phrase "seem to" makes it appear that nonbiological explanations for disparities existed in the same speculative realm as biological explanations. This clearly was not the case. The October UK report is just one of many studies powerfully demonstrating the impact of nonbiological (i.e., social) factors causing race-based health disparities. In actuality, it was the studies of purported racial biological differences that only "seem to" show some possible correlations between certain allele frequencies and higher levels of morbidity or mortality. In the end, the authors deployed these speculative terms not simply to further scientific exploration but to create a space for geneticizing racial disparities.

The above-noted study referenced African populations' differential expressions of ACE2 and TMPRSS2 and speculated that "these data suggest that a genetic component might contribute to lower numbers of reported COVID-19 cases in Africa." Yet, it also went on to note that "it remains likely that non-genetic factors such as age and comorbidities might play a more important role than host genetic elements, especially in determining disease severity and outcome in infected individuals."[47] In other words, even when identifying possible genetic contributors to disparities, the authors of the cited study carefully noted that the significance of genetic contributors likely paled in comparison to nongenetic factors. Moreover, this study was of variable "gene expression," not of genetic variation itself.

As a different study of TMPRSS2 expression and COVID-19 stated, "although this study suggests one factor that may partially contribute to COVID-19 risk . . . many additional factors are likely, especially because gene expression and race/ethnicity reflect multiple social, environmental, and geographic factors."[48] In other words, studying the impact of variable TMPRSS2 expression on COVID-19 disparities could not be separated from social and environmental factors. In the hands of those arguing for strong genetic contributions to disparities, however, such caveats tend to fall by the wayside. As Merlin Chowkwanyun and Adolph L. Reed Jr. cautioned in a May 2020 article, when such context is lost, "data in a vacuum can give rise to biologic explanations for racial health disparities."[49]

Another study looking for possible genetic bases for COVID-19 racial disparities hypothesized that certain common ion-channel-regulating genetic variants might be, "contributing to the spike in sudden deaths and racial health disparities observed in COVID-19 epicenters." These variants occurred at higher frequencies in "individuals of African origin" and had been linked to "an increased risk of ventricular arrhythmia (VA) and sudden cardiac death (SCD) in African Americans across the age spectrum." The authors of this study also deployed the common rhetorical move of opening with an acknowledgment that "this phenomenon is likely explained by the convergence of multiple cultural and socioeconomic factors" before moving on to hypothesize that "an underlying genetic susceptibility to SARS-CoV-2 infection ... could also contribute." Yet even on its own terms, after creating this space for geneticizing racial disparities, this study concluded that its hypothesis "remains to be proven" and, strikingly, "may not even be testable."[50]

In contrast, the authors of a similar study looked at possible relations between COVID-19 outcomes and purported genetic bases to racial "thrombotic outcome disparities" (i.e., negative blood clotting events) chose to subtitle their article, "Beyond Social and Economic Explanations." In this study, the authors hypothesized that "differences in mortality and thromboembolic event occurrences in COVID-19 may also be, in part, explained by important, but comparatively unrecognized, race-related disparities in intrinsic thrombogenicity."[51] The word "intrinsic" is critical here as it biologizes the disparities of the affected racialized group. By way of comparison, a systemic review of sixty-eight studies exploring possible genetic contributions to racial disparities in cardiovascular disease published in 2015 found little evidence to support any purported genetic connection, concluding that "most associations reported from genome-wide searches were small, difficult to replicate, and in no consistent direction that favored one racial group or another."[52]

It has not been just the individual researchers foregrounding genetic explanations of racial disparities, but major journals themselves. In the wake of a major controversy surrounding a *JAMA* podcast that effectively denied the existence of racism in the medical profession, former New York City Health Commissioner Dr. Mary Bassett asserted that "the biomedical literature just has not embraced racism as more than a topic of conversation, and hasn't seen it as a construct that should help guide analytic work. ... But it's not just JAMA—it's all of them." With specific reference to COVID-19, Bassett lamented the telling example of *JAMA* rejecting her analysis of COVID-19 mortality rates by race and age, while publishing another paper proposing that a racial variation in a cellular receptor for the coronavirus might be an explanation for the pandemic's disproportionate toll on Black people.[53]

The above discussion is a representative sample of the wide range of articles published in major biomedical journals. Reviewing them is important for several reasons. First, they set a narrative frame that affected understandings of and substantive responses to the observed phenomenon of COVID-19 racial disparities. Second, they played into and reinforced preexisting dynamic of racialized care with potential implications for triage and treatment in a time of public health emergency. Third, they set the stage for racializing clinical trials for vaccines in a manner that further reinforced the narrative that races are genetically distinct groups, and these differences must play a role in our response to COVID-19 (and other) health disparities.

UNPACKING THE GENETICIZATION OF RACIAL DISPARITIES IN COVID-19

Perhaps the most remarkable thing about the almost reflexive search for genetic bases to COVID-19 health disparities was the mountain of evidence that such disparities were profoundly and overwhelmingly due to historical, social, environmental, legal, and political (i.e., *not* genetic) factors. Even if some possible genetic correlations to racial disparities were to be uncovered, the chances were minimal that they would explain more than a tiny fraction of such difference relative to the impact of nongenetic factors. This must have been obvious to anyone with the slightest familiarity with the history of racial disparities in the United States. As Lance Gravlee put it: "human biology is more than the genome. Our environments, experiences and exposures have profound impacts on how our bodies develop, turning genetic potential into whole beings."[54] But if your idea of looking at gene-environment interactions is considering how methyl groups affect gene expression, perhaps you never develop a feel or appreciation for considering the power forces beyond the genome to shape health.

The nongenetic factors contributing to COVID-19 disparities were myriad. Minoritized populations already bore a disproportionate burden of underlying comorbidities that can place them at greater risk of higher mortality from COVID-19. Moreover, not everyone was in an equal position to manage their risk of encountering others with COVID-19. As cardiologist Clyde Yancy noted, social distancing is a form of privilege. Many of the conditions that have structured access to such privilege along racial lines are the result of decades-long legal and political actions, such as mortgage red-lining, employment discrimination, and urban transportation policies.[55] Such policies have led to the reality that ethnic and racial minorities are more likely to live in substandard crowded housing, be exposed to toxic environmental pollutants, be forced to rely on public transportation, and have reduced access to health care.[56] They are also more likely to work in settings that have been deemed "essential" such as healthcare facilities, farms, factories, grocery stores, and public transportation, not to mention that these populations had, and have, higher rates of incarceration in prisons where COVID-19 ran rampant.[57]

In their analysis of the legal underpinning of COVID-19 disparities, Yearby and Mohapatra noted that "structural racism in employment causes disparities in exposure; structural racism in housing causes disparities in susceptibility; and structural racism in healthcare causes disparities in treatment."[58] Moving from the structural to the interpersonal, another study published in November 2020 found a strong correlation between levels of implicit anti-Black bias among non-Hispanic Whites and higher overall COVID-19 mortality rates and larger Black-White incidence rate gaps. In that study, the authors concluded that racism may not merely aggravate disparities but may actually be "harmful for everyone's health."[59]

As a September 2020 report from the National Academies of Sciences, Engineering, and Medicine (NASEM) concluded:

> This disproportionate burden [of COVID-19 morbidity, mortality, and transmission] largely reflects the impacts of systemic racism and socioeconomic factors that are associated with increased likelihood of acquiring the infection (e.g., frontline jobs that do not allow social distancing, crowded living conditions, lack of access to personal protective equipment, inability to work from home) and of having more severe disease when infected (as a result of a higher prevalence of comorbid conditions or other factors).[60]

All of these factors combined to offer up a decidedly nongenetic menu of contributors to health disparities. Yet, the search for genetic causes was widespread and persistent.

Not only were there many studies documenting the myriad nongenetic contributors to COVID-19 disparities, but there were also studies explicitly concluding that genetics were *not* contributing to COVID-19 disparities. In addition to the UK report mentioned earlier, finding that "biological factors such as genetics are unlikely to explain the inequalities in ethnic groups from COVID-19," NASEM similarly declared in September 2020 that "currently there is little evidence that this is biologically mediated, but rather reflects the impact of systemic racism."[61] Further, longtime health and science reporter Gina Kolata authored a December 2020 article in the *New York Times* with the unequivocal title, "'Nothing to Do with Genes:' Racial Gaps in Pandemic Stem from Social Inequities, Studies Find."[62]

One of the studies discussed in Kolata's article assessed racial disparities in hospitalization and mortality in patients with COVID-19 in New York City hospitals. It found that "although Black patients were more likely than White patients to test positive for COVID-19, after hospitalization they had lower mortality, suggesting that neighborhood characteristics may explain the disproportionately higher out-of-hospital COVID-19 mortality among Black individuals." Ultimately, this led to the conclusion that "existing structural determinants pervasive in Black and Hispanic communities may explain the disproportionately higher out-of-hospital deaths due to COVID-19 infections in these populations."[63] In commenting on the study for the *New York Times* story, its lead author Dr. Gbenga Ogedegbe noted that "we hear this all the time—'Blacks are more susceptible' . . . It is all about the exposure. It is all about where people live. It has nothing to do with genes."[64]

A similar study of hospitalization and mortality in Louisiana, published in June 2020 in the *New England Journal of Medicine*, likewise found that "Black race was not associated with higher in-hospital mortality than white race." That is, once they were admitted, Black patients fared no worse than White patients. Therefore, the study concluded, differences in clinical presentation and mortality likely reflected social factors such as risk of community exposure or "longer wait to access care."[65] In a larger study of ninety-two hospitals across twelve states, Yehia and colleagues found that "there was no statistically significant difference in all-cause, in-hospital mortality between White and Black patients after adjusting for other factors."[66] This study again indicated that once Black patients actually obtained access to comparable care, the disparities in mortality disappeared.

Now, contrast these studies with those discussed earlier that looked, for example, "beyond social and economic explanations"[67] for genetic bases to explain COVID-19 disparities. The former were based on empirical observations of existing conditions informed by historical understandings of the social structures shaping those conditions. The latter tended to be abstract hypotheses of possible genetic pathways that *might* involve differing certain alleles that occur at different frequencies in certain socially identified racial groups. Such studies almost always involved trying to explain conditions that disproportionately impact social groups that have experienced a history of social dispossession and injustice.

It is quite reasonable to want to explore how genes contribute to the spread and severity of COVID-19. Such analysis was central to the development of the mRNA vaccines by Pfizer and Moderna.[68] However, looking for a genetic explanation of the structure and functioning of SARS-CoV-2 is not the same thing as using genes to explain racial disparities in morbidity and mortality in those suffering from COVID-19. As epidemiologist Jon Zelner (leader of another of the studies discussed in the Kolata article) asserted, the toll on Black and Hispanic Americans, "could easily have been ameliorated in advance of the pandemic by a less threadbare and cruel approach to social welfare and health care in the U.S."[69] Trying to find race-specific genetic differences to explain such disparities not only had little scientific basis, it diverted attention exactly from these sorts of historical and social inequities that predated the pandemic and structured its racialized impact.

HEALTH, RACE, AND ALGORITHMS

Misconceiving the relationship between race and biological difference can also negatively impact medical treatment. Racism in medical treatment and research has a long and sordid history in this country.[70] As COVID-19 was ravaging the United States in the summer of 2020, Vyas et al. published a major study in the *New England Journal of Medicine* documenting the problems created by the, "subtle insertion of race into medicine [through] diagnostic algorithms and practice guidelines that adjust or 'correct' their outputs on the basis of a patient's race or ethnicity." The authors stated, "Physicians use these algorithms to individualize risk assessment and guide clinical decisions. By embedding race into the basic data and decisions of health care, these algorithms propagate race-based medicine. Many of these race-adjusted algorithms guide decisions in ways that may direct more attention or resources to white patients than to members of racial and ethnic minorities." Such disparate allocation of attention and resources might be justified if there were a true genetic basis to the racial algorithms, but the Vyas study found that developers of such algorithms often had either no clearly articulated rationale for using race as they did, or used rationales based on faulty or outdated data.[71]

This was problematic both for immediate concerns of quality of care and also because, as the authors noted:

> Most race corrections implicitly, if not explicitly, operate on the assumption that genetic difference tracks reliably with race. If the empirical differences seen between racial groups were actually due to genetic differences, then race adjustment might be justified: different coefficients for different bodies.
>
> Such situations, however, are exceedingly unlikely. Studies of the genetic structure of human populations continue to find more variation within racial groups than between them.[72]

This study looked at race-adjusted algorithms in such areas as cardiology, nephrology, and obstetrics. This study was conducted largely before COVID-19 had taken hold in the United States, but it speaks directly to how the biologization of racial difference can affect access to and delivery of care.

As the nature of COVID-19 became better understood, doctors began looking to measure levels of oxygen in the blood using devices known as pulse oximeters to diagnose hypoxemia, or low blood oxygen, as indicative of the presence and severity of the disease. A pulse oximeter is a clamp-like device that clips onto a finger. As a Michigan study published in the *New England Journal of Medicine* in December 2020 noted: "oxygen is among the most frequently administered medical therapies, with a level that is commonly adjusted according to the reading on a pulse oximeter that measures patients' oxygen saturation."[73] Yet this study of two large cohorts determined:

> Black patients had nearly three times the frequency of occult hypoxemia that was not detected by pulse oximetry as White patients. Given the widespread use of pulse oximetry for medical decision making, these findings have some major implications, especially during the current coronavirus disease 2019 (Covid-19) pandemic. Our results suggest that reliance on pulse oximetry to triage patients and adjust supplemental oxygen levels may place Black patients at increased risk for hypoxemia.[74]

The reason for this particular technologically mediated disparity was fairly simple. Pulse oximeters work by sending two types of red lights through the finger from one arm of the clamp which is then picked up by a sensor of the other side to detect the color of your blood. The redder the blood, the more highly oxygenated it is. But pulse oximeters were developed using algorithms calibrated to detect oxygen levels in lighter-skinned people; so they often misread blood levels of darker-skinned people. This has nothing to do with the genetics of blood oxygen levels, it has to do with using a White norm to develop diagnostics.

Following the publication of the *New England Journal of Medicine* study, Senators Ron Wyden, Cory Booker, and Elizabeth Warren wrote to the acting director of the FDA requesting it, "conduct a review of the interaction between a patient's skin color and the accuracy of pulse oximetry measurements."[75] The next month, the FDA issued a public warning about the devices, acknowledging they had limitations but pointedly not using the word "race." Instead, the FDA alert cautioned that "a recent report ... suggests that the devices may be less accurate in people with dark skin pigmentation."[76] On the one hand, this alert appropriately decoupled dark skin pigmentation (a biological attribute) from race (a social attribute). On the other hand, the original report explicitly mentioned race in part because it noted that the inaccuracies in pulse oximeter readings could have racially disproportionate impacts that might aggravate already existing health disparities. This is one of the great tensions in using race in medicine—sometimes taking race out of the picture obscures the impact of racism.

The FDA notably took a step beyond the race-neutral framing of issues with pulse oximeter readings on November 16, 2023, when it published a discussion paper, "Approach for Improving the Performance Evaluation of Pulse Oximeter Devices Taking Into Consideration Skin Pigmentation, Race and Ethnicity," and solicited public feedback.[77] Presentations on the issue made to an FDA advisory committee in February 2024 were generally careful to emphasize the significance of not reflexively using race as a proxy for skin tone in studies assessing the accuracy of pulse oximeters. This is significant because it acknowledged the salience of race as a framing concepts in public discourse (and several previous medical studies) but took pains to disentangle it from the distinctively relevant (and not directly racially concordant) phenotype of skin pigmentation.[78]

A February 2021 comment published in *The Lancet Respiratory Medicine* similarly asked: "Could routine race-adjustment of spirometers exacerbate racial disparities in COVID-19 recovery?"[79] Spirometers are devices that originated in the nineteenth century and used to measure lung function. As Lundy Braun has shown in her masterful study of the development and application of spirometry, the use of the device has long been plagued by the use of highly problematic and inaccurate "race-corrections" that often lead to the misdiagnosis of Black patients.[80] The authors of the 2021 comment cautioned that "these race adjustments could potentially cause clinicians to miss important diagnoses" in Black COVID-19 patients and "influence treatment plans" to their detriment.[81]

Beyond medical treatment decisions, the biologization of racial difference in response to COVID-19 persisted and proliferated in the weird world of conspiracy theories, most notably in the person of noted anti-vaxxer, Robert F. Kennedy Jr., who in 2023 opined: "There is an argument that [COVID-19] is ethnically targeted. COVID-19 attacks certain races disproportionately. . . . COVID-19 is targeted to attack Caucasians and black people. The people who are most immune are Ashkenazi Jews and Chinese."[82] Kennedy later defended himself, saying his comments had been mischaracterized and cited one of the studies from early on in the COVID-19 pandemic (June 2020) analyzing genetic polymorphisms in ACE2 and TMPRSS2.[83] Like many of these studies, the one cited by Kennedy hypothesized about, "differential polymorphisms which may explain susceptibility and even outcome in different ethnic populations."[84] The early search for genetic bases for racial differences in response to COVID-19 affected not only public health responses during the initial outbreak, but continued to influence political discourse years later.

LAW AND RACE UNDER COVID-19

Biologized racial difference in relation to COVID-19 was also invoked under very different circumstances, for example, in trying to obtain compassionate release from prison. In such instances, courts embraced the idea of race as social rather than genetic difference in order to deny those compassionate release requests. In one federal case, a prisoner seeking COVID-19-related compassionate release from prison cited a number of physiological conditions which he alleged placed him at higher risk for severe COVID-19. Among them, simply that he was an "African-American male" and that "the virus has hit the African-American community particularly hard and is killing men at a higher rate than women." The court refused his release plea, noting that "while some have suggested that genetic factors might explain these differences, defendant concedes that the reasons remain unknown."[85] Similarly, in the case of *United States v. Alexander*, the court denied the defendant prisoner's application for compassionate release, stating:

> It is not clear that being African-American increases Defendant's risk of complications from the COVID-19, in the same manner as one's underlying medical conditions. Indeed, although African Americans are overrepresented in data regarding COVID-19 hospitalizations and deaths in America as a whole, this overrepresentation may result from other systemic economic and social issues affecting the African American community, including access to health care, higher prevalence of underlying conditions, and lack of access to health insurance.[86]

In response to a similar claim from a prisoner in Colorado, federal district court Judge Robert Blackburn cited anthropologist Lance Gravlee's

Scientific American post to support his contention that "although some researchers suggest there also may be a biological component to African Americans' observed susceptibility to the disease . . . that idea is neither proven nor uncontested. . . . It thus is entirely speculative whether Mr. Billings's race, in and of itself, increases his risks of contracting the virus or experiencing a more severe course of the virus. This consideration therefore cannot be considered compelling."[87]

In light of the legal underpinnings to COVID-19 disparities identified by Yearby and Mohapatra, the irony of federal courts embracing the idea that race is not genetic is palpable.[88] While these courts certainly got the science right, their position stands in marked contrast to the ways in which representatives from other federal agencies, pharmaceutical corporations, and academic research centers discussed race and biological difference in relation to COVID-19 during the same period of time. Specifically, those representatives, agencies, and corporations made repeated and widespread calls to "diversify" clinical trials for COVID-19 vaccines by enrolling more Black bodies as subjects. One might well wonder why it is that some agents of the federal government were so willing to embrace ideas about the social construction of race when it served to keep Black bodies in detention while under different circumstances other agents of the federal government were so ready to invoke biological racial difference as grounds for enrolling Black bodies in clinical trials. Such divergence points up the importance of examining the work that conceptualizations of race and diversity were doing in different contexts, and in whose interests such conceptualizations were being deployed.

DIVERSITY AND THE RACIALIZATION OF VACCINE TRIALS

The broad discussions of biological difference framing COVID-19 disparities created a powerful rationale for racializing emergent vaccine trials by calling for inclusion of "diverse" bodies as a scientific imperative. At the same time that discourses of biological difference were being used by some to *explain* disparities, others were using biological difference as a purported basis for *addressing* disparities. As sociologist Steven Epstein showed over a decade ago in his foundational book,

Inclusion: The Politics of Difference in Medical Research, diversifying the racial composition of clinical trials has been a growing concern of researchers and the federal government since the 1980s. A key transitional moment in this story was the passage of the 1993 National Institutes of Health (NIH) Revitalization Act, which required that women and members of "minority groups" be included as research subjects in NIH-funded studies unless a valid reason for noninclusion was articulated. The Revitalization Act also established an Office of Research on Minority Health within the NIH.[89]

Despite the passage of the Revitalization Act, some of those working on health disparities expressed discomfort at the implicit geneticization of racial differences encompassed in such calls for inclusion. Otis Brawley, then-director of the National Cancer Institute's (NCI) Office of Special Populations Research, worried that implementation of the NIH Revitalization Act guidelines, "may eventually do more harm than good for the minority populations that it hopes to benefit. The legislation's emphasis on potential racial differences fosters the racism that its creators want to abrogate by establishing government-sponsored research on the basis of the belief that there are significant biological differences among the races." Moreover, Brawley opined, it might, "distract from truly important health care issues ... [by] encourag[ing] scientists to waste time and resources looking for minute, insignificant biological differences and to ignore social and environmental influences."[90]

Yet since 1994, diversity in clinical trials has been a mantra of the industry even as racial disparities have persisted and, in some cases, deepened. In 1997, Congress passed the FDA Modernization Act, which called upon pharmaceutical companies to include racial data in their new drug approval submissions. The FDA itself followed up with official guidance papers in 1999, 2005, and 2016, each of which encouraged and provided direction for the collection of racespecific data in the drug development and approval process.[91]

This is not to say that such calls for diversity in clinical trials were nefarious or wholly without merit. Certainly, racial inclusion, when understood as related to variable social, environmental, and historical conditions along with differential rates of comorbidities, can be very relevant for clinical trials and a potential way to address issues of racial equity. However, it is very difficult, although not impossible, to maintain the use of racial categories as social variables once introduced into biomedical contexts. This is especially true where the researchers involved are generally not trained to understand or appreciate the complexities of race as a biocultural construct—that is, a social construct with very real biological implications for individuals whose bodies are being racialized. The general foundations for racializing COVID-19 clinical trials were thus broad and deep, but the proximate context and frame for racializing specific vaccine trials in 2020 was clearly conditioned by the racialization of the disease itself.

Calls for racial inclusion and diversity in clinical trials for COVID-19 vaccines came early and persistently from a wide variety of sources, ranging from industry and government to media, civil rights groups, and the biomedical community itself. As early as April 2020, a group of sixteen U.S. Democratic senators sent an open letter to the CEOs of major pharmaceutical corporations requesting that "any vaccine or therapeutic drug trials related to COVID-19 include women, minorities, and LGBTQ+ persons." The letter noted various social factors contributing to already-observed racial disparities in COVID-19 morbidity and mortality and noted that "alarming research shows that although 'African Americans represent 12% of the United States population, they make up only 5% of all clinical trial participants. Only 1% of clinical trial participants were Hispanic, though they comprise 16% of the national population.'"[92]

In July 2020, NIH Director Francis Collins and Centers for Disease Control (CDC) Director Robert Redfield appeared before a Senate subcommittee with other top government scientists where they emphasized the importance of diversifying COVID-19 vaccine trials. Redfield insisted that "the last thing we want is to be trying to recommend who gets the vaccine and we don't have any data on how the vaccine works in the population that needs it."[93] Similarly, in August, Director of the National Institute of Allergy and Infectious Diseases Dr. Anthony Fauci expressed his desire, "to see minorities enrolled in coronavirus vaccine trials at levels at least double their percentages in the population, because COVID-19 has hit those groups especially hard."[94]

By June 2020, the FDA had already issued a "Guidance for Industry on the Development and Licensure for Vaccines to Prevent COVID-19," which "encourage[d] the inclusion of diverse populations in all phases of vaccine clinical development."[95] In November, it followed up with another guidance, this time specifically on, "Enhancing the Diversity of Clinical Trial Populations," wherein it elaborated on issues of eligibility criteria, enrollment practices, and trial design and noted that the FDA had been steadily working to diversify clinical trials over the past few decades.[96]

Prominent academic and nonprofit organizations such as the Henry Ford Health System and the Kaiser Family Foundation similarly emphasized the importance of racially diversifying COVID-19 vaccine trials. They released statements purporting to explain, "Why Is Diversity So Important In Vaccine Trials," and declared that "ensuring racial and ethnic diversity in clinical trials for development of COVID-19 vaccines is particularly important."[97] Especially resonant were the calls from prominent Black leaders in the medical community, such as the National Medical Association and leaders at Historically Black Colleges and Universities (HBCUs). In June, the National Medical Association and the Alliance of Multicultural Physicians urged Congress and the FDA, "to make diversity in clinical trials a greater priority."[98] Similarly, the presidents of Dillard and Xavier Universities issued a joint statement in September 2020, declaring that "it is of the utmost importance that a significant number of black and brown subjects participate" in COVID-19 vaccine trials.[99] That same month, representatives of four historically Black medical schools (Meharry Medical College, Howard University, Morehouse School of Medicine, and Charles Drew University of Medicine and Science) stated their commitment "to the inclusion of Black, Indigenous and people of color (BIPOC) as we engage in research initiatives focused on the novel coronavirus, SARS CoV-2."[100]

Popular media reported widely on the calls for clinical trial diversity. In August, National Public Radio aired an interview with Renee Mahaffey Harris, president of the Center for Closing the Health Gap, which was titled, "More People of Color Needed in COVID-19 Vaccine Trials."[101] That same month, *STAT News* published a story noting, "Covid-19 clinical trials are failing to enroll diverse populations, despite awareness efforts."[102] In June, NBC National News ran a story on the importance of diversifying clinical trials asserting that a "COVID-19 vaccine will work only if trials include Black participants."[103] Many other major news outlets, including the *Wall Street Journal*, the *New York Times*, and the *Washington Post* published stories and opinion pieces on the importance of diversifying COVID-19 vaccine trials.[104]

The pharmaceutical industry was responsive to these calls. In May 2020, the chief diversity officer and the head of U.S. Medical Affairs for Genentech issued a "call for more inclusive COVID-19 research."[105] Testifying before a U.S. House Subcommittee in July 2020, a representative from Johnson & Johnson declared that was "is committed to robust representation of diverse populations in our studies."[106] The major trade group PhRMA issued a statement in October, recognizing that "achieving clinical trials that include diverse populations presents an ongoing challenge," and affirmed its commitment "to enhance the diversity of clinical trial participants."[107] In presenting data on its COVID-19 vaccine to the FDA for approval in December, Pfizer emphasized: "the importance of conducting the study in people of color," and asserted that "from the very start [it was] focused on targeting in recruitment from racial and ethnic minorities."[108] Moderna was so responsive to these concerns that it actually paused its clinical trial in October in order to increase minority enrollment.[109] By April 2021, the industry trade publication *Scrip* was even promoting the Twitter hashtag #ClinicalTrialsSoWhite.[110]

Concerns for clinical trial diversity were also evident throughout the FDA review process for both the Pfizer and Moderna vaccines. Pfizer reported that, of a total clinical trial population of 40,277 subjects, 3,929 or 9.8 percent were Black and 10,553 or 26.2 percent were Hispanic/Latino. Moderna similarly reported that of its 30,351 clinical trial subjects approximately 10 percent were Black and 21 percent Hispanic or Latino.[111] A number of people and organizations submitting comments to the FDA on the proposed vaccines also raised concerns about clinical trial diversity. For example, the American Academy of Pediatrics cautioned that "the population studied *must reflect the racial and ethnic diversity* of the US population"; the associate director of Health Policy for the National Consumers League requested "that the

FDA continue to prioritize vaccine clinical trial data that reflects diversity"; and the executive director of the National Women's Health Network flatly asserted that "there were not enough Black and Indigenous people included in the Moderna phase 3 trial."[112] In response, Senior Vice President of Pfizer Vaccine Clinical Research and Development Dr. William Gruber assured the FDA review committee that "we also recognized the importance of conducting the study in people of color; so we have adopted an approach that assures a diverse racial and ethnicity profile."[113]

Such calls to diversify clinical trials for COVID-19 vaccines at times came to take on the character of a sort of diversity panic. Horrible things, it seemed, would happen if clinical trials did not sufficiently represent diverse populations: the vaccine would not work; it would not be safe; it would not be trusted. Only through diversity could such catastrophes be avoided. Upon close inspection, however, such claims were often based on questionable assumptions, incomplete data, and problematic definitions of such basic concepts as "diversity" and "representation."

Calls to diversify the clinical trials for COVID-19 vaccines were repeated, widespread, and persistent. Rationales for diversifying clinical trials, and more specifically diversifying them by *race*, can be organized into three basic (and sometimes overlapping) categories: biological, social, and political. Biological rationales generally had to do with concerns about safety, efficacy, and generalizability of trial results. Social rationales revolved around addressing issues of trust and vaccine hesitancy. Political rationales involved commitments to address equity in access to care and therapeutics.

UNPACKING THE SCIENTIFIC RATIONALES FOR DIVERSE VACCINE CLINICAL TRIALS

Social and political rationales for diversifying clinical trials may be reasonable, but only if clearly articulated and justified. Too often when race is introduced into the field of biomedicine these rationales become entangled, confused, and conflated in a manner that threatens to reify race as genetic and divert attention away from the social and political bases of health inequities.

Empirical Bases for Concerns about Diversity

Calls to diversify vaccine clinical trials focused on the importance of obtaining data that showed the vaccines were safe and effective in *all* relevant racial and ethnic groups. The Democratic senators' April 20, 2020 letter to leading pharmaceutical CEOs was very clear about this, ominously quoting from a statement issued by the Johns Hopkins University Science Policy Group in 2017. The quote read, "Inequitable research can lead to dangerous outcomes for those who are not represented in clinical trials. Drugs including chemotherapeutics, antiretrovirals, antidepressants, and cardiovascular medications have been withdrawn from market due to differences in drug metabolism and toxicity across race and sex."[114] Months later, in answering "Key Questions" on "Racial Diversity within COVID-19 Vaccine Clinical Trials," researchers from the Kaiser Family Foundation asserted that "diversity within clinical trials for a COVID-19 vaccine helps ensure safety and effectiveness." Significantly, this report also noted that "diverse racial/ethnic representation in COVID-19 vaccine trials is important because drugs and vaccines can differentially affect groups reflecting variation in underlying experiences and environmental exposure."[115]

Such sensitivity to the complex relationship between race and non-genetic factors in immune response stands in marked contrast to many media reports, such as one by NBC News from June 2020 which quoted a retired pulmonologist who asserted that "genetics related to racial differences make it essential that we be involved in broad-based and diverse clinical trials of medications and vaccines." The NBC report then flatly stated this meant that a "vaccine might not work in African Americans if African Americans do not participate in the clinical trials to create the drug."[116]

Not all media reports quite so baldly geneticized race. A quote from a story in *STAT News* by UC San Francisco oncologist Hala Borno is more typical of the media reporting. Borno stated, "I think that if we do not ensure diversity in these Covid-19 clinical research studies, we may ultimately render interventions, whether it be drug or vaccines, that do not uniformly demonstrate efficacy across populations, or have side effects that we only capture later on."[117] A more recent study of inclusion in vaccine trials published in *JAMA Network Open* asserted, "Historically, clinical trials have lacked equitable inclusion of people identifying as members of racial/ethnic minority groups and female and older individuals. When people with diverse backgrounds are not adequately represented, treatments shown to be effective in trials may not be generalizable to or effective for all populations."[118] Similarly, Paulette Chandler, a primary care physician and lead of community engagement and education for COVID-19 vaccine trials at Boston's Brigham and Women's Hospital stated in September 2020 that "unless we have a diverse group of people involved in the trial, we will not be able to generalize our findings to every group."[119]

The theme was clear and consistent. There was an assumption that essential biological racial difference was somehow directly related to how a vaccine worked—either in terms of its efficacy or its side effects. References to "generalizability" are particularly concerning because, without explicitly invoking the idea of racial differences, they imply that results from trials conducted in people of one race simply could not be extrapolated to people of another race. The problem is that there was little or no evidence to support these race-specific concerns about safety, efficacy, or generalizability, and certainly not sufficient evidence to warrant major efforts to reconfigure (and perhaps delay) clinical trials in a time of pandemic emergency.

While diversifying clinical trials has long been a concern of both the federal government and an array of biomedical institutions, prior to 2020 there had been relatively little discussion of the need or importance of diversity specifically in vaccine trials. Just over two years before the COVID-19 outbreak, the World Health Organization issued a report titled, "Design of Vaccine Efficacy Trials to be Used During Public Health Emergencies—Points of Considerations and Key Principles," that made no reference to race at all, and only one reference to ethnicity in the context of considering vulnerable groups in potential need of protection from exploitation by researchers.[120] An article titled, "Design of Vaccine Efficacy Trials During Public Health Emergencies," published in *Science Translational Medicine* in July 2019 on the eve of the COVID-19 outbreak, similarly had no mention of race or ethnicity in relation to issues of safety, efficacy, of generalizability of trial results.[121] Another lengthy review article published in 2019 examining "factors that influence the immune response to vaccination" did mention studies of ethnic variability in response to vaccines in one sentence out of thirty pages of text.[122] However, the studies it references involved not large racial groups, but localized ethnic populations, such as different ethnic groups within Guatemala or a particular region of China.[123] None of the cited studies found any difference in safety or efficacy based on race or ethnicity.[124] To the contrary, instead of focusing on race, the bulk of the Zimmerman and Curtis article was devoted to examining factors such as comorbidities, behavior, nutrition, environmental exposures, and the nature of the vaccine itself and its administration.[125]

As for the three COVID-19 vaccines that were ultimately developed in the United States during 2020 and approved by the FDA—all were found to have a similar safety and efficacy profile across all racial subgroups. The FDA reported that the first approved vaccine, developed by Pfizer, showed efficacy to be "consistent across various subgroups, including racial and ethnic minorities," and its "safety profile" to be "generally similar across age groups, genders, ethnic and racial groups."[126] The FDA came to similar conclusions regarding the Moderna vaccine, finding vaccine efficacy among racial and ethnic "subgroups" to be similar to that "seen in the overall study population" and identifying "no specific safety concerns . . . in subgroup analyses by age, race, [or] ethnicity."[127] Finally, for the Johnson & Johnson vaccine, the FDA concluded that efficacy "among the subgroups (age, comorbidity, race, ethnicity) appears to be similar to the [vaccine efficacy] in the overall study population," and "there were no specific safety concerns identified in subgroup analyses by age, race, ethnicity, medical comorbidities, or prior SARS-CoV-2 infection."[128] In 2023, the National Academies of Sciences, Engineering, and Medicine formed a committee on the Use of Race, Ethnicity, and Ancestry as Population Descriptors in Genomics Research, whose final report stated quite simply: "Vaccines for coronavirus disease 2019 (COVID-19) work via the same immunological mechanisms in Peru or Poland."[129]

Given the consistency of these results (not to mention their concordance with experience from previous vaccines) it is, perhaps, worth revisiting some of the early, urgent calls to diversify the COVID-19 vaccine trials. What were their rationales and what sort of evidence did they present to support their concerns? Noteworthy in this regard is the

April 20, 2020 letter from the sixteen Democratic U.S. senators to leading pharmaceutical CEOs calling for diversity in vaccine clinical trials. Coming from such prominent and powerful politicians, relatively early in the pandemic, it set a powerful frame for conceptualizing the clinical trial process. Of particular force was the quotation, referenced above, to a statement issued by the Johns Hopkins University Science Policy Group in 2017. The full paragraph in the letter stated:

> Alarming research shows that although "African Americans represent 12% of the United States population, they make up only 5% of all clinical trial participants. Only 1% of clinical trial participants were Hispanic, though they comprise 16% of the national population." As a result, "inequitable research can lead to dangerous outcomes for those who are not represented in clinical trials. Drugs including chemotherapeutics, antiretrovirals, antidepressants, and cardiovascular medications have been withdrawn from market due to differences in drug metabolism and toxicity across race and sex."[130]

Let us unpack this reference and the work it did in the senators' letter. First, even on its face, the quoted passage refers to drugs, not vaccines. More specifically, it refers to drug metabolism and toxicity. The biological mechanisms underlying drug metabolism and immune response, while often related, are distinct.[131] Variance in drug metabolism is common and is affected by many things, including diet, other medications, comorbidities, and genetics. For any given drug there will be a standard or normal rate of metabolization, and there will also be a certain range of individuals who metabolize the drug more quickly or more slowly than the norm. This can be particularly important for determining proper dosage. Rapid metabolizers might require a higher dose for the drug to be effective, while slower metabolizers might need a lower dose because the drug stays in their system longer. When slow metabolizers receive normal or high doses, this can cause toxic reactions. Several drugs' labels mention race or ethnicity as a factor in determining whether someone is likely a rapid or slow metabolizer, but this is a crude proxy for prediction of the rate of metabolization.[132] It is important to note that there is no clear concordance of race and drug metabolization.

Instead, in some cases, there is merely an observation that for some drugs certain racial groups may have, on average, a higher frequency of rapid or slow metabolizers than others. In any event, drug metabolization is not directly related to the immune response provoked by a vaccine.[133]

Second, the Johns Hopkins statement cited by the senators is itself only a blog post that, in turn, cited a study from the *Journal of Women's Health* to support its statement.[134] This study does *not* support the Johns Hopkins assertion that many drugs, "have been withdrawn from market due to differences in drug metabolism and toxicity across race and sex." Rather, it states that "several drugs have been withdrawn from the market over the last two decades because of *sex-based* adverse events," and then goes on to merely observe that "with regard to race and ethnicity, a number of studies have found variations in drug metabolism and toxicity" in various drugs. Again, this was without reference to any vaccines.[135] This is not an insignificant difference. It refers only to withdrawing drugs for sex-based differences, not race. For race, it simply restates the widely observed phenomenon that for certain drugs there may be different frequencies of rapid and slow metabolizers in different ethnic groups.

This second study itself referenced a third study to support its statement about sex-based drug withdrawals.[136] That study, published in the *European Journal of Clinical Pharmacology*, observed that "several publications indicate that the female gender experiences a higher incidence of adverse drug reactions (ADRs) than does the male gender. *The reasons, however, remain unclear.* Gender-specific differences in the pharmacokinetic and pharmacodynamic behaviour of drugs could not be identified as an explanation." The study concluded that while, "our data confirm the higher risk of ADRs among female subjects compared with a male cohort . . . no single risk factor could be identified."[137] This study, while confirming gender-specific differences in drug response, made no mention of withdrawing drugs from the market.

This is not to say the claim regarding FDA concerns with sex-based differences in drug response is wholly without merit. A 2001 General Accounting Office study (not cited by the Johns Hopkins statement) did find that eight of the ten drugs withdrawn from the market between 1997 and 2000 posed higher risks for women.[138] Even so, there is a marked difference between finding that the drugs withdrawn *happened* to have a higher risk for women, and determining that the drugs were withdrawn *because* they had a higher risk for women. Still, none of this speaks to alleged race-based differences in safety or efficacy as grounds for drug withdrawal, and also does not speak to race-based differences in vaccine response. In short, despite what the letters from the senators claimed, existing evidence supported the conclusion of bioethicists Angela Ballantyne and Agomoni Ganguli-Mitra that "on balance there is no biological imperative to achieve representative recruitment of minoritized populations in COVID vaccine trials."[139]

The example of the senators' letter illustrates how the casual blurring of boundaries and elision of distinctions can lead to the reification of racial difference as genetic. Moreover, as is made evident by the 2019 review of factors influencing vaccine response referenced above, nongenetic behavioral, environmental, and social factors certainly dwarf any possible importance of race for assessing variable vaccine response.[140] A vaccine cannot elicit any response in someone who is either unable to access it or unwilling take it.

The questionable empirical basis for concerns about racial variance in vaccine efficacy or safety is only the first issue that needs to be unpacked in the realm of biological rationales for diversifying clinical trials. The second issue is what exactly was meant by diversity when diversity was called for in clinical trials for COVID-19 vaccines.

What Was Meant by Diversity in Vaccine Trials?

Calls to diversify the COVID-19 vaccine clinical trials typically invoked, in a commonsense fashion, the basic census categories of race and ethnicity (i.e., White, Black, Asian/Pacific Islander, American Indian/Alaska Native, Hispanic/Non-Hispanic) that have their foundation in a 1997 directive from the Office of Management and Budget (OMB).[141] These categories structured the 1993 NIH Revitalization Act directive to increase diversity in clinical trials, and have since become the default categories for much of biomedical research and regulation.[142] They are also generally the same categories that were used to identify and trace racial disparities in the impact of COVID-19.

But why the focus on race in the first place? Often in studies arguing for the relevance of race as a genetic category in biomedical research, the authors argue that race is an apt proxy for continental genetic ancestry.[143] This sort of conceptualization of diversity was evident in some of the studies of the possible genetic underpinnings to COVID-19 disparities, particularly those involving the global distribution of HLA haplotypes.[144] Yet, such efforts to capture ancestral genetic diversity were not prominent in discussions of COVID-19 vaccine trials. Presentations before the FDA during the review processes for the various vaccines generally did not engage issues of genetic diversity but simply focused on the demographic census categories.

Apparently, then, when it came to calls to diversify vaccine trials, clinical trial designers simply assumed, without evidence, that demographic categories of race had some biologic relevance to vaccine performance. One might just as easily have called for diversifying vaccine trials to ensure adequate representation of left-handed people. Left-handedness as a phenotype is arguably more closely related to genetic difference than is race.[145] In the United States about 13 percent of the population is left-handed.[146] Roughly the same percentage of the population is Black; thus, if researchers are concerned with genetically diverse representation, why was nobody talking about this or any number of other possible population group differences? Perhaps because we do not have a centuries-old biomedical tradition of trying to discern biological differences between "righties" and "lefties" as a means to justify social hierarchy or explain away disparities (Of course, historically, being left-handed has also been somewhat stigmatized, but the consequences of such stigmatization have always been understood to be social in character and not due to any underlying genetic difference). Hence, there is no need to enroll lefties in clinical trials in proportion to their representation in the population. Black people, however, are another story. As the marked disparities in COVID-19's impact were attributed "in part" to some presumed, if unidentified, genetic differences—that is, as the consequences of the stigmatized characteristic were attributed to biology instead of society—there emerged a purported biological rationale to single out race, among a wide range of possible demographic characteristics, as essential for representation in vaccine trials.

As we look more closely at invocations of the need for diversity, we see further refinements of the basic census categories that elaborate the biological rationales for the call. For example, in its November 2020 guidance for industry on, "Enhancing the Diversity of Clinical Trial Populations," the FDA emphasized the importance of enrolling subjects who will better reflect the population most likely to use the drug, i.e., those affected with the condition the drug is meant to treat. However, the same guidance also made it clear that the "sponsors should enroll participants who reflect the characteristics of clinically relevant populations with regard to age, sex, race, and ethnicity . . . [because] analyzing data on race and ethnicity may assist in identifying population-specific signals."[147] This echoes concerns about possible racial differences in response to some drugs; however, the COVID-19 pandemic was not a drug, nor were we dealing with a "population most likely to use the drug." Rather, we were dealing with a vaccine, and the population most likely to use it was, ideally, everyone.

More specific to the COVID-19 vaccine trials was the idea that the trials had to be representative of communities bearing the heaviest disease burden or at the highest risk of contracting the disease. These categories certainly had a public health logic to them, but they can also be moving targets during a period of emergent crisis. Certainly, as in the case of COVID-19 with its well-documented disparate impact in the United States, they may loosely map onto racial groupings. Thus, for example, in its June 2020 guidance to industry on "Development and Licensure of Vaccines to Prevent COVID-19," the FDA, when urging the "inclusion of diverse populations" in clinical trials, "strongly encourage[d] the enrollment of populations most affected by COVID-19, specifically racial and ethnic minorities." In order to evaluate "vaccine safety and efficacy," the guidance encouraged, "adequate representation of elderly individuals and individuals with medical comorbidities."[148] The distinctions between the two categories of inclusion are significant. This is the only place race was mentioned in the guidance, and it was *not* directly connected to concerns of safety or efficacy. The rationale for racial inclusion in the guidance was that it was a marker of populations "most affected" by COVID-19. Similarly, Dr. Ann Falsey of the University of Rochester School of Medicine further contextualized diversity as a function more of *risk* than of race when she told a journalist at *JAMA*: "we are thinking very hard about not only how to get a diverse population that reflects the US population but also people at high risk—postal workers, home health workers, you name it."[149]

In comments submitted to the FDA regarding its review of the Pfizer vaccine, the Infectious Diseases Society of America emphasized that "COVID-19 vaccines should be adequately studied in populations that have been disproportionately impacted by the pandemic and who face disparities in care, including the elderly; individuals with chronic conditions; and Black/African American, Indigenous, Latinx, and other communities of color."[150] Here disparate impact was connected with access to care. There was, in other words, no explicit biological rationale for racial inclusion. Rather, in these examples the rationale for using race was being grounded in concepts of risk or concerns for equity. It was thus possible to make appeals for diversity in ways that did not assert or imply that racial difference was genetic. Nonetheless, given that at this time many studies were being published arguing that genetic susceptibility to COVID-19 varied by race, it is possible that this framing also reflected those biologized understandings of racial disparities in COVID-19.

Other calls for diversity in clinical trials more directly geneticized race. In a blog posting by Henry Ford Health System, Dr. Paul Kilgore, coprincipal investigator for its Johnson & Johnson COVID-19 vaccine trial, declared: "When people have different genetic and biologic makeup, their bodies can produce antibodies differently. This means to ensure a vaccine will protect people of all ethnic groups, we need to make sure everyone is fully represented in clinical trials."[151] We also see an interest in disparate impact blur into presumptions of essential biological difference in the concerns expressed by Dr. Anna Durbin, principal investigator at the Johns Hopkins Center for Immunization Research, who insisted that racial diversity was necessary "to make sure it works in the groups most affected by COVID-19."[152] The focus on equity is laudable, and concerns for safety and efficacy are reasonable, but it is difficult to keep these issues separate and distinct. In juxtaposing the two in relation to race, the reification of race as genetic becomes almost inevitable, especially in a biomedical culture suffused with understandings of race as an essential biological category.[153]

Calls to diversify COVID-19 vaccine trials also extended to hiring the health workforce responsible for carrying out the trials. The general idea was that upstream representation of racial minorities in the health professions would increase downstream willingness to participate in clinical trials. For example, in discussing COVID-19 vaccine trials on National Public Radio, President of the Center for Closing the Health Gap Renee Mahaffey Harris noted that "due to the fact that COVID-19 has had a disproportionate impact to Black and Latino people across this country, it is more paramount than ever that the trials be reflective of a bigger proportion of Black and Latino people." She then went on to assert the importance of having more Black and Latino doctors and researchers represented at "the early part of creating the trial."[154] In October 2020, the Pharmaceutical Research and Manufacturers of America (PhRMA) issued, "Principles on Conduct of Clinical Trials Communication of Clinical Trial Results," in which it declared that "enhancing meaningful representation of diverse participants in clinical trials would help provide information about drug response and measures of safety and efficacy in populations that have been historically under-represented and under-studied, in particular Black and Brown people." PhRMA went on to assert that to achieve this goal it was necessary to "enhance[] diversity among clinical trial investigators."[155] This shows both the flexibility and ambiguity of the concept of diversity. In relation to clinical trial subjects, diversity was often used in a way that presumed essential biological differences among racial groups. Yet it was often invoked by the same actors to apply in a wholly social sense to the need to diversify the workforce. The issue here is not that such uses of diversity are wrong or incorrect, but that it is a slippery concept that ranges across social and biological domains. The concept of diversity needs to be employed carefully in contexts where the conflation or confusion of social and biological conceptions of racial difference are likely to occur.

How Much Diversity is Enough?

Once you determine that you want or need representation from diverse racial groups in clinical trials, the next question becomes just how much diversity is enough? In terms of biomedical concerns, one generally wants to enroll "enough" people to ensure statistically robust results showing safety and efficacy in any given "population." This is, in theory at least, what has been driving calls for inclusion going back to the 1993 NIH Revitalization Act.[156] When it comes to racially marked populations, concepts of what constitutes adequate representation vary significantly. Such variance may be due, in part, to the particular rationale for diversity one is seeking to serve. It may also simply be due to a lack of attention as to why racial diversity may or may not matter in a vaccine trial. Clarifying such issues matters because it allows us to differentiate between biological, social, and political rationales for diversity and ensure that sociopolitical concerns for diversity do not become entangled with or transformed into reified biological understandings of race.

The FDA never specified how much minority representation it wanted to see in COVID-19 vaccine trials, and NIH Director Francis Collins acknowledged in July 2020 that there was no general agreement on the percentage of minorities the trials should include. While some scientists were arguing that percentages should be representative of the distribution of the current burden of disease, others were arguing for more straightforward demographic representation equivalent to each group's percentage of the overall use population—approximately 13 percent for Blacks and 18 percent for non-White Hispanics.[157]

In August 2020, Director Fauci told CNN that he, "wanted to see minorities enrolled in coronavirus vaccine trials at levels at least double their percentages in the population, because Covid-19 has hit those groups especially hard."[158] That would equate to roughly 26 percent representation for Blacks and 36 percent for non-White Hispanics, amounting to 62 percent of the entire trial population just for those two groups. The final numbers reported by Pfizer and Moderna were nowhere near this, but they did approach (although fall short of) proportional demographic representation. In December 2020, shortly after the FDA had approved the Pfizer and Moderna vaccines, the National Medical Association reviewed the trial enrollment data and concluded that the roughly 10 percent of Black people enrolled in each trial was "sufficient to have confidence in health outcomes of the clinical trials."[159]

What does this mean in terms of absolute numbers? In any clinical trial, if there is a population of interest, enough of those subjects should be enrolled to provide statistically robust data about that population. A typical Phase 3 drug trial enrolls between 300 and 3,000 subjects. Phase

Epilogue 299

3 trials demonstrate whether or not a product offers a treatment benefit to a specific population and immediately precedes submission for FDA approval.[160] Vaccine trials are often much larger, usually in the tens of thousands, in part because they are typically administered to large numbers of otherwise healthy individuals, and so concerns for safety are higher.[161]

For example, trials for the recently developed HPV (human papilloma virus) vaccine ranged from 5,500 to 18,500 subjects.[162] One review of trials for thirteen vaccines conducted between 2000 and 2010 found that enrollment in Phase 3 trials ranged from 2,358 to 80,427 with a mean of 29,844 and a median of 22,938.[163] As noted above, Pfizer enrolled 40,477 subjects and Moderna 30,351 in their COVID-19 trials, thus falling within a fairly standard range, somewhat above both the mean and the median for recent vaccine trials. While the considerations in designing a vaccine trial differ somewhat from those of designing a drug trial, it is worth considering that between them, in absolute numbers, Pfizer and Moderna enrolled 6,633 Black subjects.[164] This number alone would be larger than the total enrollment for many Phase 3 drug trials. Indeed, it is over six times the number enrolled in the Phase 3 trial for the race-specific heart failure Drug, BiDil, which enrolled *only* self-identified African American subjects.[165] In such cases 6,663 subjects would clearly be deemed "enough" representation to address issues of safety and efficacy. Moreover, each trial separately enrolled numbers of Black subjects comparable to the 3,000 subjects typically enrolled in a Phase 3 drug trial.

Vaccine trials typically try to enroll larger numbers. Consider the total of 6,633 as compared with the median and mean numbers from the study mentioned above. The 6,633 would amount to approximately 22 percent of the mean enrollment of 29,844 and 29 percent of the median enrolment of 22,938, coming very close to the enriched representation numbers suggested by Dr. Fauci. These are very crude calculations and not directly applicable to the complexities of evaluating the adequacy of a given trial design. Nonetheless, they are instructive for thinking about how "representation" was being conceptualized in relation to COVID-19 vaccine trials and just what sorts of consideration might be used in considering how much diversity was enough. These figures also bring into relief the sort of diversity panic that seemed to swirl around the trial enrollment process—a panic, it seems, which cannot be wholly explained by biomedical considerations of safety, efficacy, and generalizability. Rather, the widespread calls for diversity in clinical trials gained much of their urgency by combining such biomedical concerns with broader social concerns relating to trust and vaccine hesitancy, and political concerns for equity.

UNPACKING THE SOCIAL RATIONALES FOR DIVERSE VACCINE CLINICAL TRIALS

Foremost among the social rationales for increasing vaccine trial diversity was a concern to address issues of trust and vaccine hesitancy in minority, particularly Black, communities.[166] In 2015, the Strategic Advisory Group of Experts on Immunization (SAGE) Working Group on Vaccine Hesitancy defined vaccine hesitancy as, "delay in acceptance or refusal of vaccination despite availability of vaccination services." The report also noted that it is, "complex and context specific, varying across time, place and vaccines. It is influenced by factors such as complacency, convenience and confidence."[167] Vaccine hesitancy, in short, is neither a static nor monolithic phenomenon. It can range from unyielding "anti-vaxxers" who will not get any vaccine under any circumstance to those who simply do not wish to be first in line, but rather, want to wait and see how any given vaccine rollout progresses. It could also include many intermediate attitudes. As one review put it, "not all mistrust is created equal." Vaccine hesitancy among African Americans is distinctively situated in a history of past and ongoing racist encounters with biomedical institutions, and in many cases is quite different in character and tone from vaccine mistrust expressed by White people.[168] In the context of COVID-19, such differences have become particularly pronounced over time, as Black rates of vaccine hesitancy steadily declined while rates of hesitancy among White Republicans (particularly White male Republicans) remained consistently and stubbornly high.[169]

Before COVID-19 there was very little, if any, discussion of the importance of diversifying vaccine trials to address issues of trust among minorities. One 2020 study looking at past vaccination hesitancy experiences in order to develop, "a Social and Behavioral Research

Agenda to Facilitate COVID-19 Vaccine Uptake in the United States," focused on the importance of improving experiences of health care delivery and listed many strategies for improving "transparency and community engagement," but nowhere did it consider diversifying vaccine trials as a means to improve trust and increase vaccine acceptance.[170] A major 2016 review of forty-three studies of vaccine hesitancy listed twenty-three different determinants affecting vaccine uptake; diversifying clinical trials was not among them, nor was it even mentioned in the article. Beyond this, the article concluded that "most interventions to increase vaccine acceptance have shown little or no effect."[171] The SAGE Working Group of Vaccine Hesitancy similarly found that "despite extensive literature searching, there are (1) few existing strategies that have been explicitly designed to address vaccine hesitancy; and (2) even fewer strategies that have quantified the impact of the intervention."[172]

In light of such findings, it is perhaps surprising that so many biomedical professionals, academics, and policymakers assumed without evidence that diversifying clinical trials (a strategy that had not even been the subject of previous studies) would reduce vaccine hesitancy among affected minority groups. As an illustrative example, consider a 2021 statement from the Kaiser Family Foundation on, "Racial Diversity within COVID-19 Vaccine Clinical Trials: Key Questions and Answers," which declared without citation that "diversity within clinical trials for a COVID-19 vaccine helps ensure safety and effectiveness across populations and may increase confidence in getting the vaccine among people of color."[173] A 2021 study published in *JAMA Network Open* on diversity in clinical trials similarly asserted that "improving racial/ethnic diversity in clinical trials is important because enrollment *may* be associated with vaccination rates in minority groups. Efforts to improve inclusion *may* help to address vaccine hesitancy, provide education, and counter safety concerns about vaccines by ensuring equitable representation in definitive clinical trials." Note the equivocal use of the term "may" to qualify these claims. This was perhaps wise because, while the authors here did provide three citations to support this claim, none of the cited studies actually discussed diversifying clinical trials as a means to address vaccine hesitancy; instead, they simply explored the phenomenon generally in certain minority populations.[174] As Ballantyne and Ganguli-Mitra note, "greater participation of minoritized groups in trials may lead to greater trust in the vaccine products—but not necessarily."[175]

Those wringing their hands about potential vaccine hesitancy among minority populations might have done well to consider such studies as the one conducted in 2014 by the CDC's Office of Minority Health and Health Equity. The CDC study found that the congressionally authorized Vaccines for Children program had effectively reduced racial and ethnic disparities in vaccination coverage for the Measles, Mumps, and Rubella (MMR) and Polio vaccines by focusing on providing practical access.[176] Addressing structural issues is thus perhaps more relevant to addressing disparities than speculating about changing trust attitudes through diversifying clinical trials. Such a conclusion comports well with the findings of the Tuskegee Legacy Project, which addressed a range of issues related to the recruitment and retention of Black people and other minorities in biomedical research studies in the early 2000s. The Tuskegee Legacy Project found that "although Blacks self-report having a higher fear of participation, they are just as likely as Whites to self-report willingness to participate in biomedical research."[177] In these contexts, wariness or mistrust did not translate directly into a refusal to participate.

There is also an important distinction to be made between vaccine hesitancy and trial hesitancy. At an outreach session to Black Americans conducted at Meharry Medical College in the summer of 2020, one participant declared: "The word 'vaccination' don't scare me ... the word 'trial' do."[178] Similarly, there was pushback when the presidents of Dillard and Xavier Universities (both HBCUs) urged community members to participate in clinical trials. "Our children are not lab rats for drug companies. I cannot believe that Xavier is participating in this," wrote one parent on Xavier's Facebook page.[179]

Such sentiments raise the concern that overzealous calls to diversify clinical trials could create a backlash in terms of both trial and vaccine hesitancy. First, as noted by Rachel Hardiman of the Center for Antiracism Research for Health Equity at the University of Minnesota, overenergetic outreach to Black communities could increase wariness, rather than alleviate it.[180] Those targeted may simply, and for good reason, feel they are being exploited—like lab rats—and hence decline to participate. Second, if the trials themselves do not meet the diversity goals they or others have set, then taking the proffered biological rationales for diversity at face value, members of minority groups might reasonably conclude that the vaccines have not been proven safe or effective for their group. In calling for increasing diverse enrollment in trials in a *New York Times* op-ed, leaders from several major HBCU medical centers fed into this dynamic, asserting that "without significant participation in clinical trials, there will be no proof that our patients should trust the vaccine."[181]

Specific differences among types of vaccine hesitancy are evident when comparing the attitudes of Black Americans vs. White Republicans. Throughout 2020, polls consistently showed Black Americans expressing greater degrees of hesitancy than Whites, but they also showed Republicans expressing more hesitancy than Democrats.[182] Polls conducted by the Pew Research Center showed the percentage of Black Americans saying they would "definitely or probably" get a vaccine falling from 54 percent in May to below 40 percent in September and then climbing back to 42 percent in December. These numbers were consistently close to 20 percent below those for Whites. During this same period, the percentages for Republicans began at 65 percent in May, falling well below 50 percent in September before climbing back to 50 percent in December. These numbers were between 15–19 percent below those for Democrats.[183]

Given such numbers, the widespread concern about vaccine hesitancy among Blacks during this time is certainly understandable. However, there was no comparable concern being expressed about Republican hesitancy. Why is it then that when it came to getting more subjects for clinical trials, trust only became an issue for enrolling more Black bodies instead of more Republican bodies? Clearly, there is no definitive answer, but one cannot help but be concerned that the difference lies, at least in part, in an implicit understanding or belief—a frame—that Black bodies were biologically different from White bodies while Republican bodies (quite understandably) were never conceived of as being biologically different from Democratic bodies.

Partisan divides in vaccine hesitancy had been well documented for years before the outbreak of COVID-19. Back in 2009, in response to the outbreak of H1N1 (swine flu), a poll conducted by the Pew Research Center found only 41 percent of Republicans said they would get vaccinated, compared to 60 percent of Democrats.[184] Conservative commentator Glenn Beck told his roughly three million followers in 2009, "if you have some idiot government official demanding, telling me I must take this vaccine, I'll never take it."[185] There was invariably an aspect of racialization informing the rising perception that a government led by a Black man, President Obama, was incompetent and a threat to freedom.

As Matt Nisbet noted in a 2016 article on "Partisan Pandemics": "trust in government was ultimately the key driver of decisions to be vaccinated. In contrast to their Democratic counterparts, Republicans . . . were less likely to say that they were willing to take the vaccine."[186] Similarly, a 2018 study found that Republicans were far more likely than Democrats to have inaccurate beliefs about vaccines and attributed that difference, at least in part, to "Democrats' greater support for government programs and science"; a finding echoed in another 2018 study concluding that political ideology had a powerful effect on attitudes toward vaccines that "is mediated by trust in government medical experts."[187]

Awareness of the partisan divide was not confined to the pages of academic journals. In May 2019, Politico published an article titled, "How the Anti-Vaccine Movement Crept into the GOP Mainstream." In the story Executive Director of the Colorado Children's Immunization Coalition Stephanie Wasserman noted that "the antivax messaging has shifted from a focus on questions of safety to things like parental rights and data privacy, and those messages resonate more with conservative lawmakers and play to the GOP political base."[188] The shift in focus from safety to freedom is notable both because it presaged much Republican resistance to COVID-19 vaccines and because it contrasts markedly with the type of mistrust expressed by Blacks.

The racialization of trust runs powerfully through recent responses to vaccines. In January 2021, Lauren Bunch reflected upon current issues in vaccine hesitancy in response to COVID-19 by discussing a 2016 study which found that while Blacks' mistrust was rooted in concerns over governmental motives, White mistrust tended to focus more on issues of competence.[189] One might reframe this as Blacks' concern being rooted in experiences of exploitation and Whites' more in an ideology of individualistic meritocracy and skepticism toward scientific expertise. This difference perhaps accounts for the radical divergence between Black and Republican vaccine hesitancy as the COVID-19 vaccines were actually rolled out in 2021. In December 2020, Black hesitancy was still quite high, with only 42 percent saying they would "definitely or probably" get vaccinated, while the number for Republicans was 50 percent. After this point, the numbers began to change dramatically. By February 2021, Pew was finding that 61 percent of Black people were saying they would "definitely or probably" get vaccinated—only 8 percent below the number for Whites; in contrast, only 56 percent of Republicans said they would "definitely or probably" get vaccinated. In two months' time the numbers had shifted by 19 percent for Blacks but only 6 percent for Republicans.[190]

In March, a series of polls conducted by Civiqs showed the following trend from November 9, 2020 to March 28, 2021: Blacks went from 41 percent saying they would not get vaccinated and 23 percent unsure (for a total of 64 percent hesitant) to 11 percent no and 10 percent unsure (for a total of 21 percent hesitant); Republicans went from 41 percent saying they would not get vaccinated and 23 percent unsure (for a total of 64 percent hesitant) to 42 percent no and 11 percent unsure (for a total of 53 percent hesitant).[191] By April, a poll conducted by Monmouth University found 43 percent of Republicans declaring they "likely will never get" vaccinated, in contrast to only 20 percent of Black people.[192] Such trends illustrate how vaccine hesitancy is neither static nor monolithic, and throws into relief differences between vaccine wariness and outright refusal. Blacks' concerns over exploitation were apparently susceptible to amelioration through experience. For Republicans, however, anti-government attitudes and suspicion of expertise remained relatively intransigent. The polling trends would seem to indicate that Blacks were taking a wait-and-see attitude toward the vaccine and steadily moved toward acceptance. Republicans, in contrast, demonstrated remarkably consistent and stubbornly high levels of outright refusal in their stance toward vaccination. By January 2022, the Kaiser Family Foundation was reporting that 26 percent of Republicans and 27 percent of White evangelical Christians (groups with significant overlap) were reporting that they would "definitely not" get vaccinated; while the similar number for Black adults was 8 percent. Indeed, overall resistance to vaccines by all White adults was twice that of Black adults, coming in at 16 percent; while Democrats as a group came in with the lowest overall number at 5 percent.[193]

There is no evidence that this shift in Black attitudes was attributable to diversifying clinical trials. There is evidence, however, from the studies discussed above; first, previous attempts to address vaccine hesitancy generally had little effect; and second, Blacks, while initially more hesitant than Whites, would get vaccinated in similar numbers if simply given access.

Despite the well-documented partisan differences in vaccine hesitancy going back over a decade, it was not until such trends became evident in early 2021 that biomedical professionals, commentators, and policymakers started wringing their hands over Republican vaccine hesitancy—this despite the fact that the overall response to the pandemic had become highly politicized and polarized over the course of 2020.[194] While major initiatives were taken to address Black vaccine hesitancy by seeking to enroll more Black bodies in vaccine trials, no one was calling for demographically representative samples of Republicans to be enrolled. Moderna did not pause its trial so that it could enroll more Republicans. From all indications, partisan affiliation of those enrolling in clinical trials was never tracked. It certainly was not presented in the demographic breakdown of trial results to the FDA by vaccine developers.

In other words, as COVID-19 ravaged through the United States during 2020 and vaccine developers were frantically seeking to enroll people in clinical trials, somehow White mistrust, specifically White Republican mistrust, was largely ignored—that is until the clinical trials were completed, the vaccines came online, and it became evident that White Republican vaccine refusal presented a much greater challenge than Black vaccine wariness.

THE RACIALIZATION OF TRUST

How specifically, then, did trust become racialized during the COVID-19 pandemic and what sort of work did this racialization do? Black mistrust of biomedical institutions is a widely recognized and much-discussed phenomenon. It has been part of the debates around increasing inclusion in clinical trials going back to the NIH Revitalization Act of 1993.[195] Many calls for inclusion invoked the legacy of the Tuskegee syphilis experiments where, for decades, representatives from the U.S. Department of Public Health withheld treatments from Black men in order to study the course of a disease that they thought functioned somehow differently in Black bodies.[196]

Typical of such framings is a report from National Public Radio that opined: "A lingering mistrust of the medical system makes some Black Americans more hesitant to sign up for COVID-19 vaccines.... The mistrust is rooted in history, including the infamous U.S. study of syphilis that left Black men in Tuskegee, Ala., to suffer from the disease."[197] Notably, this report aired in mid-February 2021, well after polls had begun to show a dramatic shift in Black attitudes toward vaccine acceptance, thus testifying to the power of this narrative. One problem with framing the issue in this manner is that there is little evidence to support it. Back in 2006, the Tuskegee Legacy Project found that there was no association between knowledge of Tuskegee and actual willingness to participate in research.[198] Reflecting back on the project during the height of the COVID-19 pandemic Director of the National Center for Bioethics in Research and Health Care at Tuskegee University Dr. Reuben Warren declared that the association of Tuskegee with Black vaccine refusal, "was a false assumption.... The hesitancy is there, but the refusal is not. And that's an important difference."[199]

Much more powerful in driving contemporary Black hesitancy has been the Black community's ongoing everyday lived experiences of racism in encounters with the health care system. As Dr. Lauren Nephew wrote in January 2021:

> As a Black woman, I have borne witness to the very system that says it is ready to protect me with a vaccine, systematically disempower my community, putting many at risk of comorbidity and death.... Unfortunately, this lack of trust is not just the result of residual pain from past atrocities like the Tuskegee experiments.... Many patients of color, including myself, can describe present-day experiences in the health care system where we have been discounted, ignored, and devalued.[200]

Dr. Nephew also cited the case of Dr. Susan Moore, a Black physician, whose video of experiencing racism while being treated for COVID-19 went viral after she died from the disease.[201] "'He made me feel like a drug addict,' Dr. Moore said, accusing a white doctor of downplaying her complaints of pain and suggesting she should be discharged." As the *New York Times* noted, "Dr. Moore's experience highlighted what many Black professionals said they regularly encountered. Education cannot protect them from mistreatment, they say, whether in a hospital or other settings."[202]

The focus on Black trust did the work of diverting attention away from the structural and institutional underpinnings of COVID-19 racial disparities and allowed society to concentrate instead on the subjective attitudes of Black people. Focusing on the symptom of vaccine hesitancy allowed policymakers and biomedical professionals to avoid addressing the underlying structural causes of the mistrust. "It's a scapegoat," said Karen Lincoln, a professor of Social Work at the University of Southern California. "It's an excuse. If you continue to use it as a way of explaining why many African Americans are hesitant, it almost absolves you of having to learn more, do more, involve other people—admit that racism is actually a thing today."[203] As historian of Tuskegee, Susan Reverby, put it, "the news media's focus on mistrust or seemingly ridiculous conspiracies . . . ignores the racist structures that shape economic, political, and social realities that lead to health disparities. The alarming statistics on who is getting the vaccines, and who is not, should shift our attention away from mere mistrust in communities of color and toward the structures of racism that cause that mistrust."[204]

During 2020, a focus on Black mistrust became a cheap and easy way to superficially address concerns about the manifest problem of racial disparities. This focus may explain why White Republican mistrust was of no major concern until 2021. White Republicans were not yet experiencing a disproportionate burden of COVID-19 morbidity and mortality, so there was no need to divert attention from any deeper structural issues giving rise to the mistrust. White mistrust was not perceived to be a problem until 2021 because it was not related to issues of racial equity but rather to concerns about herd immunity. This problem did not become manifest until 2021, and so it was not until then

Epilogue 309

that White mistrust became a concern among policymakers and the broader public health community.

UNPACKING THE POLITICAL RATIONALES FOR DIVERSE VACCINE CLINICAL TRIALS

Concerns for racial equity lie at the heart of the political rationales for diversifying COVID-19 vaccine trials. In her August 2020 letter to Director Collins and Director Fauci, Representative Nanette Diaz Barragán (D-Calif) directly connected her call to diversify trials to equitable "access to the full range of treatments" for COVID-19.[205] A similar letter sent by twenty-two members of Congress to FDA Commissioner Stephen Hahn and Secretary of Health and Human Services Alex Azar in June 2020 emphasized that diverse clinical trials were essential "to ensure that no demographic group is left behind."[206] The working assumption was that diverse representation in trials was a prerequisite to racially equitable access to safe and effective treatments.

In this scheme, what were the burdens of diversity and who bore them? The dangers of COVID-19 were linked to concepts of risk—risk of contracting the virus, risk of morbidity and mortality from contracting the virus, and risk of "being left behind" or not having access to vaccines and other care. As Paul Slovic has argued, "danger is real, but risk is socially constructed.... Whoever controls the definition of risk controls the rational solution to the problem at hand."[207] In 2020, there was a recognized and very real danger that COVID-19 was having a disproportionate impact on racial minorities. In addressing this danger, calls to diversify clinical trials defined risk in terms of representation of Black and Brown bodies. If there was not a sufficiently diverse representation of subjects in vaccine trials then we risked the possibility that the vaccines might not be safe and effective for all groups, thereby exacerbating disparities. Thus, researchers had to make concerted efforts to reach out to communities of color and convince them to enroll. Even if a given vaccine were proven safe and effective in all groups, there was the additional risk that communities of color, particularly Black Americans, might be mistrustful and hesitate to get vaccinated. Diversifying clinical trials was cast as a means to address this risk.

These constructions of risk placed the burdens of addressing disparities directly on the minds and bodies of Black people. Some vaccine advocates focused on broader structural issues such as the need to develop a better community health infrastructure to ensure equitable access to vaccines and related health care, or easing intellectual property protection for therapeutics developed with federal funds.[208] In contrast, the focus on diversifying clinical trials foregrounded access as a function of the willingness of Black people to make their bodies available for medical experimentation—which, after all is what a "trial" is. Highlighting trust as a barrier to vaccine uptake made the problem of potential disparities in vaccine uptake a function of addressing attitudes in Black minds. In this framing, the problem was not the ongoing practices and structures of racism; rather, it was the attitudes of Black people. What needed to be changed here was Black minds, not White institutions. Moreover, emphasizing the legacy of Tuskegee made it seem like the racism responsible for fostering such mistrust was largely a thing of the past, marginalizing the significance of ongoing lived experiences of racism in the health care system.

Similarly, calling for diversity also in the biomedical workforce in order help encourage minority enrollment or mitigate disparities similarly placed the burdens on those diverse hires.[209] The implication of such calls was that the mere act of hiring Black and Brown professionals would address the problem of trust in the community. Beyond perhaps taking a course in "cultural competency" (which has been happening at medical schools and medical centers for years), White professionals would not have to confront their own racist behavior, change their practices, or redistribute resources. All they had to do was diversify the workforce a bit and the rest would take care of itself. Or rather, the newly hired diverse professionals would take care of diversifying enrollment for them.

Given that the studies finding that most interventions to address vaccine hesitancy had little or no effect, why did so many policymakers and biomedical professionals think that increasing diversity of enrollment was necessary?[210] The answer may be that it was easier to focus on changing Black minds than to improve the infrastructure of vaccine delivery. If diverting attention from larger, more difficult structural reforms involved getting more bodies enrolled in clinical trials—so much the better.

I do not mean to assert that there was some sort of nefarious conspiracy to exploit Black bodies and divert attention from deeper structural issues. I do not think calls to diversify clinical trials necessarily involved the sort of "predatory inclusion" identified by Keeanga-Yamahtta Taylor in her study of real estate practices and mortgage pricing in the African American community.[211] Nonetheless, the fact remains that the rationales for including more Black bodies in the COVID-19 vaccine trials were tenuous; further, they dangerously threatened to reify race as genetic, thereby potentially inadvertently worsening problems of the racial disparities they were trying to address. Calls for inclusion are appealing because they are comparatively easy. Such calls do not involve conflict or competition for limited resources. They merely expand the pie—and in this case, the pie is simply the number of available test subjects.

In this framing, inclusion became a substitute for substantive equity. Without evidence, this frame assumed that, downstream, inclusion would reduce vaccine hesitancy and improve disparities. Increasing enrollment, in whatever form, serves the interests of those conducting the research—i.e., pharmaceutical corporations. Focusing on equity in trial enrollment diverts attention not only from structural conditions causing disparities, it also diverts attention away from economic issues specific to vaccine equity, such as intellectual property protection.[212]

The intellectual property issue had been raised as early as April 2020, when a group of intellectual property scholars came together to propose an "Open COVID Pledge" in response to President Trump's emergency orders on COVID-19. "The Open COVID Pledge call[ed] upon organizations around the world to make their patents and copyrights freely available in the fight against the pandemic."[213] In October 2020, India and South Africa petitioned the World Trade Organization (WTO) to waive temporarily certain intellectual property protections to increase production of and access to vaccines.[214] A year after the initial Open COVID Pledge was proposed in April 2021, a group of Democratic senators sent a letter to pharmaceutical executives calling upon them to consider loosening up access to their vaccine-related intellectual property holdings.[215] By this time, Pfizer was reporting that it had already taken in $3.5 billion in vaccine revenues just in the first quarter of 2021.[216] Shortly after the Pfizer announcement, the Biden administration proclaimed its support for the WTO proposal that had been submitted by India and South Africa to make it easier for countries that permit compulsory licensing to allow a manufacturer to export vaccines.[217] While this step was certainly important, the call to waive patents rights, at least in the short term, was likely more symbolic than substantive; it would still take months of international negotiation before the proposal would take effect, if at all.[218] As journalist Melody Schreiber noted soon after the Biden administration announcement, "freeing vaccine patents is just the first step. Next, companies need to share how to make the vaccines—known as technology transfer—and governments need to provide the resources, from raw materials to production capacity, to ramp up global vaccine production in a matter of weeks instead of years."[219] Time matters in a pandemic; it also matters in market share. Given the choice, what corporation would not rather have had senators sending them letters about diversifying clinical trials instead of letters calling upon them to vitiate their patent protection?

CONCLUSION

Two years before Derrick Bell wrote his critique of diversity as deployed by the Supreme Court in *Grutter v. Bollinger* in 2003, Charles Lawrence III expressed similar concerns over what he characterized as "the liberal defense of affirmative action." Lawrence declared himself to be an "unambivalent advocate for affirmative action," but he, like Bell, was profoundly uneasy with the ways in which "liberal supporters of affirmative action have used the diversity argument to defend affirmative action at elite universities and law schools without questioning the ways that traditional admissions criteria continue to perpetuate race and class privilege."[220] I share a similar unease with respect to the racialized response to COVID-19. On the one hand, I am an unambivalent advocate of the need to take race-conscious measures to address the deep, persistent, and pervasive health disparities in this country. On the other hand, I am intensely wary of attempts to do so in ways that either biologize race or divert attention from deeper structural issues of racism—or both. The otherwise well-meaning liberal concern to take race seriously in the face of a global pandemic is laudable, but like the liberal approach to affirmative action, it might win certain discrete battles (as it did in *Grutter v. Bollinger*) while losing the larger war of challenging race-based privilege (as it ultimately did in *SFFA v. Harvard*).

In the case of COVID-19, I am concerned that some of the well-meaning, short-term *means* chosen to address racial disparities might end up placing the *end* of racial health justice further out of reach. In 2001, Charles Lawrence III argued that "as diversity has emerged as the dominant defense of affirmative action in the university setting, it has pushed other, more radical substantive defenses to the background. These more radical arguments focus on the need to remedy past discrimination, address present discriminatory practices, and reexamine traditional notions of merit and the role of universities in the reproduction of elites."[221] In the story of liberal responses to COVID-19, the general desire to address issues of racial justice was certainly justified. But proximate concerns about the safety, efficacy, and uptake of vaccines across racial groups were largely based on weak empirical data and faulty assumptions about biological difference and the sources of Black mistrust. As a result, the means taken to address those concerns, while functioning to signal a symbolic concern for racial equity, may actually have been reinforcing pernicious ideas about essential biological difference among the race while "push[ing] other, more radical substantive" approaches to addressing the structural bases of health disparities "to the background."[222] This tends to be how the discourse of diversity works, whether in affirmative action or in health disparities. In the case of health disparities, this discourse has the added danger of confusing and conflating sociopolitical concepts of diversity with genetic concepts of diversity. Even in the face of the victory for affirmative action in *Grutter v. Bollinger*, Bell lamented: "These are difficult times for those working for racial equity, and there seemed reason for declaring victory after a years-long litigation that many, including this writer, predicted would result in the invalidation of any use of race in the admissions process. I fear, though, that further events—even in the short term—will render this latest civil rights victory, like so many before it, hard to distinguish from defeat."[223]

There are no easy solutions to the issues I have identified here. Over the years, many suggestions have been presented to guide the responsible use of racial categories in biomedicine, yet these problems persist.[224] Given the complex and dynamic nature of how race is understood and deployed in biomedical contexts and the persistent controversies around recognizing and addressing the historical and structural manifestations of racism in health and healthcare, I am hesitant to proffer any definitive actions that must be taken to avoid the dangers of geneticizing both race and racial disparities. The most important thing, perhaps, is simply to remain attentive to the dynamics I have discussed here and demand that those who are using racial categories provide more complete and clearer justifications for how they are choosing to use race in particular situations. This is even more important under emergent exigent circumstances such as a pandemic. Bearing this in mind, I offer the following modest suggestions for researchers, clinicians, and policymakers to consider:

1. Adopt a skeptical attitude toward any hypothesized but not yet proven causal link between genetics and disparities. Note that this is different from looking at genetic contributions to specific conditions, for example, high blood pressure (or COVID-19), that disproportionately impact minoritized communities. Understanding genetic contributions to disease is quite distinct from understanding genetic contributions to disease disparities. Do not confuse or conflate the two.
2. When social, historical, legal, and economic factors are clearly shown to be significant contributors to an observed disparity, adopt a null hypothesis approach assuming that such disparities are NOT driven by genetics until proven otherwise.
3. When seeking to diversify clinical trials, particularly under emergent circumstances, clearly distinguish between social or political reasons for inclusion—such as encouraging trust or increasing equitable access to therapies—versus biological or genetic reasons, such

as concerns about generalizability or genetically based differential responses to safety or efficacy.
4. When seeking to diversify trials to address such social concerns as trust or vaccine hesitancy, do not simply assume that diversity in trials will address the problem. Do not use this as a rationale to address issues of trust or hesitancy unless you have evidence to support it.
5. When seeking to diversify trials to address biological concerns, especially with vaccines, again assume a null hypothesis: do not assume racial or ethnic diversity is necessary to ensure cross-racial/ethnic safety and efficacy unless you have sound evidence supporting the assumption. The mere fact that some drugs have on average shown differential rates of efficacy across racial groups on their own is not enough, as drug response can be affected by myriad social and environmental factors. With respect to vaccines, the fact that certain alleles appear to occur at different frequencies in different populations is not enough to justify calls for racial diversity unless you can show that those alleles have a direct relation to vaccine response *and* that the difference in frequencies is significant enough in absolute terms to merit substantial race-based diversification in the trial population. If you are using racial categories as proxies for certain environmental or social factors that might affect vaccine response, be explicit about this, and maintain clarity about the use of race as a proxy throughout the process.
6. Do not assume that the mere substitution of terms such as "genetic ancestry" for race will solve any of these problems. In practice, genetic ancestry all too readily becomes continental ancestry which, in turn, blurs back into racial classifications. If you use terms such as "genetic ancestry," be specific about precisely what you mean by this term and how it is being used to construct groupings among present-day populations.

Reflecting back upon the early years of COVID-19, we can see that in the short term, vaccine developers did a decent job of enrolling minorities in their clinical trials. In the short term, all of the vaccines have proven to have the same safety and efficacy across races. And in the short term, Black hesitancy did not prove to be a significant obstacle to vaccination. This is all to the good; but let us be careful as we move forward to ensure that the stories we tell of these short-term successes do not further contribute to the narrative that the best way to promote health equity is to focus on purported genetic differences in Black bodies or allegedly misguided attitudes in Black minds. Let us not reflect back upon these victories and find them hard to distinguish from defeat.

Notes

INTRODUCTION

1. Annie Massa, Katherine Burton, and Amanda Gordon, "Bill Ackman 'Just Fixing Things' with Moves Against University Leaders and Diversity Programs: 'That's What I Do,'" *Forbes*, January 13, 2024, https://fortune.com/2024/01/13/bill-ackman-fixing-things-university-leaders-diversity-programs-dei/.
2. Lena Sun and Juliet Eilperin, "CDC Gets List of Forbidden Words: Fetus, Transgender, Diversity," *Washington Post*, December 15, 2017, https://www.washingtonpost.com/national/health-science/cdc-gets-list-of-forbidden-words-fetus-transgender-diversity/2017/12/15/f503837a-e1cf-11e7-89e8-edec16379010_story.html.
3. Theresa M. Duello, Shawna Rivedal, Colton Wickland, and Annika Weller, "Race and Genetics Versus 'Race' in Genetics: A Systematic Review of the Use of African Ancestry in Genetic Studies," *Evolution, Medicine, and Public Health* 9, no. 1 (2021): 232–45, 232.
4. Adam P. Van Arsdale, "Population Demography, Ancestry, and the Biological Concept of Race," *Annual Review of Anthropology* 48 (2019): 227–41, 227. https://www.annualreviews.org/content/journals/10.1146/annurev-anthro-102218-011154.
5. United States Census Bureau, "2020 Census Frequently Asked Questions About Race and Ethnicity," August 21, 2021, https://www.census.gov/programs-surveys/decennial-census/decade/2020/planning-management/release/faqs-race-ethnicity.html.

6. Joseph L. Graves Jr. and Alan H. Goodman, *Racism, Not Race: Answers to Frequently Asked Questions* (New York: Columbia University Press, 2021), 4.
7. The White House, "Remarks by the President, Prime Minister Tony Blair of England (via satellite), Dr. Francis Collins, Director of the National Human Genome Research Institute, and Dr. Craig Venter, President and Chief Scientific Officer, Celera Genomics Corporation, on the Completion of the First Survey of the Entire Human Genome Project," last modified June 26, 2000, https://clintonwhitehouse4.archives.gov/textonly/WH/EOP/OSTP/html/00628_2.html.
8. Michael Omi and Howard Winant, *Racial Formation in the United States: From the 1960s to the 1990s*, 2nd ed. (New York: Routledge, 1994), 55.
9. Graves and Goodman, *Racism, Not Race*, 3.
10. Krieger, "If 'Race' Is the Answer," accessed February 20, 2019, http://raceandgenomics.ssrc.org/Krieger/#31.
11. 438 U.S. 265 (1978).
12. Kathy Hudson, Rick Lifton, B. Patrick-Lake, E. Gonzalez Burchard, T. Coles, R. Collins, and A. Conrad, "The Precision Medicine Initiative Cohort Program-Building a Research Foundation for 21st-Century Medicine," *Precision Medicine Initiative (PMI) Working Group Report to the Advisory Committee to the Director*, 2015, https://acd.od.nih.gov/documents/reports/PMI_WG_report_2015-09-17-Final.pdf.
13. "Data Snapshots," All of Us Research Program, accessed January 20, 2024, https://researchallofus.org/data-tools/data-snapshots/.
14. National Institutes of Health, "Precision Medicine Initiative (PMI) Working Group," https://acd.od.nih.gov/working-groups/pmi.html.
15. Hudson et al., "The Precision Medicine Initiative."
16. Hudson et al., "The Precision Medicine Initiative."
17. Esteban González Burchard, Elad Ziv, Natasha Coyle, Scarlett Lin Gomez, Hua Tang, Andrew J. Karter, Joanna L. Mountain, Eliseo J. Pérez-Stable, Dean Sheppard, and Neil Risch, "The Importance of Race and Ethnic Background in Biomedical Research and Clinical Practice," *New England Journal of Medicine* 348, no. 12 (2003): 1170–75.
18. Richard S. Cooper, Jay S. Kaufman, and Ryk Ward, "Race and Genomics," *New England Journal of Medicine* 348, no. 12 (2003): 1166–70, 1169.
19. Sandford Levinson, "1999 Owen J. Roberts Memorial Lecture: Diversity," *University of Pennsylvania Journal of Constitutional Law* 2.3 (2000): 573–607, 579.
20. Peter Wood, *Diversity: The Invention of a Concept* (New York: Encounter Books, 2003), 5.
21. Peter H. Schuck, *Diversity in America: Keeping Government at a Safe Distance* (Cambridge, MA: Harvard University Press, 2003), 19–29.
22. Schuck, *Diversity in America*, 31.

23. Richard J. Herrnstein and Charles Murray, *The Bell Curve: Intelligence and Class Structure in American Life* (New York: Free Press, 1994).
24. *Grutter v. Bollinger*, 539 U.S. 306 (2003).
25. Jonathan Kahn, *Race in a Bottle: The Story of BiDil and Racialized Medicine in a Post-Genomic Age* (New York: Columbia University Press, 2012).
26. Prabarna Ganguly, "Putting Diversity Front and Center," *National Human Genome Research Institute*, June 19, 2019, https://www.genome.gov/news/news-release/Putting-diversity-front-and-center.
27. 600 U.S. 181 (2023).
28. Just as the early eugenics movement in the United States was often driven by some of the most progressive, "scientific" thinkers of the day. See, e.g., Michael Freeden, "Eugenics and Progressive Thought: A Study in Ideological Affinity," *The Historical Journal* 22, no. 3 (1979): 645–71; and more generally, Daniel J. Kevles, *In the Name of Eugenics: Genetics and the Uses of Human Heredity.* (Cambridge, MA: Harvard University Press, 1995), 95.
29. Richard C. Lewontin, "The Apportionment of Human Diversity," *Evolutionary Biology* (New York: Springer, 1972): 381–98.
30. See, e.g., Michael Yudell, *Race Unmasked: Biology and Race in the Twentieth Century* (New York: Columbia University Press, 2014), 198; Charles C. Roseman, "Lewontin Did not Commit Lewontin's Fallacy, His Critics Do: Why Racial Taxonomy is not Useful for the Scientific Study of Human Variation," *BioEssays* 43, no. 12 (2021): 2100204; and Adam P. Van Arsdale, "Population Demography, Ancestry, and the Biological Concept of Race," *Annual Review of Anthropology* 48 (2019): 227–41.
31. *DeFunis v. Odegaard*, 416 U.S. 312 (1974).
32. Proposition 209. Prohibition Against Discrimination or Preferential Treatment by State and Other Public Entities. Initiative Constitutional Amendment, https://vigarchive.sos.ca.gov/1996/general/pamphlet/209.htm.
33. *Hopwood v. Texas*, 78 F.3d 932 (5th Cir. 1996), 945
34. See, generally, Steven Epstein, *Inclusion: The Politics of Difference in Medical Research* (Chicago: University of Chicago Press, 2008).
35. Jonathan K. Pritchard, Matthew Stephens, and Peter Donnelly, "Inference of Population Structure Using Multilocus Genotype Data," *Genetics* 155, no. 2 (2000): 945–59.
36. Jo C. Phelan, Bruce G. Link, Sarah Zeiner, and Lawrence H. Yang, "Direct-to-Consumer Racial Admixture Tests and Beliefs About Essential Racial Differences," *Social Psychology Quarterly* 77, no. 3 (2014): 296–318.
37. In re African-American Slave Descendants Litig., 375 F.Supp.2d 721 (N.D. Ill. 2005) (affirm'd in part as modified, rev'd in part by *In re* African-American Slave Descendants Litig., 471 F. 3d 754 (7th Cir. 2006).

38. Amy Harmon, "Seeking Ancestry in DNA Ties Uncovered by Tests," *New York Times*, April 12, 2006, https://www.nytimes.com/2006/04/12/us/seeking-ancestry-in-dna-ties-uncovered-by-tests.html.
39. Deborah A. Bolnick, "Individual Ancestry Inference and the Reification of Race as a Biological Phenomenon," in *Revisiting Race in a Genomic Age*, ed. Barbara A. Koenig, Sandra Soo-Jin Lee, and Sarah S. Richardson (New Brunswick, NJ: Rutgers University Press, 2008): 70–85, 77. See also Rob DeSalle and Ian Tattersall, *Troublesome Science: The Misuse of Genetics and Genomics in Understanding Race* (New York: Columbia University Press, 2018),143–48.
40. Yahoo News Video, "'America First 2.0:' Vivek Ramaswamy's Presidential Campaign Explained," August 11, 2023, https://uk.news.yahoo.com/america-first-2-0-vivek-130035358.html.
41. *Students for Fair Admissions, Inc. v. President and Fellows of Harvard University v. University of North Carolina, et al.*, 143 S.Ct. 2141 (2023).

1. THE ROOTS OF MODERN DIVERSITIES

1. Michael R. Dietrich, "Richard C. Lewontin (1929–2021)," *Nature* 595, no. 7868 (2021): 489.
2. Richard C. Lewontin, "The Apportionment of Human Diversity," *Evolutionary Biology* (New York: Springer, 1972): 381–98, 396
3. Michael Yudell, *Race Unmasked: Biology and Race in the Twentieth Century* (New York: Columbia University Press, 2014), 198; see also Charles C. Roseman, "Lewontin Did Not Commit Lewontin's Fallacy, His Critics Do: Why Racial Taxonomy Is Not Useful for the Scientific Study of Human Variation," *BioEssays* 43, no. 12 (2021): 2100204; and Adam P. Van Arsdale, "Population Ancestry and the Biological Concept of Race," *Annual Review of Anthropology* 48 (2019): 227–41, 230.
4. See, e.g., Armand Marie Leroi, "A Family Tree in Every Gene," *New York Times*, March 14, 2005, https://www.nytimes.com/2005/03/14/opinion/a-family-tree-in-every-gene.html.
5. Charles Darwin, *The Origin of Species* (New York: P. F. Collier & Son, 1909), 25, 30.
6. Jenny Reardon, *Race to the Finish* (Princeton, NJ: Princeton University Press, 2009), 29.
7. Reardon, *Race to the Finish*, 29–31; see also Angela Saini, *Superior: The Return of Race Science* (Boston: Beacon Press, 2019), 56–59.
8. Frank B. Livingston, "On the Non-Existence of Human Races," *Current Anthropology* 3, no. 3 (1962): 279.

9. Arsdale, "Population Ancestry," 227–241, 228.
10. "Collections: Human Variation," Coriell Institute, accessed April 18, 2023, https://www.coriell.org/1/NIGMS/Collections/Human-Variation.
11. Jean Heller, "AP WAS THERE: Black Men Untreated in Tuskegee Syphilis Study," *AP News*, May 10, 2017, https://apnews.com/article/business-science-health-race-and-ethnicity-syphilis-e9dd07eaa4e74052878a68132cd3803a.
12. Steven Epstein, *Inclusion: The Politics of Difference in Medical Research* (Chicago: University of Chicago Press), 43–44; and see, generally, Susan M. Reverby, ed. *Tuskegee's Truths: Rethinking the Tuskegee Syphilis Study* (Chapel Hill: University of North Carolina Press, 2012).
13. Epstein, *Inclusion*, 44.
14. Juan Siliezar, "A Pioneering Geneticist and Renaissance Man of Parts," *The Harvard Gazette*, July 27, 2021, https://news.harvard.edu/gazette/story/2021/07/dick-lewontin-remembered-as-loyal-mentor-and-friend/.
15. Richard Herrnstein "I.Q." *Atlantic* 228, no. 3 (1971): 44–64.
16. Arthur Jensen, "How Much Can We Boost IQ and Scholastic Achievement?," *Harvard Educational Review* 39 (1969): 1–123.
17. Jensen, "How Much Can We Boost IQ and Scholastic Achievement," 1–123, 15, 123
18. Richard C. Lewontin, "Race and Intelligence," *Bulletin of the Atomic Scientists* 26, no. 3 (1970): 2–8, 7.
19. Lewontin, "Race and Intelligence," 8.
20. David J. Depew, "Richard Lewontin and the Argument from Ethos," *Poroi: An Interdisciplinary Journal of Rhetorical Analysis & Invention* 13, no. 2 (2018), 19.
21. Depew, "Richard Lewontin and the Argument from Ethos," 19.
22. Richard J. Herrnstein and Charles Murray, *The Bell Curve: Intelligence and Class Structure in American Life* (New York: Free Press, 1994).
23. Charles Murray, *Human Diversity: The Biology of Gender, Race, and Class* (New York: Twelve, 2020).
24. Edward O. Wilson, *Sociobiology: The New Synthesis* (Cambridge. MA: Harvard University Press, 1975).
25. Aaron Panofsky, *Misbehaving Science* (Chicago: University of Chicago Press, 2021), 130.
26. Elizabeth Allen, Barbara Beckwith, Jon Beckwith, Steven Chorover, David Culver et al., "Against 'Sociobiology,'" *New York Review of Books*, November 13, 1975, https://www.nybooks.com/articles/1975/11/13/against-sociobiology/.
27. David B. Oppenheimer, "Archibald Cox and the Diversity Justification for Affirmative Action," *Virginia Journal of Social Policy & the Law* (2018): 158–204, 165.
28. *DeFunis v. Odegaard*, 416 U.S. 312 (1974), 314

29. Archibald Cox, "Brief of the President and Fellows of Harvard College, Amicus Curiae," in *DeFunis v. Odegaard* (submitted February 4, 1974), appended to David Oppenheimer, "Archibald Cox and the Diversity Justification for Affirmative Action," *Virginia Journal of Social Policy & Law* 25 (2018–2019): 208 ([i]-52).
30. *DeFunis v. Odegaard*, 319–20.
31. See *Regents of the University of California v. Bakke*, 438 U.S. 265 (1978).
32. See, e.g., J. Harvie Wilkinson, *From Brown to Bakke: The Supreme Court and School Integration: 1954–1978* (New York: Oxford University Press, 1981).
33. *Regents of the Univ. of Cal. v. Bakke*, 438 U.S. 265, 314 (1978), 317.
34. *Regents of the Univ. of Cal. v. Bakke* .438 U.S. 265 at 359 (citing *Califano v. Webster*, 430 U.S. 313, 317 (1977), quoting *Craig v. Boren*, 429 U.S. 190, 197 (1976).
35. *Regents of the Univ. of Cal. v. Bakke*, 418 (1978).
36. *Regents of the Univ. of Cal. v. Bakke*, 320 (1978).
37. Oppenheimer, "Archibald Cox and the Diversity Justification," 158–204.
38. Oppenheimer, "Archibald Cox and the Diversity Justification," 158–204; see also Ellen Berrey, *The Enigma of Diversity: The Language of Race and the Limits of Racial Justice* (Chicago: University of Chicago Press, 2015), 30:

> Although the keyword *diversity* emerged in the late 1960s, it was neither a cornerstone of civil rights reforms nor a principle of decisions such as *Griggs* and *Gautreaux*. Its earliest supporters were white college administrators and community activists trying to encourage black integration into exclusively white universities and neighborhoods. These leaders derived the notion of diversity to make sense of those experiences, deflect legal opposition mounted by conservatives, and—for the universities—to prevent federal interference in their admissions processes.

39. Oppenheimer, "Archibald Cox and The Diversity Justification," 158–204, 158.
40. Oppenheimer, "Archibald Cox and The Diversity Justification," 158–204, 165; Cox's earlier elaborations of the arguments for the value of diversity in his *DeFunis* brief, and indeed the existence of the Harvard plan itself, undermines legal scholar Peter Wood's contention that "Diversity, in sum, was a small, intellectual unelaborated part of the University of California's case before the Supreme Court." Peter Wood, *Diversity: The Invention of a Concept* (San Francisco: Encounter Books:, 2003), 108.
41. Oppenheimer, "Archibald Cox and the Diversity Justification, 158–204, 191; see also Anthony S. Chen and Lisa M. Stulberg, "Before *Bakke*: The Hidden History of the Diversity Rationale," *University of Chicago Law Review Online* (2020): 78. https://lawreviewblog.uchicago.edu/2020/10/30/aa-chen-stulberg/, noting:

Harvard especially "took pride in the diversity of its student body" as early as the turn of the twentieth century. This sensibility was manifest in President Charles W. Eliot's welcome to new students in 1900. "The majority [of our students] are of moderate means," Eliot said, "and it is this diversity of condition that makes the experience of meeting men here so valuable."

42. Oppenheimer, "Archibald Cox and The Diversity Justification," 158–204, 165.
43. Chen and Stulberg, "Before *Bakke*," 78.
44. Peter Schmidt, " 'Bakke' Set a New Path to Diversity for Colleges 30 Years After the Ruling, Academe Still Grapples with Race in Admissions," *Chronicle of Higher Education* (2008), https://perma.cc/P5EZ-ZYVJ.
45. Chen and Stulberg, "Before *Bakke*." 78.
46. Brief Amicus Curiae of the Law School Admission Council in Support of Petitioner the Regents of the University of California, June 7, 1977. 1977 WL 188017, 23.
47. Brief Amicus Curiae, 23, 52.
48. An amicus brief submitted by Cleveland State University's chapter of the Black American Law Students Association, similarly dismissed Jensen's work in noting that admissions programs based solely on test scores had embedded cultural biases that disadvantaged Blacks—"unless one subscribes to the view of . . . Jensen." Brief of Amicus Curiae Cleveland State University Chapter of the Black American Law Students Association, April 7, 1977. 1977 WL 187973 at 22.
49. Office of Management and Budget (OMB), *Directive 15: Race and Ethnic Standards for Federal Statistics and Administrative Reporting* (as adopted on May 12, 1977), accessed on March 15, 2022, https://wonder.cdc.gov/wonder/help/populations/bridged-race/Directive15.html; Michael Omi, "Racial Identity and the State: The Dilemmas of Classification," *Law & Inequality* 15 (1997): 7–23, 10.
50. Melissa Nobles, *Shades of Citizenship: Race and Census in Modern Politics* (Stanford, CA: Stanford University Press, 2000), 75–79.
51. U.S. Census Bureau, *Racial and Ethnic Classifications Used in Census 2000 and Beyond*, http://www.census.gov/population/www/socdemo/race/racefactcb/html.
52. Alice Robbin, "The Politics of Representation in the US National Statistical System: Origins of Minority Population Interest Group Participation," *Journal of Government Information* 27 (2000): 431, 435.
53. Michael Omi, *Racial Identity and the State: Contesting the Federal Standards for Classification* (New York: Routledge, 1999), 7; Alice Robbin, "Classifying Racial and Ethnic Group Data in the United States: The Politics of Negotiation and Accommodation," *Journal of Government Information* 27, no. 2 (2000): 148–50.

54. Office of Management and Budget (OMB), Directive 15.
55. See, e.g., Ellen Berrey, *The Enigma of Diversity: The Language of Race and the Limits of Racial Justice* (Chicago: University of Chicago Press, 2015), 29. See also Jonathan Kahn, *Race in a Bottle: The Story of BiDil and Racialized Medicine in a Post-Genomic Age* (New York: Columbia University Press, 2012), 27–47; Epstein, *Inclusion*.
56. *DeFunis v. Odegaard*, 438 U.S. 265, 314 (1978), 315, 317.
57. *DeFunis v. Odegaard*, 315, 317.
58. "1972 Democratic Party Platform," The American Presidency Project, accessed March 16, 2022, https://www.presidency.ucsb.edu/documents/1972-democratic-party-platform.
59. See, e.g., Anthony M. Champagne, "The Segregation Academy and the Law," *The Journal of Negro Education* 42, no. 1 (1973): 58–66; Marilyn Grady and Sharon C. Hoffman, "Segregation Academies Then and School Choice Configurations Today in Deep South States," *Contemporary Issues in Educational Leadership* 2 (2018): 1–24.
60. "1972 Democratic Party Platform."
61. Charlene Galarneau, "Getting King's Words Right," *Journal of Health Care for the Poor and Underserved* 29, no. 1 (2018): 5–8, 5.
62. J. Nadine Gracia, "Remembering Margaret Heckler's Commitment to Advancing Minority Health," *Health Affairs Blog* (2018), https://www.healthaffairs.org/do/10.1377/forefront.20181115.296624/full/.
63. Margaret Heckler, "The Report of the Secretary's Task Force on Black & Minority Health," *U.S. Department of Health and Human Services* (1985), 5, 7, 32, https://archive.org/details/reportofsecretaroousde/page/n1/mode/2up.
64. Catherine Lee and John D. Skrentny, "Race Categorization and the Regulation of Business and Science," *Law & Society Review* 44, no. 3–4 (2010): 617–50, 631.
65. Heckler, "The Report of the Secretary's Task Force," 24.
66. Heckler, "The Report of the Secretary's Task Force," 18–19.
67. Heckler, "The Report of the Secretary's Task Force," 56, 101.
68. Epstein, *Inclusion*, 126.
69. Thomas Byrne Edsall and Mary D. Edsall, *Chain Reaction: The Impact of Race, Rights, and Taxes on American Politics* (New York: Norton, 1992), 186–97.
70. Edsall and Edsall, *Chain Reaction*, 191.
71. See, e.g., Kahn, *Race in a Bottle*, 193–224.
72. Department of Energy, *Human Genome Project Information Archive, 1990–2003*, https://web.ornl.gov/sci/techresources/Human_Genome/project/timeline.shtml.
73. Department of Energy, Office of Health and Environmental Research, *Sequencing the Genome: Summary Report of the Santa Fe Workshop* (1986), 8, 9, 10, https://web.ornl.gov/sci/techresources/Human_Genome/publicat/1986santafereport.pdf.

74. Charles DeLisi, "Santa Fe 1986: Human Genome Baby-Steps," *Nature* 455, no. 7215 (2008): 876–77; Walter Fred Bodmer and Robin McKie, *The Book of Man: The Human Genome Project and the Quest to Discover Our Genetic Heritage* (New York: Oxford University Press, 1997), 224–25.
75. Francis S. Collins, "Medical and Societal Consequences of the Human Genome Project," *New England Journal of Medicine* 341, no. 1 (1999): 28–37, 28, https://www.nejm.org/doi/full/10.1056/NEJM199907013410106; Leslie Fink, "Five-Year Plan Goes to Capitol Hill," *Human Genome News* 2, no. 1 (1990).
76. See, generally, *Saint Francis College v. Al-Khazraji*, 481 U.S. 604 (1987).
77. *Saint Francis College v. Al-Khazraji*, 609, 613.
78. *Saint Francis College v. Al-Khazraji*, 612.
79. *Saint Francis College v. Al-Khazraji*, fn 4.
80. *Saint Francis College v. Al-Khazraji*, fn 4.
81. *Saint Francis College v. Al-Khazraji*, fn 4.
82. Khiara M. Bridges, "The Dangerous Law of Biological Race," *Fordham Law Review* 82 (2013): 21–80, 54, quoting *Saint Francis College v. Al-Khazraji*, 481 U.S. 604, 613 (1987).
83. Bridges, "The Dangerous Law," 54.
84. *Saint Francis College v. Al-Khazraji*, 613.
85. Bridges, "The Dangerous Law," 53–55.

2. MODERN DIVERSITIES TAKING SHAPE

1. See, e.g., Anna Lorbiecki and Gavin Jack, "Critical Turns in the Evolution of Diversity Management," *British Journal of Management* 11, no. s1 (2000): S17–S31, S20; Frederick R. Lynch, *The Diversity Machine: The Drive to Change the White Male Workplace* (New York: Free Press, 1997), 9–10; Pushkala Prasad and A. Mills, "Understanding the Dilemmas of Managing Workplace Diversity," in *Managing the Organizational Melting Pot: Dilemmas of Workplace Diversity* (Thousand Oaks, CA: SAGE, 1997): 3–27, 3–4; Lauren B. Edelman, Sally Riggs Fuller, and Iona Mara-Drita, "Diversity Rhetoric and the Managerialization of Law," *American Journal of Sociology* 106, no. 6 (2001): 1589–1641, 1612–13.
2. William B. Johnston and Arnold E. Packer, *Workforce 2000: Work and Workers for the 21st Century* (New York: Hudson Institute, 1987), 95
3. Peter Wood, *Diversity: The Invention of a Concept* (San Francisco: Encounter Books, 2003), 204–5
4. Judith J. Friedman and Nancy DiTomaso, "Myths About Diversity: What Managers Need to Know About Changes in the US Labor Force," *California Management Review* 38, no. 4 (1996): 54–77, 58.
5. Wood, *Diversity*, 207–8.

6. Edelman et al., "Diversity Rhetoric," 1589–1641, 1613–14.
7. See, generally, *Richmond v. Croson*, 488 U.S. 469 (1989).
8. As Justice Marshall lamented in his dissent, "Today, for the first time, a majority of this Court has adopted strict scrutiny as its standard of Equal Protection Clause review of race-conscious remedial measures," 488 U.S. 469, 551 (1989) (Marshall, J., dissenting).
9. Peter H. Schuck, *Diversity in America: Keeping Government at a Safe Distance* (Cambridge, MA: Harvard University Press, 2006), 161.
10. Schuck, *Diversity in America*, 161–62; see also Scott Palmer, "Diversity and Affirmative Action: Evolving Principles and Continuing Legal Battles," in *Diversity Challenged: Evidence on the Impact of Affirmative Action*, ed. Gary Orfield and Michal Kurlaender (Cambridge, MA: Harvard Educational Review, 2001): 81–98.
11. Oppenheimer, "Disappearance of Voluntary Affirmative Action From the U.S. Workplace," *Journal of Poverty and Social Justice* 24: 37–50, 45.
12. R. Roosevelt Thomas Jr., "From Affirmative Action to Affirming Diversity," *Harvard Business Review* 68, no. 2 (1990): 107–17, 107, 108, 117, https://hbr.org/1990/03/from-affirmative-action-to-affirming-diversity%202.
13. Thomas, "From Affirmative Action," 109, 112
14. Richard J. Herrnstein, "Still an American Dilemma," *Public Interest* 98 (1990): 3–17.
15. National Research Council, "A Common Destiny: Blacks and American Society" (Washington, D.C.: National Academies Press, 1990.)
16. Herrnstein, "Still an American Dilemma," 4, 5–6.
17. Vidu Soni, "A Twenty-First-Century Reception for Diversity in the Public Sector: A Case Study," *Public Administration Review* 60, no. 5 (September–October 2000): 395–408, 395. Fredrick Lynch, an avowed critic of the diversity management industry, noted that "In the take-off year of 1991, diversity management books fairly exploded onto bookstore library shelves." Lynch, *The Diversity Machine*, 93–94.
18. Terry L. Cross, Barbara Bazron, Karl W. Dennis, and Mareasa R. Isaacs, *Towards a Culturally Competent System of Care: A Monograph on Effective Services for Minority Children Who are Severely Emotionally Disturbed* (Washington, D.C.: CASSP Technical Assistance Center, Georgetown University Child Development Center, 1989), https://files.eric.ed.gov/fulltext/ED330171.pdf; On the emergence of cultural competence at this time, see Kwame McKenzie, "A Historical Perspective of Cultural Competence," *Ethnicity and Inequalities in Health and Social Care* 1 (2008): 5.
19. National Quality Improvement Center, Terry L. Cross, https://qiclgbtq2s.org/meet-the-team/terry-l-cross/.

20. Cross et al., "Towards a Culturally Competent System of Care," iv.
21. Cross et al., "Towards a Culturally Competent System of Care," iv, v.
22. Mareasa R. Isaacs and Marva P. Benjamin, *Towards a Culturally Competent System of Care, Volume II: Programs Which Utilize Culturally Competent Principles*, Washington, D.C.: CASSP Technical Assistance Center, Georgetown University Child Development Center, 1991, 6. https://files.eric.ed.gov/fulltext/ED345393.pdf, discussing Taylor H. Cox and Stacy Blake. "Managing Cultural Diversity: Implications for Organizational Competitiveness," *Academy of Management Perspectives* 5, no. 3 (1991): 45–56
23. See, e.g., Harriet A. Washington, *Medical Apartheid: The Dark History of Medical Experimentation on Black Americans from Colonial Times to the Present* (New York: Doubleday, 2006).
24. Lundy Braun, *Breathing Race into the Machine: The Surprising Career of the Spirometer from Plantation to Genetics* (Minneapolis: University of Minnesota Press, 2014), xvii.
25. James H. Jones, *Bad Blood* (New York: Simon & Schuster, 1993), 106.
26. Jean-Paul Lallemand-Stempak, "What Flows Between Us: Blood Donation and Transfusion in the United States (Nineteenth to Twentieth Centuries)," in *Giving Blood: The Institutional Making of Altruism*, ed. Johanne Charbonneau and André Smith (New York: Routledge, 2015): 37–51, 45–46.
27. Darshali A. Vyas, Leo G. Eisenstein, and David S. Jones, "Hidden in Plain Sight—Reconsidering the Use of Race Correction in Clinical Algorithms," *New England Journal of Medicine* 383, no. 9 (2020): 874–82.
28. Jeffrey T. Berger and Dana Ribeiro Miller, "Health Disparities, Systemic Racism, and Failures of Cultural Competence," *American Journal of Bioethics* 21, no. 9 (2021): 4–10, 8
29. Catherine Bliss, "Translating Racial Genomics: Passages in and Beyond the Lab," *Qualitative Sociology* 36, no. 4 (2013): 423–43, 423.
30. U.S. Department of Energy, Office of Health and Environmental Research, *Human Genome: 198990 Program Report*, March 1990, 17, https://web.ornl.gov/sci/techresources/Human_Genome/publicat/89report/hgreport89_90_img.pdf
31. U.S. Department of Health and Human Services, U.S. Department of Energy, *Understanding Our Genetic Inheritance: The U.S. Human Genome Project: The First Five Years FY 1991–1995*, https://web.ornl.gov/sci/techresources/Human_Genome/project/5yrplan/firstfiveyears.pdf
32. Jenny Reardon, *Race to the Finish: Identity and Governance in the Age of Genomics* (Princeton, NJ: Princeton University Press, 2009), 1.
33. Henry T. Greely, "The Control of Genetic Research: Involving the Groups Between," *Houston Law Review* 33, no. 5 (1997): 1397–1430, 1415.

34. Human Genome Diversity Project, Human Genome Diversity *Workshop*, 1992, 1, https://www.osti.gov/servlets/purl/505330.
35. Anita Bernstein, "Diversity May be Justified," *Hastings Law Journal* 64 (2012): 201–255, 245–46.
36. Edward O. Wilson, *The Diversity of Life* (Cambridge, MA: Harvard University Press, 1992).
37. "1992 Democratic Party Platform," The American Presidency Project, accessed March 16, 2022, https://www.presidency.ucsb.edu/documents/1992-democratic-party-platform.
38. Deborah R. Litvin, "The Discourse of Diversity: From Biology to Management," *Organization* 4, no. 2 (1997): 187–209, 194–95.
39. Steven Epstein, *Inclusion: The Politics of Difference in Medical Research* (Chicago: University of Chicago Press, 2008), 73.
40. Henry T. Greely, "Human Genome Diversity: What About the Other Human Genome Project?," *Nature Reviews Genetics* 2, no. 3 (2001): 222–227, 224.
41. Reardon, *Race to the Finish*, 76.
42. David B. Resnik, "The Human Genome Diversity Project: Ethical Problems and Solutions," *Politics and the Life Sciences* 18, no. 1 (1999): 15–23, 15.
43. Reardon, *Race to the Finish*, 1; Jonathan Marks, "The Human Genome Diversity Project," in *Bioethics*, 4th edition, ed., Bruce Jennings (New York: Gale Publishing, 2014): 1578–82, 1579.
44. Litvin, "Discourse of Diversity," 187–209, 197.
45. Reardon, *Race to the Finish*, 98–126.
46. Miranda Oshige McGowan, "Diversity of What?," *Representations* 55 (1996): 129–38, 132.
47. See, e.g., Dina G. Okamoto, *Redefining Race: Asian American Panethnicity and Shifting Ethnic Boundaries* (New York: Russell Sage Foundation, 2014).
48. Debra Harry and Jonathan Marks, "Human Population Genetics Versus the HGDP," *Politics and the Life Sciences* 18, no. 2 (1999): 303–5, 303.
49. Eric T. Juengst, "Group Identity and Human Diversity: Keeping Biology Straight from Culture," *The American Journal of Human Genetics* 63, no. 3 (1998): 673–77, 674, 677.
50. See, e.g., Luigi Luca Cavalli-Sforza, "Origin and Differentiation of Human Races," *Proceedings of the Royal Anthropological Institute of Great Britain and Ireland* 1972 (1972): 15–25. "Most of the difficulties for taxonomy of human races arise from the fact that no matter how races are defined almost no measurable trait shows really sharp discontinuities from one 'race' to the next," Cavalli-Sforza, "Origin and Differentiation," 15.
51. Henry T. Greely, "The Control of Genetic Research: Involving the 'Groups Between,'" *Houston Law Review* 33 (1996): 1397, 1414.

52. Resnik, "Human Genome Diversity Project," 15–23, 16.
53. Greely, "Control of Genetic Research," 1397–1430, 1414–15.
54. Margaret Lock, "The HGDP and the Politics of Bioethics," *Politics and the Life Sciences* 18, no. 2 (1999): 323–25.
55. Joseph S. Alper and Jon Beckwith, "Racism: A Central Problem for the Human Genome Diversity Project," *Politics and the Life Sciences* 18, no. 2 (1999): 285–88, 285–86.
56. Reardon, *Race to the Finish*, 98–108.
57. Lock, "The HGDP," 323–25, 323.
58. Reardon, *Race to the Finish*, 98.
59. Luigi Luca Cavalli-Sforza, "The Human Genome Diversity Project: Past, Present and Future," *Nature Reviews Genetics* 6, no. 4 (2005): 333–340; 333.
60. J. Philippe Rushton, *Race, Evolution, and Behavior: A Life History Perspective* (New Brunswick, NJ: Transaction Books, 1994).
61. See Michael Schulson, "New Evidence Revives Old Questions About E. O. Wilson and Race," *Undark*, February 26, 2022, https://undark.org/2022/02/16/new-evidence-revives-old-questions-about-e-o-wilson-and-race/.
62. Richard Herrnstein and Charles Murray, *The Bell Curve: Intelligence and Class Structure in American Life* (New York: Free Press, 1994).
63. Herrnstein and Murray, *The Bell Curve*, 564.
64. Richard Herrnstein and Charles Murray, "Race and IQ: An Apologia with Responses," *The New Republic* 211, no. 18 (1994).
65. See, e.g., Richard Jacoby and Naomi Glauberman, eds., *The Bell Curve Debate: History, Documents, Opinions* (New York: Times Books, 1995); Steven Fraser, ed., *The Bell Curve Wars: Race, Intelligence, and the Future of America* (New York: Basic Books, 2008); Alan Davis, "The Bell Curve and Its Critical Progeny: A Review," *Journal of Education for Students Placed at Risk* 2, no. 2 (1997): 183–89; Bernie Devlin, Stephen E. Fienberg, Daniel P. Resnick, and Kathryn Roeder, eds., *Intelligence, Genes, and Success: Scientists Respond to* The Bell Curve (New York: Springer Science & Business Media, 1997).
66. Aaron Panofsky, *Misbehaving Science: Controversy and the Development of Behavior Genetics* (Chicago: University of Chicago Press, 2014), 1–4. Thirty years later reflecting back on these controversies in an article titled, "There's Still No Good Reason to Believe Black-White IQ Differences are Due to Genes," psychologists Eric Turkheimer, Kathryn Paige Harden, and Richard E. Nisbet noted, for example, that Herrnstein and Murray: "acknowledge all the reasons why the heritability of intelligence doesn't necessarily mean that group differences are due to genes. They then proceed to draw their conclusions as if those reasons don't really matter." *Vox*, June 17, 2017, https://www.vox.com/the-big-idea/2017/6/15/15797120/race-black-white-iq-response-critics.

67. American Anthropological Association, *Statement on "Race" and Intelligence*, December 1994, https://americananthro.org/about/policies/american-anthropological-association-statement-on-race-and-intelligence/.
68. As the eminent geneticist David Botstein said, "People keep asking me why I do not rebut *The Bell Curve*. The answer is because it is so stupid that it is not rebuttable." Aaron Panofsky, "Rethinking Scientific Authority: Behavior Genetics and Race Controversies," *American Journal of Cultural Sociology* 6, no. 2 (2018): 322–58, 329.
69. Michael Lind. "Brave New Right," *The New Republic*, October 31, 1994. https://newrepublic.com/article/120890/responses-new-republics-bell-curve-excerpt.
70. Dorothy Nelkin and M. Susan Lindee, *The DNA Mystique: The Gene as a Cultural Icon* (New York: Freeman, 1995).
71. A. Allen, B. Anderson, L. Andrews, J. Beckwith, J. Bowman, R. Cook-Deegan, D. Cox, T. Duster, R. Eisenberg, B. Fine, N. Holtzman, P. King, P. Kitcher, J. McInerney, V. McKusick, J. Mulvihill, J. Murray, R. Murray, T. Murray, D. Nelkin, R. Rapp, M. Saxton, and N. Wexler, "The Bell Curve: Statement by the NIH-DOE Joint Working Group on the Ethical, Legal, and Social Implications of Human Genome Research," *American Journal of Human Genetics* 59, no. 2 (1996): 487.
72. M. W. Browne, "Chinyelou, M. (1995), Debunking the Bell Curve and Scientific Racism, Vol 1," in *Intelligence, Genes, and Success: Scientists Respond to The Bell Curve*, ed. Bernie Devlin, Stephen E. Fienberg, Daniel P. Resnick, and Kathryn Roeder (New York: Springer Verlag, 1997); R. G. Newby, "The Bell Curve: Laying Bare the Resurgence of Scientific Racism [Special issue]," *American Behavioral Scientist* 39, no. 1 (1995).
73. Herrnstein and Murray, *The Bell Curve*, 459, 535.
74. Thomas, "From Affirmative Action," 107–17, 112.
75. Charles Murray and Richard Herrnstein, "Race, Genes, and IQ-An Apologia," *New Republic* 211, no. 18 (1994): 27–37.
76. Jonathan Marks, *Human Biodiversity: Genes, Race, and History* (Berlin: Aldine de Gruyter, 1995), 271–73, 274
77. Aaron Panofsky, Kushan Dasgupta, and Nicole Iturriaga, "How White Nationalists Mobilize Genetics: From Genetic Ancestry and Human Biodiversity to Counterscience and Metapolitics," *American Journal of Physical Anthropology* 175, no. 2 (2021): 387–98.
78. Ari Feldman, "Human Biodiversity: the Pseudoscientific Racism of the Alt-Right," *The Forward*, August 5, 2016, https://forward.com/opinion/346533/human-biodiversity-the-pseudoscientific-racism-of-the-alt-right/.
79. Public Law 103–43 (1993)

80. Epstein, *Inclusion*, 82–83; the Act also established an Office of Research on Women's Health.
81. Epstein, *Inclusion*, 82–83, 148.
82. Office of Management and Budget, "Revisions to the Standards for the Classification of Federal Data on Race and Ethnicity," 58782 F.R. 62, No. 210. October 30, 1997, accessed April 6, 2022, https://www.govinfo.gov/content/pkg/FR-1997-10-30/pdf/97-28653.pdf.
83. Jonathan Kahn, *Race in a Bottle: The Story of BiDil and Racialized Medicine in a Post-Genomic Age* (New York: Columbia University Press, 2012), 27–29.
84. Office of Management and Budget, "Directive 15: Race and Ethnic Standards for Federal Statistics and Administrative Reporting," (as adopted on May 12, 1977), accessed April 6, 2022, https://wonder.cdc.gov/wonder/help/populations/bridged-race/Directive15.html.
85. Epstein, *Inclusion*, 96–97.
86. Epstein, *Inclusion*, 72–73.
87. Epstein, *Inclusion*, 102–3.
88. See, generally, *Adarand Constructors, Inc. v. Peña*, 515 U.S. 200 (1995).
89. *Adarand Constructors, Inc. v. Peña*, 207.
90. See, generally, *Metro Broadcasting, Inc. v. FCC*, 497 U.S. 547 (1990), 566–70, 556, 569, 570.
91. Thomas, "From Affirmative Action," 107–17.
92. *Adarand Constructors, Inc. v. Peña*, 258.
93. Lynch, *The Diversity Machine*, 7.
94. Anna Lorbiecki and Gavin Jack, "Critical Turns in the Evolution of Diversity Management," *British Journal of Management* 11, no. s1 (2000): S17-S31, S22.
95. Proposition 209, Prohibition Against Discrimination or Preferential Treatment by State and Other Public Entities. Initiative Constitutional Amendment. https://vigarchive.sos.ca.gov/1996/general/pamphlet/209.htm.
96. *Resolution of the University of California Board of Regents Adopting a Policy, "Ensuring Equal Treatment" of Admissions*, July 20, 1995, reprinted in, *Representations*, no. 55 (Summer 1996), Appendix SP-1: 184–186.
97. Robert Post, "Introduction: After Bakke," *Representations* 55 (1996): 1–12.
98. *Resolution of the University of California Board of Regents Adopting a Policy*, Appendix: SP-1, 186.
99. See, generally, *Hopwood v. Texas*, 78 F.3d 932 (5th Cir. 1996), 945
100. Khiara M. Bridges, "The Dangerous Law of Biological Race," *Fordham Law Review* 82 (2013): 21–80, 54.
101. See, e.g., David A. Hollinger, "Amalgamation and Hypodescent: The Question of Ethnoracial Mixture in the History of the United States," *American*

Historical Review 108, no. 5 (2003): 1363–90; and Marvin Harris, *Patterns of Race in the Americas*, vol. 1. (New York: Walker, 1964).

102. Richard C. Lewontin, "The Apportionment of Human Diversity," *Evolutionary biology* 6 (1972): 381–98.
103. Lynch, *The Diversity Machine*, 7; see also Republican Party Platform, 1996, https://www.presidency.ucsb.edu/documents/republican-party-platform-1996, stating, "The diversity of our nation is reflected in this platform. . . . We view this diversity of views as a source of strength," and, "America's ethnic diversity within a shared national culture is one of our country's greatest strengths."
104. Democratic Party Platform, 1996. https://www.presidency.ucsb.edu/documents/1996-democratic-party-platform.
105. "Report on the Joint NIH/DOE Committee to Evaluate–the Ethical, Legal, and Social Implications Program—of the Human Genome Project," National Institutes of Health, 1996, https://www.genome.gov/10001745/elsi-evaluation-report.
106. National Research Council, "Evaluating Human Genetic Diversity" (National Academies Press, 1997), 14, 25, https://nap.nationalacademies.org/download/5955.
107. National Research Council, Evaluating Human Genetic Diversity, 37, 48
108. Marks, "The Human Genome Diversity Project," 1578–82, 1581.
109. A single nucleotide polymorphism (abbreviated SNP, pronounced snip) is a genomic variant at a single base position in the DNA. Scientists study if and how SNPs in a genome influence health, disease, drug response, and other traits. "Single Nucleotide Polymorphisms (SNPs)," National Human Genome Research Institute, https://www.genome.gov/genetics-glossary/Single-Nucleotide-Polymorphisms
110. Ramya M. Rajagopalan and Joan H. Fujimura, "Variations on a Chip: Technologies of Difference in Human Genetics Research," *Journal of the History of Biology* 51, no. 4 (2018): 841–73, 841, 850.
111. Walter Fred Bodmer and Robin McKie, *The Book of Man: the Human Genome Project and the Quest to Discover Our Genetic Heritage* (New York: Oxford University Press, 1997), 70–78.
112. Rajagopalan and Fujimura, "Variations on a Chip," 841–873, 867.
113. Catherine Bliss, "Translating Racial Genomics: Passages in and Beyond the Lab," *Qualitative Sociology* 36, no. 4 (2013): 428.
114. Epstein, *Inclusion*.
115. Urs A. Meyers, "Pharmacogenetics–Five Decades of Therapeutic Lessons from Genetic Diversity," *Nature Reviews Genetics* 5 (2004): 669–767, 670.
116. Adam Hedgecoe, *The Politics of Personalised Medicine: Pharmacogenetics at the Clinic* (Cambridge: Cambridge University Press, 2004), 1.

117. Epstein, *Inclusion*.
118. *Food and Drug Administration Modernization Act of 1997*, Pub. L. No. 105–115, 111 Stat. 2296, § 115 (1997).
119. Catherine Lee and John D. Skrentny, "Race Categorization and the Regulation of Business and Science," *Law & Society Review* 44, no. 3/4 (2010): 617–50, 634.
120. Food and Drug Administration, "Guidance for Industry: Population Pharmacokinetics," 1999, https://web.archive.org/web/20060113070421/http://www.fda.gov/cder/guidance/1852fnl.pdf.
121. See, generally, Kahn, *Race in a Bottle*, 36–39.
122. Melissa Nobles, *Shades of Citizenship: Race and the Census in Modern Politics* (Stanford, CA: Stanford University Press, 2000), 78, 82.
123. Michael Omi, "Racial Identity and the State: The Dilemmas of Classification," *Journal of Law and Inequality* 15 (1997): 23.
124. Nobles, *Shades of Citizenship*, 82.
125. National Research Council, "Spotlight on Heterogeneity: The Federal Standards for Racial and Ethnic Classification," Washington, D.C: The National Academies Press, 1996, 18 n11, https://doi.org/10.17226/9060.
126. Francis Collins, "Medical and Societal Consequences of the Human Genome Project," *New England Journal of Medicine* 341, no. 1 (1999): 28–37, 33.
127. Francis Collins, *The Language of Life: DNA and the Revolution in Personalized Medicine* (New York: Harper Perennial, 2010), 231.
128. Collins, "Medical and Societal Consequences," 30, 35.
129. Alan H. Goodman and Thomas Leland Leatherman, eds., *Building a New Biocultural Synthesis: Political-Economic Perspectives on Human Biology* (Ann Arbor: University of Michigan Press, 2010); Thomas Leatherman and Alan H. Goodman, "Building on the Biocultural Syntheses: 20 Years and Still Expanding," *American Journal of Human Biology: The Official Journal of the Human Biology Council* 32, no. 4 (2019): e23360–e23360.

3. FORENSIC DIVERSITY

1. The following discussion of forensic DNA evidence draws from Jonathan Kahn, "Race, Genes, and Justice: A Call to Reform the Presentation of Forensic DNA Evidence in Criminal Trials," *Brooklyn Law Review* 74, no. 2 (2008): 325–75, 331–43.
2. David Kaye, "DNA Evidence: Probability, Population Genetics and the Courts," *Harvard Journal of Law & Technology* 7 (1993): 101, 104.
3. Mildred Cho and Pamela Sankar, "Forensic Genetics and Ethical, Legal and Social Implications Beyond the Clinic," *Nature Genetics* 36 (2004): S9.

4. John M. Butler, *Forensic DNA Typing: Biology, Technology, and Genetics of STR Makers*, 2nd ed. (Boca Raton, FL: CRC Press, 2005), 2–6.
5. Older analyses typically put the figure at 99.9 percent but a more recent study indicates that 99.5 percent may be a more accurate finding. See Rick Weiss, "Mom's Genes or Dad's? Map Can Tell," *Washington Post*, September 4, 2007, A01, http://www.washingtonpost.com/wp-dyn/content/article/2007/09/03/AR2007090301106.html.
6. Butler, *Forensic DNA Typing*, 2–6.
7. Alec J. Jeffreys, Victoria Wilson, and Swee Lay Thein, "Hypervariable 'Minisatellite' Regions in Human DNA," *Nature* 312 (1985), 367; see also National Institutes of Justice (NIJ), *The Future of Forensic DNA Testing* (Lightning Source, 2002), 14–15.
8. Butler, *Forensic DNA Typing*, 2–3; Richard Lewontin and Daniel Hartl, "Population Genetics in Forensic DNA Typing," *Science* (1991),1745.
9. Butler, *Forensic DNA Typing*, 2–3.
10. National Research Council, *DNA Technology in Forensic Science* 10 (1992).
11. Richard Lewontin and Daniel Hartl, "Population Genetics in Forensic DNA Typing," *Science* (1991): 1745–46; Cho and Sankar, *Forensic Genetics*, S9.
12. Richard Lempert defines the product rule as follows:

   > According to the product rule, the probability of two independent events equals the probability of the first event times the probability of the second; with n independent events the separate probabilities of each of the n events are multiplied together to give the probability of their joint occurrence. Thus if the probability that a person had allele A = 1/10 and the probability that he had allele B = 1/10 and the probability that he had allele C = 1/10, and if the probability that the person had one of these alleles was not affected by whether or not he had either or both of the others, the probability that the person would have alleles A, B, and C would be 1/10 × 1/10 × 1/10, or 1/1000.

   Richard Lempert, "The Suspect Population and DNA Identification," *Jurimetrics* 4, no. 3 (1993): 1–2.
13. Butler, "Forensic DNA Typing," 481, 486; Lewontin and Hartl, "Population Genetics," 1746. That the individual be "unrelated" is significant because related individuals will have a higher likelihood of sharing a greater percentage of DNA, hence altering the probabilities of a random match.
14. Lewontin and Hartl, "Population Genetics," 1746.
15. Butler, "Forensic DNA Typing," 3.
16. Tracey Maclin, "Is Obtaining an Arrestee's DNA a Valid Special Needs Search Under the Fourth Amendment?," *Journal of Law, Medicine & Ethics* 34 (2006): 165.
17. National Institutes of Justice, *Future of DNA Testing*, 14–15.

18. Lempert, "Suspect Population," 1–3; Cho and Sankar, *Forensic Genetics*, S9.
19. Lempert, "Suspect Population"; see also David H. Kaye, "Logical Relevance: Problems with the Reference Population and DNA Mixtures in People v. Pizarro," *Law, Probability and Risk* 3 (2004): 211–215.
20. This latter area of concern was brought front and center in 1995 in the highly publicized murder trial of O. J. Simpson where defense lawyers undermined apparently airtight evidence connecting Simpson to the crime by calling into question the methods (or lack thereof) employed by the Los Angeles Police Department in the collecting and handling of relevant DNA samples. See, e.g., David Lazer, "Introduction," in *DNA and Criminal Justice System*, ed. David Lazer (Cambridge, MA: MIT Press, 2004), 4; and Sheila Jasanoff, "DNA's Identity Crisis," reprinted in *DNA and the Criminal Justice System*, 340–45.
21. See, generally, *People v. Gastro*, 545 N.Y.S.2d 985 (N.Y. Crim. Ct. 1989); see, generally, *Schwartz v. State*, 447 N.W.2d 422 (Minn. 1989).
22. Jasanoff, "DNA's Identity Crisis," 339–40.
23. See National Research Council, Committee on DNA Technology in Forensic Science, *DNA Technology in Forensic Science* (Washington, D.C.: National Academy Press, 1992).
24. For an excellent discussion of the NRC I report and the subsequent controversies about its findings, see Jay Aronson, *Genetic Witness: Science, Law, and Controversy in the Making of DNA Profiling* (New Brunswick, NJ: Rutgers University Press, 2007), 153–71.
25. National Research Council, *DNA Technology*, 10–13, 75–82; Kaye, *DNA Evidence*, 106–08.
26. See Kaye, *DNA Evidence*, 12.
27. National Research Council. *DNA Technology*, 75–82.
28. National Research Council, *DNA Technology*, 75–82; Lewontin and Hartl, "Population Genetics," 1746.
29. National Research Council, *DNA Technology*, 75–85; Cho and Sankar, *Forensic Genetics*, S9.
30. See, generally, Butler, *Forensic DNA Typing*, 455–519.
31. Aronson, *Genetic Witness*, 120–71, 120–46; Pilar N. Ossorio, "About Face: Forensic Genetic Testing for Race and Visible Traits," *Journal of Law, Medicine & Ethics* 34, no. 2 (2006): 277–92; Cho and Sankar, Forensic Genetics, S8–12; Robert F. Oldt and Sreetharan Kanthaswamy, "Expanded CODIS STR Frequencies–Evidence for the Irrelevance of Race-Based DNA Databases," *Legal Medicine* 42 (2020): 101642.
32. United States Office of Management and Budget, "Revisions to the Standards for the Classification of Federal Data on Race and Ethnicity," https://obamawhitehouse.archives.gov/omb/fedreg_1997standards.

33. See, e.g., Neil Risch, Esteban Burchard, Elad Ziv, and Hua Tang, "Categorization of Humans in Biomedical Research: Genes, Race, and Disease," *Genome Biology* 3 (2002): 1.
34. Lewontin and Hartl, *Population Genetics*; see also National Research Council, *DNA Technology*, 10–13.
35. Aronson, *Genetic Witness*, 120–171, 120–45.
36. Cho and Sankar, *Forensic Genetics*, S9.
37. Aronson, *Genetic Witness*, 139–71, 139–45.
38. Lewontin and Hartl, *Population Genetics*, 1747.
39. Richard C. Lewontin, "The Apportionment of Human Diversity," *Evolutionary Biology* 6 (1972): 381.
40. Lewontin and Hartl, *Population Genetics*, 1748.
41. Lewontin and Hartl, *Population Genetics*, 1747, 1749.
42. Ranajit Chakraborty and Kenneth Kidd, "The Utility of DNA Typing in Forensic Work," *Science* 20 (1991): 1735.
43. Jay Aronson, *Genetic Witness*, 153–171, 153–58.
44. National Research Council, *DNA Technology*, 13, 80, 82.
45. Kaye, *DNA Evidence*, 102–3.
46. *People v. Barney*, 10 Cal.Rptr.2d 731, 743 (Cal. Ct. App. 1992). David Kaye notes that this case was followed in *People v. Wallace*, 17 Cal.Rptr.2d 721 (Cal. Ct. App. 1993); *Commonwealth v. Lanigan*, 596 N.E.2d 311 (Mass. 1992) (finding that product rule calculation method not prescribed by NRC panel for calculating frequency of DNA pattern is not generally accepted among population geneticists); *State v. Vandebogart*, 616 A.2d 483 (N.H. 1929) (same); *State v. Cauthron*, 846 P. 2d 502 (Wash. 1993) (finding error in allowing expert to testify that defendant was the source of the incriminating DNA and yet excluding testimony of frequency of the DNA pattern given that the NRC panel had proposed a generally accepted method of calculation); *cf. State v. Bible*, 858 P. 2d 1152 (Ariz. 1993) (holding method as applied to 1988 database not generally accepted); *Springfield v. State*, 860 P. 2d 435 (Wyo. 1993) (holding frequency recalculated with "the most conservative" NRC method admissible under relevance standard); *People v. Atoigue*, DCA No. CR 91–95A (Guam Dist. Ct. App. Div. 1992) (method not generally accepted among population geneticists). Kaye, *DNA Evidence*, 103, n.20.
47. Aronson, *Genetic Witness*, 168–71.
48. Kaye, *DNA Evidence*, 397.
49. Eric Lander and Bruce Budowle, "DNA Fingerprinting Dispute Laid to Rest," *Nature* 371 (1994): 735.
50. Lander and Budowle, "DNA Fingerprinting," 735.
51. Lander and Budowle, "DNA Fingerprinting," 738.

52. Lander and Budowle, "DNA Fingerprinting," 737–38.
53. See, e.g., Herman Trautman, "Logical or Legal Relevancy—A Conflict in Theory," *Vanderbilt Law Review* 5 (1952): 385.
54. Kaye, *DNA Evidence*, 104, 127–28.
55. Lander and Budowle, "DNA Fingerprinting," 735.
56. National Research Council, *The Evaluation of Forensic DNA Evidence* (National Academy Press 1996).
57. National Research Council, *Evaluation*, 22, 33, 34.
58. National Research Council, *Evaluation*, 56.
59. National Research Council, *Evaluation*, 57.
60. National Research Council, *Evaluation*, 34, 56, 57.
61. See, e.g., National Research Council, *DNA Technology*.
62. National Research Council, *DNA Technology*, 83.
63. National Academies of Sciences, *The Evaluation of Forensic DNA Evidence* (Washington, D.C.: The National Academies Press, 1996), 35.
64. National Academies of Sciences, *Evaluation*, 57–58.
65. National Academies of Sciences, *Evaluation*, 57–58.
66. Aronson, *Genetic Witness*, 153–71, 168–71.
67. National Research Council, *DNA Technology*, 5.
68. Khiara M. Bridges, "The Dangerous Law of Biological Race," *Fordham Law Review* 82 (2013): 21–80, 54.
69. Butler, *Forensic DNA Typing*, 63.
70. Butler, *Forensic DNA Typing*, 85.
71. Butler, *Forensic DNA Typing*, 13, 85, 94.
72. Butler, *Forensic DNA Typing*, 95; Bruce Budowle, Ranajit Chakraborty, George Carmody, and Keith L. Monson, "Source Attribution of a Forensic DNA Profile," *Forensic Science Communication* 2 (2000): 3, https://www.researchgate.net/profile/Keith-Monson/publication/267683506_Source_Attribution_of_a_Forensic_DNA_Profile_Forensic_Science_Communications_July_2000/links/55269be50cf2d0000c7fc2d13/Source-Attribution-of-a-Forensic-DNA-Profile-Forensic-Science-Co.
73. Butler, *Forensic DNA Typing*, 13, 253.
74. *DNA Identification Act of 1994*, Pub. L. No. 103–322, 108 Stat. 2068 (1994), codified in 42 U.S.C. § 14131.
75. More information on the Federal Bureau of Investigation, *NDIS Statistics*, is available at https://le.fbi.gov/science-and-lab/biometrics-and-fingerprints/codis/codis-ndis-statistics. Last Accessed February 20, 2024.
76. Butler, *Forensic DNA Typing*, 439. The database is also used to aid investigations in identifying human remains.

77. Butler, *Forensic DNA Typing*, 439.
78. Bruce Budowle, Brendan Shea, Stephen Niezgoda, and Ranajit Chakraborty, "CODIS STR Loci Data from 41 Sample Populations," *Journal of Forensic Sciences* 46 (2001): 453.
79. Butler, *Forensic DNA Typing*, 439.
80. See, e.g., *People v. Wilson*, 136 P. 3d 864, 866 (Cal. 2006).
81. Peter M. Vallone, Amy E. Decker, and John M. Butler, "Allele Frequencies for 70 Autosomal SNP Loci with U.S. Caucasian, African-American, and Hispanic Samples," *Forensic Science International* 149 (2005): 279; Other similar treatments of both race and technique in forensic DNA analysis can be found in John M. Butler, Richard Schoske, Peter M. Vallone, Janette W. Redman, and Margaret C. Kline, "Allele Frequencies for 15 Autosomal STR Loci on U.S. Caucasian, African American, and Hispanic Populations," *Journal of Forensic Sciences* 48, no. 1(2003): 1; Bruce Budowle et al., "CODIS STR," 453; Bruce Budowle and Tamyra Moretti, "Genotype Profiles for Six Population Groups at the 13 CODIS Short Tandem Repeat Core Loci and Other PCRBBased Loci," *Forensic Science Communications* 1 (1999): 2; The Human Identity Project at the NIST has been funded by the National Institute of Justice to improve forensic DNA testing methods.
82. Vallone et al., "Allele Frequencies," 279.
83. This lack of comparable care is not restricted to the arena of forensics. For example, a recent survey of biomedical studies using race as a variable found that 72 percent of 268 reports analyzed did not explain their methods of assigning race or ethnicity as independent variables. Hasan Shanawani et al., "Non-Reporting and Inconsistent Reporting of Race and Ethnicity in Articles that Claim Associations Among Genotype, Outcome, and Race or Ethnicity," *Journal of Medical Ethics* 32 (2006): 724.
84. Lewontin, "Apportionment," 381–84.
85. I am indebted to Mike Fortun for the concept of "care of the data."
86. U.S. Office of Management and Budget (OMB), *Revisions to the Standards for the Classification of Federal Data on Race and Ethnicity*, 1997, https://www.whitehouse.gov/wp-content/uploads/2017/11/Revisions-to-the-Standards-for-the-Classification-of-Federal-Data-on-Race-and-Ethnicity-October30-1997.pdf.
87. Sandra Soo-Jin Lee, Joanna Mountain, Barbara Koenig, "The Meaning of 'Race' in the New Genomics: Implications for Health Disparities Research," *Yale Journal of Health Policy, Law, and Ethics* 1 (2001): 33, 55.
88. Michael Omi, "Racial Identity and the State: The Dilemmas of Classification," *Journal of Law & Inequality* 7 (1997): 7.
89. Budowle et al., "Source Attribution of a Forensic DNA Profile," 3.
90. Lewontin and Hartl, *Population Genetics*, 1745.
91. *People v. Wilson*, 136 P. 3d 864, 866 (Cal. 2006).

92. Ranajit Chakraborty and Kenneth Kidd, "The Utility of DNA Typing in Forensic Work," *Science* 254 (1991): 1735.
93. National Institute of Justice, *The Future of Forensic DNA Testing: Predictions on the Research and Development* Working Group, National Commission on the Future of DNA Evidence, 2002, 27 (emphasis added), http://www.ncjrs.gov/pdffiles1/nij/183697.pdf.
94. "Remarks by the President, Prime Minister Tony Blair of England (via satellite), Dr. Francis Collins, Director of the National Human Genome Research Institute, and Dr. Craig Venter, President and Chief Scientific Officer, Celera Genomics Corporation, on the Completion of the First Survey of the Entire Human Genome Project," The White House Office of the Press Secretary, June 26, 2000, accessed 9/11/2023, http://www.ornl.gov/sci/techresources/Human_Genome/project/clinton2.shtml.
95. See, e.g., Jo C. Phelan, Bruce G. Link, Sarah Zelner, and Lawrence H. Yang, "Direct-to-Consumer Racial Admixture Tests and Beliefs About Essential Racial Differences," *Social Psychology Quarterly* 77, no. 3 (2014): 296–318; Ann Morning, *The Nature of Race: How Scientists Think and Teach About Human Difference* (Oakland: University of California Press, 2011); Sandra Lee, "Biobanks of a 'Racial Kind': Mining for Difference in the New Genetics," *Patterns of Prejudice* 40 (2006): 443, 447–48.
96. Lee, "Biobanks," 448.
97. Troy Duster, "The Molecular Reinscription of Race: Unanticipated Issues in Biotechnology and Forensic Science," *Patterns of Prejudice* 40 (2006): 427.
98. See, e.g., Jonathan Kahn, "From Disparity to Difference: How Race-Specific Medicines May Undermine Policies to Address Inequalities in Health Care," *Southern California Interdisciplinary Law Journal* 15 (2005): 105–6, and "Race-ing Patents/Patenting Race: An Emerging Political Geography of Intellectual Property in Biotechnology," *Iowa Law Review* 92 (2007): 353; Deborah Bolnick et al., "The Science and Business of Genetic Ancestry Tracing," *Science* 318 (2007): 399.
99. National Institute of Standards and Technology, "Population Survey," https://strbase-archive.nist.gov/population/PopSurvey.htm.
100. *People v. Wilson*, 866.
101. *People v. Wilson*, 868; Answer Brief on the Merits, *People v. Wilson*, No. S130157 (Cal. Nov. 18, 2005).
102. Michael K. Brown, Martin Carnoy, Elliott Currie, Troy Duster, David B. Oppenheimer, Marjorie M. Schultz, and David Wellman, *Whitewashing Race: The Myth of a Color-Blind Society* (Berkeley: University of California Press, 2003).
103. Karen Rothenberg and Alice Wang, "The Scarlet Gene: Behavioral Genetics, Criminal law, and Racial and Ethnic Stigma," *Law and Contemporary Problems* 69 (2004): 343, 352.

104. Simon Cole, "How Much Justice Can Technology Afford? The Impact of DNA Technology on Equal Criminal Justice," *Science and Public Policy* 34 (2007): 95, 98.
105. Brown et al., *Whitewashing Race*, 149–51, 151
106. Dorothy Roberts, "The Social and Moral Cost of Mass Incarceration in African American Communities," *Stanford Law Review* 56 (2004): 1271–1305, 1297; see also Rose Brewer and Nancy Heitzeg, "The Racialization of Crime and Punishment," *American Behavioral Scientist* 51 (2008): 625.
107. See *People v. Soto*, 21 Cal.4th 512, 534 (1999).
108. Oldt and Kanthaswamy, "Expanded CODIS," 101642.

4. DIVERSITY AFFIRMED IN THE NEW MILLENIUM

1. "The White House Office of the Press Secretary, Remarks by the President, Prime Minister Tony Blair of England (via satellite), Dr. Francis Collins, Director of the National Human Genome Research Institute, and Dr. Craig Venter, President and Chief Scientific Officer, Celera Genomics Corporation, on the Completion of the First Survey of the Entire Human Genome Project," June 26, 2000, accessed September 11, 2023, http://www.ornl.gov/sci/techresources/Human_Genome/project/clinton2.shtml.
2. See *Grutter v. Bollinger*, 137 F. Supp. 2d 821, 823 (E.D. Michigan, 2001).
3. *Grutter v. Bollinger*, 137 F. Supp. 2d, 847–49, 849
4. See, generally, *Grutter v. Bollinger*, 288 F.3d 732 (6th Cir. 2002).
5. See, generally, *Grutter v. Bollinger*, 539 U.S. 306 (2003).
6. *Grutter v. Bollinger*, 539 U.S., 343.
7. *2000 Democratic Party Platform*, https://www.presidency.ucsb.edu/documents/2000-democratic-party-platform.
8. *2000 Democratic Party Platform*.
9. Celeste Condit, Deirdre Moira Condit, and Paul J. Achter. "Human Equality, Affirmative Action, and Genetic Models of Human Variation," *Rhetoric & Public Affairs* 4, no. 1 (2001): 85–108, 86, 90, 102, 104.
10. Quinetta M. Roberson, "Disentangling the Meanings of Diversity and Inclusion in Organizations," *Group & Organization Management* 31, no. 2 (2006): 212–36, 212.
11. Sanford Levinson, "1999 Owen J. Roberts Memorial Lecture: Diversity," *University of Pennsylvania Journal of Constitutional Law* 2, no. 3 (2000): 573, 578.
12. Frederick R. Lynch, *The Diversity Machine: The Drive to Change the White Male Workplace* (New York: Free Press, 1997), 7, 18.
13. Lauren B. Edelman, Sally Riggs Fuller, and Iona Mara-Drita. "Diversity Rhetoric and the Managerialization of Law," *American Journal of Sociology* 106, no. 6 (2001): 1589–1641, 1589, 1618, 1626.

14. María C. Ledesma, "Public Discourse Versus Public Policy: Latinas/os, Affirmative Action, and the Court of Public Opinion," *Association of Mexican American Educators Journal* 9, no. 1 (2015), 59. Or, as the conservative Fifth Circuit Court of Appeals put it in striking down the University of Texas's affirmative action program, "To believe that a person's race controls his point of view is to stereotype him." *Hopwood v. Texas*, 78 F.3d 932, 945 (5th Cir. 1996).
15. William G. Bowen and Derek Bok, *The Shape of the River* (Princeton, NJ: Princeton University Press, 1998); W. G. Tierney and J. K. Chung, "Review of *The Shape of the River: Long-Term Consequences of Considering Race in College and University Admissions* and *Chilling Admissions: The Affirmative Action Crisis and the Search for Alternatives*," *Journal of Higher Education* 71, no. 2: 247–55, 247.
16. *Grutter v. Bollinger*, 539 U.S., 330.
17. Bowen and Bok, *Shape of the River*, xxxii, 10, 225.
18. Gary Orfield and Michal Kurlaender, eds., *Diversity Challenged: Evidence on the Impact of Affirmative Action* (Cambridge, MA: Harvard Educational Review, 2001), 7.
19. Scott Palmer, "A Policy Framework for Reconceptualizing the Legal Debate Concerning Affirmative Action in Higher Education," in *Diversity Challenged: Evidence on the Impact of Affirmative Action*, ed. Gary Orfield and Michal Kurlaender (Cambridge, MA: Harvard Educational Review, 2001), 49–80, 55 (quoting Tanya Y. Murphy, "An Argument for Diversity Based Affirmative Action in Higher Education," *Annual Survey of American Law* 95 [1996]: 515).
20. Quoted in Ellen Berrey, *The Enigma of Diversity: The Language of Race and the Limits of Racial Justice* (Chicago: University of Chicago Press, 2015), 93.
21. See *Grutter v. Bollinger*, 539 U.S. 306 (2003).
22. *Gratz v. Bollinger*, 539 U.S. 244 (2003).
23. *Grutter v. Bollinger*, 539 U.S., 315–25, 323, 325.
24. Ofra Bloch, "Diversity Gone Wrong: A Historical Inquiry into the Evolving Meaning of Diversity from Bakke to Fisher," *University of Pennsylvania Journal of Constitutional Law* 20, no. 5 (May 2018): 1145–1210, 1178–79.
25. Patrick S. Shin and Mitu Gulati, "Showcasing diversity," *North Carolina Law Review* 89 (2010): 1017, 1046.
26. *Grutter v. Bollinger*, 539 U.S., 34.
27. Berrey, *Enigma of Diversity*, 113.
28. Roberson, "Disentangling the Meanings of Diversity," 212.
29. Peter H. Schuck, *Diversity in America: Keeping Government at a Safe Distance* (Cambridge, MA: Harvard University Press, 2003), 7, 324.
30. Schuck, *Diversity in America*, 199.
31. *The Civil Rights Cases*, 109 U.S. 3, 25 (1883).
32. See *Shelby County v. Holder*, 570 U.S. 529, 547 (2013).

33. Robert Post, "Affirmative Action and Higher Education: The View from Somewhere," *Yale Law & Policy Review* 23 (2005): 25–32, 26
34. Peter Wood, *Diversity: The Invention of a Concept* (San Francisco: Encounter Books, 2003), 7, 312, 313.
35. Derrick Bell, "Diversity's Distractions," *Columbia Law Review* 103 (2003): 1622–33, 1622.
36. Bell, "Diversity's Distractions," 1622–33, 1624.
37. Bell, "Diversity's Distractions," 1622–33, 1632.
38. Charles R. Lawrence III, "Two Views of the River: A Critique of the Liberal Defense of Affirmative Action," *Columbia Law Review* 101, no. 4 (May 2001): 928–75, 928, 940–41.
39. Lawrence, "Two Views of the River," 952.
40. See *Grutter v. Bollinger*, 539 U.S., 330–332; see also, e.g., Cynthia L. Estlund, "Putting Grutter to Work: Diversity, Integration, and Affirmative Action in the Workplace," *Berkeley Journal of Employment and Labor Law* 26 (2005): 1–39, 21–23; and Bradley Jones and Roopali Mukherjee, "From California to Michigan: Race, Rationality, and Neoliberal Governmentality," *Communication and Critical/Cultural Studies* 7, no. 4 (2010): 401–22.
41. See *Grutter v. Bollinger*, 539 U.S., 330–32; see also, e.g., Estlund, "Putting Grutter to Work," 1–39, 21–23; and Jones and Mukherjee, "From California to Michigan," 401–22.
42. Gary Gerstle, *The Rise and Fall of the Neoliberal Order: America and the World in the Free Market Era* (New York: Oxford University Press, 2022).
43. Mason Adams, "The Powell Memo: Did this Document Put Big Business in the Political Driver's Seat?," *Virginia Business*, July 30, 2021, https://www.virginiabusiness.com/article/did-this-document-put-big-business-in-the-political-drivers-seat/.
44. Gerstle, *The Rise and Fall of the Neoliberal Order*, 2–3, 108.
45. See *Grutter v. Bollinger*, 539 U.S., 330.
46. As succinctly defined by David Harvey:

> Neoliberalism is in the first instance a theory of political economic practices that proposes that human well-being can best be advanced by liberating individual entrepreneurial freedoms and skills within an institutional framework characterized by strong private property rights, free markets and free trade. The role of the state is to create and preserve an institutional framework appropriate to such practices. The state has to guarantee, for example, the quality and integrity of money. It must also set up those military, defence, police and legal structures and functions required to secure private property rights and to guarantee, by force, if need be, the proper functioning of

markets. Furthermore, if markets do not exist (in areas such as land, water, education, health care, social security, or environmental pollution) then they must be created, by state action if necessary. But beyond these tasks the state should not venture. State interventions in markets (once created) must be kept to a bare minimum because, according to the theory, the state cannot possibly possess enough information to second guess market signals (prices) and because powerful interest groups will inevitably distort and bias state interventions (particularly in democracies) for their own benefit.

David Harvey, *A Brief History of Neoliberalism* (New York: Oxford University Press, 2007), 2.

47. Recent scholarship on this concept generally finds its roots in Cedric Robinson's influential 1983 book, *Black Marxism: The Making of the Black Radical Tradition*, revised and updated 3rd ed. (Chapel Hill: University of North Carolina Press, 2020).

48. Nancy Leong, "Racial Capitalism," *Harvard Law Review* 126 (2012): 2151–2226, 2151, 2192–93.

49. By the mid-1990s some diversity consultants cost an average of $2,000 per day. J. Ryan, J. Hawdon, and A. Branick, "The Political Economy of Diversity: Diversity Programs in Fortune 500 companies," *Sociological Research Online* 7, no. 1 (2002). http://www.socresonline.org.uk/7/1/ryan.html.

50. Quoted in Celia Lury, "'The United Colors of Diversity: Essential and Inessential Culture,'" in *Global Nature, Global Culture*, ed. S. Franklin, C. Lury and J. Stacey (London: Sage, 2000), 146–87, 147.

51. *Grutter v. Bollinger*, Brief of Amicus Curiae Association of American Law Schools in Support of Respondents, 2003 WL 399076 (2003), 17 (fn. 14).

52. See e.g., Marcus W. Feldman and R. C. Lewontin, "The Heritability Hang-Up," *Science* 190 (1975): 1163–68; Marcus W. Feldman and R. C. Lewontin, "Race, Ancestry, and Medicine," in *Revisiting Race in a Genomic Age*, ed. S. Lee, B. Koening, and S. S. Richardson (New Brunswick, NJ: Rutgers University Press, 2008), 89–101.

53. Marcus W. Feldman, "The Meaning of Race: Genes, Environments & Affirmative Action," *La Raza Law Journal* 12 (2015): 365–71, 366; Expert report submitted on behalf of Intervening Defendants (Student Intervenors), *Grutter v. Bollinger*, 137 F. Supp. 2d 821 [E.D. Mich. March 27, 2001] [No. 97–75928].

54. Schuck, *Diversity in America*, 9.

55. "The White House Office of the Press Secretary, Remarks by the President," accessed June 26, 2000. https://clintonwhitehouse5.archives.gov/WH/New/html/genome-20000626.html.

56. Schuck, *Diversity in America*, 31.

57. Schuck, *Diversity in America*, 31–32.

58. Schuck, *Diversity in America*, 144.
59. Wood, *Diversity*, 5, 29, 34, 120, 312.
60. Larry Alexander and Maimon Schwarzschild, "Grutter or Otherwise: Racial Preferences and Higher Education," *Constitutional Commentary* 21 (2004): 3, 6.
61. Alexander and Schwarzschild, "Grutter or Otherwise," 3, 6. Both Alexander and Schwarzschild are listed as "contributors" in events sponsored by the conservative Federalist Society. Prof. Lawrence Alexander, *The Federalist Society*, https://fedsoc.org/contributors/lawrence-alexander; Prof. Maimon Schwarzschild, *The Federalist Society*, https://fedsoc.org/contributors/maimon-schwarzschild.

5. DIVERSITY MACHINES: RACE, BIOLOGY, AND THE SWEET SPOT OF RACIALIZED TIME

1. See *Grutter v. Bollinger*, 539 U.S. 306, 343 (2003)
2. See *Shelby County v. Holder*, 570 U.S. 529, 547 (2013).
3. Troy Duster, "A Post-Genomic Surprise. The Molecular Reinscription of Race in Science, Law, and Medicine," *British Journal of Sociology* 66, no. 1 (2015): 1–27, 8.
4. Deborah A. Bolnick, Duana Fullwiley, Troy Duster, Richard S. Cooper, Joan H. Fujimura, Jonathan Kahn, Jay S. Kaufman, Jonathan Marks, Ann Morning, Alondra Nelson, Pilar Ossorio, Jenny Reardon, Susan M. Reverby, and Kimberley TallBear, "The Science and Business of Genetic Ancestry Testing," *Science* 318, no. 5849 (2007): 399–400.
5. Bolnick et al., "Genetic Ancestry Testing," 399–400.
6. Charmaine D. Royal, John Novembre, Stephanie M. Fullerton, David B. Goldstein, Jeffrey C. Long, Michael J. Bamshad, and Andrew G. Clark, "Inferring Genetic Ancestry: Opportunities, Challenges, and Implications," *American Journal of Human Genetics* 86, no. 5 (2010): 661–73, 666.
7. For example, as Bolnick et al. noted, the test from one of the earlier genetic ancestry companies, AncestryByDNA™:

> Examined AIMs selected to differentiate between four "parental" populations (Africans, Europeans, East Asians, and Native Americans). However, these AIMs are not found in all peoples who would be classed together as a given "parental" population. The AIMs that characterize "Africans," for example, were chosen on the basis of a sample of West Africans. Dark-skinned East Africans might be omitted from the AIMs reference panel of "Africans" because they exhibit different gene variants. (Bolnick et al., "Genetic Ancestry Testing," 399–400)

8. Anna C. F. Lewis, Santiago J. Molina, Paul S. Appelbaum, Bege Dauda, Anna Di Rienzo, Agustin Fuentes, Stephanie M. Fullerton, Nanibaa' A. Garrison, Nayanika Ghosh, Evelynn M. Hammonds, David S. Jones, Eimear E. Kenny, Peter Kraft, Sandra S. -J. Lee, Madelyn Mauro, John Novembre, Aaron Panofsky, Mashaal Sohail, Benjamin M. Neale, and Danielle S. Allen, "Getting Genetic Ancestry Right for Science and Society," *Science* 376, no. 6590 (2022): 250–52.
9. Lewis et al., "Getting Genetic Ancestry Right," 250–52.
10. Jonathan Kahn, "'When Are You From?': Time, Space, and Capital in the Molecular Reinscription of Race," *British Journal of Sociology* 66, no. 1 (2015): 68–75. As a statement from the American Society for Human Genetics noted in 2008:

> Ancestry can be assessed at a number of different levels. The concept of "ancestry" is least ambiguous when we speak of our closest ancestors such as our parents or grandparents, or when we speak of our most distant ancestors, such as the earliest hominids or the first modern Homo sapiens. . . . Genetic ancestry assessment often addresses the intermediate levels of ancestry that are usually imprecisely defined and identified.

American Society of Human Genetics, *Ancestry Testing Statement*, November 13, 2008, 2. https://www.ashg.org/wp-content/uploads/2008/11/Statement-20081311-ASHGAncestryTesting.pdf.
11. Bolnick et al., "Genetic Ancestry Testing," 399–400.
12. See, generally, Alondra Nelson, *The Social Life of DNA: Race, Reparations, and Reconciliation After the Genome* (Boston: Beacon Press, 2016); Jo C. Phelan, Bruce G. Link, Sarah Zelner, and Lawrence H. Yang, "Direct-to-Consumer Racial Admixture Tests and Beliefs About Essential Racial Differences," *Social Psychology Quarterly* 77, no. 3 (2014): 296–318; and Bolnick et al., "Genetic Ancestry Testing," 399–400.
13. The American Society of Human Genetics, *Ancestry Testing Statement*.
14. Ruth Padawer, "Sigrid Johnson Was Black. A DNA Test Said She Wasn't," *New York Times* November 11, 2018, https://www.nytimes.com/2018/11/19/magazine/dna-test-black-family.html.
15. Royal et al., "Inferring Genetic Ancestry," 661–73, 668.
16. Padawer, "Sigrid Johnson Was Black," 19.
17. See *Regents of University of California v. Bakke*, 438 U.S. 265, 318 (1978); *Grutter v. Bollinger*, 539 U.S., 315–16.
18. Amy Harmon, "Seeking Ancestry in DNA Ties Uncovered by Tests," *New York Times*, April 12, 2006, https://www.nytimes.com/2006/04/12/us/seeking-ancestry-in-dna-ties-uncovered-by-tests.html.

19. Harmon, "Seeking Ancestry in DNA."
20. Harmon, "Seeking Ancestry in DNA." Indeed, as recently as 2018, the college admissions counseling company, Ivy Coach, cautioned against trying to use DNA tests to bolster one's chances, noting that "college admissions officers weren't born yesterday. They know that there are students and parents out there hoping to game the system." *DNA Testing in College Admissions*, Ivy Coach, May 7, 2018, https://www.ivycoach.com/the-ivy-coach-blog/college-admissions/dna-testing-college-admissions/. Nonetheless, the odd case of Nicole Katchur indicates the continued possibility of using DNA tests to game the system. In 2018, Katchur (who identifies as "Caucasian") filed a racial discrimination suit against the Thomas Jefferson University School of Medicine, alleging that the school's admission officer suggested she take a DNA ancestry test to see if she could qualify as Native American to garner better chances of being accepted. Katchur did not take a test and was denied admission. Here the DNA test mattered in its absence but indicates the ongoing potential for using DNA tests to game the system and provided a basis for challenging admissions practices. See *Katchur v. Thomas Jefferson University*, 354 F. Supp. 3d 655, 659 (E.D. Penn. 2019); Scott Jaschik, "DNA Testing, Race and an Admissions Lawsuit," *Inside Higher Education*, January 28, 2019, https://www.insidehighered.com/admissions/article/2019/01/28/lawsuit-raises-questions-about-dna-testing-race-and-admissions.
21. Jonah Goldberg, "Are You 11% Genetic Victim?," *National Review*, April 12, 2006, https://www.nationalreview.com/corner/are-you-11-genetic-victim-jonah-goldberg/.
22. See, generally, Nelson, *The Social Life of DNA*, 107–40.
23. See, e.g., Charles P. Henry, *Long Overdue: The Politics of Racial Reparations* (Chicago: University of Chicago Press, 2009); Boris I. Bittker, *The Case for Black Reparations* (Cambridge, MA: Harvard University Press, 2003); Ta-Nehisi Coates, "The Case for Reparations," *The Atlantic*, June 15, 2014, 54–71.
24. Nelson, *Social Life of DNA*, 125–30.
25. In re *African-American Slave Descendants Litig.*, 304 F. Supp. 2d 1027, 1048 (N.D. Ill. 2004).
26. Nelson, *Social Life of DNA*, 130–31, 131.
27. Complaint with Jury Demand, *Farmer-Paellmann v. FleetBoston Fin. Corp.*, 2004 WL 5400675 *16 (S.D.N.Y. Mar. 29, 2004).
28. *Farmer-Paellmann v. FleetBoston Fin. Corp.*, 2004 WL 5400675 *27 (S.D.N.Y. Mar. 29, 2004).
29. In re *African-American Slave Descendants Litig.*, 375 F. Supp. 2d 721 (N.D. Ill. 2005) (affirmed in part as modified, reversed in part by In re *African-American Slave Descendants Litig.*, 471 F. 3d 754 (7th Cir. 2006).

30. *African-American Slave Descendants Litig.*, 375 F. Supp. 2d, 733, 747.
31. *African-American Slave Descendants Litig.*, 375 F. Supp. 2d, 733–34.
32. NIH News Advisory, "International Consortium Launches Genetic Variation Mapping Project," October 2002, https://www.genome.gov/10005336/2002-release-genetic-variation-mapping-launch.
33. International HapMap Consortium, "Integrating Ethics and Science in the International HapMap Project," *Nature Reviews Genetics* 5, no. 6 (2004): 467–75, 469.
34. International HapMap Consortium, "A Haplotype Map of the Human Genome," *Nature* 437(2005): 1300.
35. NIH News Advisory, "International Consortium."
36. International HapMap Consortium, "Haplotype Map," 1316.
37. International HapMap Consortium, "Haplotype Map," 1316.
38. International HapMap Consortium, "Integrating Ethics," 467–75.
39. "Gene Map Points to Personal Drugs," *BBC*, October 26, 2005, http://news.bbc.co.uk/1/hi/health/4378624.stm.
40. Patricia Reaney, "Scientists Complete Map of Human Genetic Variation," *Boston Globe*, October 26, 2005.
41. Ronald Kotulak, "Genetic Map Offers Insight Into Disease," *Baltimore Sun*, October 27, 2005.
42. J.C. Mueller, Jakob C Mueller, Elin Lõhmussaar, Reedik Mägi, Maido Remm, Thomas Bettecken, Peter Lichtner, Saskia Biskup, Thomas Illig, Arne Pfeufer, Jan Luedemann, Stefan Schreiber, Peter Pramstaller, Irene Pichler, Giovanni Romeo, Anthony Gaddi, Alessandra Testa, Heinz-Erich Wichmann, Andres Metspalu, Thomas Meitinger, "Linkage Disequilibrium Patterns and tagSNP Transferability Among European Populations," *American Journal of Human Genetics* 76 (2005): 388.
43. International HapMap, "Integrating Ethics."
44. L. B. Jorde, W. S. Watkins, M. J. Bamshad, M. E. Dixon, C. E. Ricker, M. T. Seielstad, and M. A. Batzer, "The Distribution of Human Genetic Diversity: A Comparison of Mitochondrial, Autosomal, and Y-Chromosome Data," *American Journal of Human Genetics* 66 (2000): 983.
45. International HapMap Consortium. "The International HapMap Project," *Nature* 426, no. 6968 (2003): 789–96.
46. As the Consortium stated in a follow-up 2005 article in *Nature*, "Recent experience bears out the hypothesis that common variants have an important role in disease," International HapMap Consortium. "Haplotype map," 1299. See also Nicholas J. Schork, Sarah S. Murray, Kelly A. Frazer, and Eric J. Topol, "Common vs. Rare Allele Hypotheses for Complex Diseases," *Current Opinion in Genetics & Development* 19, no. 3 (2009): 212–19.

47. David E. Reich and Eric S. Lander, "On the Allelic Spectrum of Human Disease," *TRENDS in Genetics* 17, no. 9 (2001): 502–10.
48. International HapMap Consortium, "International HapMap," 789.
49. Sandra Soo-Jin Lee, "Racial Realism and the Discourse of Responsibility for Health Disparities in a Genomic Age," in *Revisiting Race in a Genomic Age*, eds. Barbara A. Koening, Sandra Soo-Jin Lee, and Sandra S. Richardson (New Brunswick, NJ: Rutgers University Press, 2008), 342–358, 344, 345.
50. J. K. Pritchard, M. Stephens, and P. Donnelly, "Inference of Population Structure Using Multilocus Genotype Data," *Genetics* 155, no. 2 (2000): 945–59.
51. Noah A. Rosenberg, Jonathan K. Pritchard, James L. Weber, Howard M. Cann, Kenneth K. Kidd, Lev A. Zhivotovsky, and Marcus W. Feldman, "Genetic Structure of Human Populations," *Science* 298 (2002): 2381–85, 2381.
52. Bolnick, "Individual Ancestry Inference," 77; see also Rob DeSalle and Ian Tattersall, *Troublesome Science: The Misuse of Genetics and Genomics in Understanding Race* (New York: Columbia University Press, 2018), 143–48.
53. Nature Biotechnology editors, "Slicing Soup," *Nature Biotechnology* 20, no. 7 (2002): 637.
54. Charles Murray, *Human Diversity: The Biology of Gender, Race, and Class*. (New York: Twelve, 2020), 178.
55. National Public Radio, "Conservative Advocate," May 25, 2001, https://www.npr.org/templates/story/story.php?storyId=1123439.
56. See, e.g., Marcus Feldman and Richard Lewontin, "The Heritability Hang-Up," *Science* 190 (1975): 1163–68; Marcus Feldman and Richard Lewontin, "Race, Ancestry, and Medicine." in *Revisiting Race in a Genomic Age*, 89–101.
57. Feldman, "Meaning of Race," 365–71, 366 (2015). Expert report submitted on behalf of Intervening Defendants (Student Intervenors), *Grutter v. Bollinger*, 137 F. Supp. 2d 821 (E.D. Mich. March 27, 2001) (No. 97–75928.
58. Lewis et al., "Getting Genetic Ancestry Right," 250–52.
59. Lewis et al., "Getting Genetic Ancestry Right," 250–51.
60. Esteban González Burchard, Elad Ziv, Natasha Coyle, Scarlett Lin Gomez, Hua Tang, Andrew J. Karter, Joanna L. Mountain, Eliseo J. Pérez-Stable, Dean Sheppard, and Neil Risch, "The Importance of Race and Ethnic Background in Biomedical Research and Clinical Practice," *New England Journal of Medicine* 348, no. 12 (2003): 1170–75.
61. Neil Risch, Esteban Burchard, Elad Ziv, and Hua Tang, "Categorization of Humans in Biomedical Research: Genes, Race, and Disease," *Genome Biology* 3, no. 7 (2002): 1–12, 3, 4.
62. Celeste Condit, "How Culture and Science Make Race 'Genetic': Motives and Strategies for Discrete Categorization of the Continuous and Heterogeneous," *Literature and Medicine*, 26, no. 1 (2007): 240–68, 255, 257.

63. Jonathan Kahn, *Race in a Bottle: The Story of BiDil and Racialized Medicine in a Post-Genomic Age* (New York: Columbia University Press, 2012), 34–37
64. Burchard et al., "The importance of Race," 1170.
65. Burchard et al, "Importance of Race," 1171–72.
66. Jonathan Kahn, "From Disparity to Difference: How Race-Specific Medicines May Undermine Policies to Address Inequalities in Health Care," *Southern California Interdisciplinary Law Journal* 15 (2005): 105.
67. See, e.g., Robert Weisbrot, *Freedom Bound: A History of America's Civil Rights Movement* (New York: Norton, 1991).
68. Sharon E. Nobles, *Shades of Citizenship: Race and the Census in Modern Politics* (Stanford, CA: Stanford University Press, 2000), 14–22.
69. See, e.g., Nobles, *Shades of Citizenship*; and Michael Omi, "Racial Identity and the State: Dilemmas of Classification," *Law and Inequality* 15 (1997): 7–23.
70. United States, Office of Minority Health, "About OMH," accessed March 25, 2005, https://minorityhealth.hhs.gov/about-office-minority-health.
71. Public Law 106–525.
72. A Bill to Improve the Health of Health Disparity Populations. S. 2217, 108th Congress, 2d sess. (2004), https://www.govinfo.gov/content/pkg/BILLS-108s2217is/pdf/BILLS-108s2217is.pdf.
73. See, e.g., Institute of Medicine, *Unequal Treatment: Confronting Racial and Ethnic Disparities in Health Care*, Smedley BD, Stith AY, Nelson AR, eds. (Washington, D.C.: National Academies Press, 2003.)
74. Alan Goodman, "Why Genes Don't Count (for Racial Differences in Health)," *America Journal of Public Health* 90 (2000):1699–1702.
75. Institute of Medicine, *Unequal Treatment*.
76. Maxwell Gregg Bloche, "Health Care Disparities–Science, Politics, and Race," *New England Journal of Medicine* 350 (2004): 1568.
77. See Shankar Vedantam, "Racial Disparities Played Down," *Washington Post*, January 14, 2004, A17.
78. Bloche, "Health Care Disparities," 1568.
79. George Ellison and Ian Ress Jones, "Social Identities and the 'New Genetics': Scientific and Social Consequences," *Critical Public* Health 12 (2002): 265–82, 267.
80. Burchard et al., "The Importance of Race," 1170–75.
81. I have dealt with this case study extensively in my book, *Race in a Bottle*. Here I merely recount the barest outlines of the story as indicative of this broader dynamic.
82. Sula Mazimba and Pamela N. Peterson, "JAHA Spotlight on Racial and Ethnic Disparities in Cardiovascular Disease," *Journal of the American Heart Association* 10, no. 17 (2021): e023650.

83. Food and Drug Administration, *Guidance for Industry: Collection of Race and Ethnicity Data in Clinical Trials* (2005), https://web.archive.org/web/20051020130952/http://www.fda.gov/cder/guidance/5656fnl.htm.
84. Dorothy E. Roberts and Oliver Rollins, "Why Sociology Matters to Race and Biosocial Science," *Annual Review of Sociology* 46 (2020): 195–214, 197.

6. GETTING BODIES: WHEN DIVERSITY DIDN'T MATTER

1. This stood in marked contrast to areas such as drug development or public health, where political activism had driven calls for the inclusion of more diverse cohorts into clinical studies. See Steven Epstein, *Inclusion: The Politics of Difference in Medical Research* (Chicago: University of Chicago Press, 2008); Jonathan Kahn, *Race in a Bottle: The Story of BiDil and Racialized Medicine in a Post-Genomic Age* (New York: Columbia University Press, 2012).
2. N. A. Rosenberg, M. D. Edge, J. K. Pritchard, and M. W. Feldman, "Interpreting Polygenic Scores, Polygenic Adaptation, and Human Phenotypic Differences," *Evolution, Medicine, and Public Health* 2019, no. 1 (2019): 26–34, 27; Catherine Heeney, "Problems and Promises: How to Tell the Story of a Genome-Wide Association Study?," *Studies in History and Philosophy of Science Part A* 89 (2021): 1–10, 1.
3. Heeney, "Problems and Promises," 1.
4. National Center for Human Genome Research (U.S.) et al., *Understanding Our Genetic Inheritance: The U.S. Human Genome Project: The First Five Years, FY 1991–1995* (1990), vii.
5. National Human Genome Research Institute, *A Brief Guide to Genomics* (2011), http://www.genome.gov/18016863.
6. Since the HGP there has been a steady succession of new, grand promises being made on behalf of new technologies in a march toward an ever-receding horizon of biotechnological Nirvana. As the initial promises from the HGP failed to materialize, successive new rounds of hype followed: Stem cell therapies would make the blind see and the lame walk; pharmacogenomics would provide individualized therapies to tailor medicines directly to your personal genetic profile; Genome Wide Association Studies (GWAS) would unravel the mysteries of common complex diseases such as diabetes; new initiatives, such as the Personal Genome Project would provide the sort of information we originally thought to glean from the HGP; the epigenome would provide the answers to how the genome really worked; synthetic biology would allow us to construct new biological entities by snapping together molecular "bricks" like so many Legos; on so on, and so on. See, e.g., Jonathan Kahn, "Synthetic Hype: A

Skeptical View of the Promise of Synthetic Biology," *Valparaiso University Law Review* 45 (2011): 1343, 1346–53.

7. Francis S. Collins, "The Case for a U.S. Prospective Cohort Study of Genes and Environment," *Nature* 429 (2004): 475, 477.
8. Teri A. Manolio, Joan E Bailey-Wilson, and Francis S Collins, "Genes, Environment and the Value of Prospective Cohort Studies," *Nature Reviews Genetics* 7 (2006): 812.
9. Framingham Heart Study, "About the Framingham Heart Study," 2013, https://www.framinghamheartstudy.org/fhs-about/.
10. Jackson Heart Study, "About Us," 2024, https://www.jacksonheartstudy.org/About.
11. National Human Genome Research Institute, "Design Consideration for a Potential United States Population-Based Cohort to Determine the Relationships Among Genes, Environment, and Health: Recommendations of an Expert Panel," http://www.genome.gov/Pages/About/OD/ReportsPublications/PotentialUSCohort.pdf; see also Manolio et al., "Genes, Environment and the Value of Prospective Cohort Studies," 812.
12. Richard Tutton, "Constructing Participation in Genetic Databases: Citizenship, Governance and Ambivalence," *Science, Technology, & Human Values Science* 32 (2007): 172. A report from a public meeting of the UK Biobank Ethics and Governance Council provides a useful, succinct description of a "biobank":

> Originally, and often still now, the term "biobanks" refers to collections of biospecimens which are available for some dispersive use (as compared with archival reference use). More recently the term has come to include collections of biospecimens along with related health and/or social information to be used in research. Often these biobanks are accumulated in the course of clinical care and are often closely held by those who created the collection. The most robust contemporary definition of "biobanks" is "rich collections of data plus biospecimens, specifically developed as resources for research."

UK Biobank Ethics and Governance Council, *Report: Public Meeting of the U.K. Biobank Ethics and Governance Council 11th June 2007 Nowgen, Manchester*, https://web.archive.org/web/20130702023924/www.egcukbiobank.org.uk/assets/wtx041249.pdf.

13. National Human Genome Research Institute, *Design Considerations for a Potential United States Population-Based Cohort to Determine the Relationships Among Genes, Environment, and Health: Recommendations of an Expert Panel*, 2, http://www.genome.gov/Pages/About/OD/ReportsPublications/PotentialUSCohort.pdf.

14. Johns Hopkins University, Genetics & Public Policy Center, *Making Every Voice Count: Public Consultation on Genetics, Environment, and Health* (2010), https://elsihub.org/grant-abstract/making-every-voice-count-public-consultation-genetics-environment-and-health.
15. Steven Epstein, "The Rise of 'Recruitmentology': Clinical Research, Racial Knowledge, and the Politics of Inclusion and Difference," *Social Studies of Science* 38 (2008): 801, 802, 803.
16. Johns Hopkins University, Genetics & Public Policy Center, *Making Every Voice Count*.
17. Shawna Williams, Joan Scott, Juli Murphy, David Kaufman, Rick Borchelt, and Kathy Hudson, *The Genetic Town Hall: Public Opinion About Research on Genes Environment and Health* (Johns Hopkins University, Genetics & Public Policy Center, 2009), 3, https://web.archive.org/web/20111203155507/http://www.dnapolicy.org/images/reportpdfs/2009PCPTownHalls.pdf.
18. Williams et al., *The Genetic Town Hall*.
19. *The Public Forum Institute* (Entrepreneurship.org, 2012), https://web.archive.org/web/20111231131557/http://www.entrepreneurship.org/en/Biz-Connect/Listings/T/The-Public-Forum-Institute.aspx.
20. Williams et al. *The Genetic Town Hall*.
21. Williams et al. *The Genetic Town Hall*.
22. Johns Hopkins University, Genetics & Public Policy Center, *Video-The Proposed Study*, https://web.archive.org/web/20090604145903/http://www.dnapolicy.org/policy.pub.biobanks.html.
23. Mayo Clinic, *Mayo Clinic Biobank* (2013), https://www.mayo.edu/research/centers-programs/mayo-clinic-biobank/overview; D. M. Roden, J. M. Pulley, M. A. Basford, G. R. Bernard, E. W. Clayton, J. R. Balser, and D. R. Masys, "Development of a Large-Scale De-Identified DNA Biobank to Enable Personalized Medicine," *Clinical Pharmacology and Therapeutics* 84, no. 3 (2008): 362.
24. Epstein, "Recruimentology," 810.
25. Johns Hopkins University, Genetics & Public Policy Center, *The Proposed Study*.
26. See, e.g., Adam Hedgecoe, *The Politics of Personalised Medicine: Pharmacogenetics in the Clinic* (Cambridge: Cambridge University Press, 2004), 9–28; Michael Fortun, *Promising Genomics: Iceland and DeCODE Genetics in a World of Speculation* (Oakland: University of California Press, 2008); Michael Arribas-Ayllon, Srikant Sarangi, and Angus Clarke, "Promissory Accounts of Personalisation in the Commercialisation of Genomic Knowledge," *Communication & Medicine* 8, no. 1 (2011): 53 (noting that in the context of direct to consumer genetic test marketing, "promising information that will empower prevention of common complex diseases and ensure better quality of life is

conflated with promising greater access to personal information"; on the concept of potential in the life sciences, see Karen-Sue Taussig, Klaus Hoeyer, and Stefan Helmreich, "The Anthropology of Potentiality in Biomedicine: An Introduction to Supplement 7," *Current Anthropology* 54 (2013): S3-S14.
27. Fortun, "Promising Genomics," 8, 10.
28. Johns Hopkins University, Genetics & Public Policy Center, *The Proposed Study*.
29. Johns Hopkins University, Genetics & Public Policy Center, *Making Every Voice Count*.
30. Johns Hopkins University, Genetics & Public Policy Center, *Making Every Voice Count*.
31. Johns Hopkins University, Genetics & Public Policy Center, *Public Consultation and Engagement* (2010), https://web.archive.org/web/20090604150103/http://www.dnapolicy.org/policy.consult.php.
32. See, e.g., Jon Elster, ed., *Deliberative Democracy* (New York: Cambridge University Press, 1998); James S. Fishkin, *When the People Speak: Deliberative Democracy and Public Consultation* (New York: Oxford University Press, 2011); Amy Gutmann and Dennis F. Thompson, *Why Deliberative Democracy?* (Princeton, NJ: Princeton University Press, 2004).
33. Williams et al., *Genetic Town Hall*.
34. Richard Tutton, "Constructing Participation in Genetic Databases: Citizenship, Governance and Ambivalence," *Science, Technology, & Human Values* 32, no. 2 (2007): 172, 175.
35. Tutton, "Constructing Participation," 188.
36. Kieran O'Doherty and Alice Hawkins, "Structuring Public Engagement for Effective Input in Policy Development on Human Tissue Biobanking," (2010), http://www.ncbi.nlm.nih.gov/pmc/articles/PMC2874727/.
37. O'Doherty and Hawkins, "Structuring Public Engagement"; see also Kieran O'Doherty and M. M. Burgess, "Engaging the Public on Biobanks: Outcomes of the BC Biobank Deliberation," *Public Health Genomics* 12 (2009): 203. Similar efforts have also been undertaken by the Mayo Clinic, in Rochester, Minnesota.
38. O'Doherty and Hawkins, "Structuring Public Engagement."
39. O'Doherty and Burgess, "Engaging the Public on Biobanks," 203.
40. Karen-Sue Taussig, "Fantasies of Human Perfectibility: Conceptualizing Potentiality and the Molecular Medical Toolkit," paper presented at the annual meeting of the American Anthropological Association, November 22, 2008 (manuscript on file with author).
41. Taussig, "Fantasies of Human Perfectibility."
42. Williams et al., *Genetic Town Hall*, 8, 11.
43. Williams et al., *Genetic Town Hall*, 5.

44. David Kaufman, Juli Murphy, Joan Scott, and Kathy Hudson, "Subjects Matter: A Survey of Public Opinions About a Large Genetic Cohort Study," *Genetics in Medicine* 10, no. 12 (2008): 831. 838.
45. U.S. Congress, Public Law 110–233, Stat. 881 (2008); Williams et al., *Genetic Town Hall*, 14.
46. Kathy L. Hudson, "Genomics, Health Care, and Society," *New England Journal of Medicine* 365, no. 11 (2011): 1033, 1039.
47. Genetic Information Nondiscrimination Act, S. Rep. Np. 110–48, 13 (1st Sess. 2007); Genetic Alliance, *The History of GINA* (2012), http://www.geneticalliance.org/ginaresource.history#2.
48. The Genetic Information Nondiscrimination Act: Hearing on H.R. 493 Before the Subcommittee on Health of the Committee on Energy and Commerce, 110th Cong. 110–15, 12–13 (2007) (statement of Francis Collins, M.D., Director, National Human Genome Research Institute, National Institutes of Health, Department of Health and Human Services).
49. The Genetic Information Nondiscrimination Act: Hearing on H.R. 493, 112 (testimony of Kathy Hudson).
50. Kathy L. Hudson, "Prohibiting Genetic Discrimination," *New England Journal of Medicine* 356, no. 19 (2007): 2021, 2022 (emphasis added).
51. Johns Hopkins University, Genetics & Public Policy Center, *Discussion Guide for Clinicians* (2012), https://web.archive.org/web/20121115054153/http://www.dnapolicy.org/resources/GINAfinal-discussionguide-3June10.pdf.
52. Gregory Petsko, "Herding CATS," *Science Translational Medicine* 3 (2011): 97.
53. Gardiner Harris, "Federal Research Center Will Help Develop Medicines," *New York Times*, January 22, 2011, http://www.nytimes.com/2011/01/23/health/policy/23drug.html?_r=1&pagewanted=all.

7. BRINGING DIVERSITY BACK IN

1. Rob DeSalle and Ian Tattersall, *Troublesome Science: The Misuse of Genetics and Genomics in Understanding Race* (New York: Columbia University Press, 2018), 8; 1000 Genome Project Consortium, "A Global Reference for Human Genetic Variation," *Nature* 526 (2015): 68.
2. 1000 Genome Project Consortium, "A Global Reference for Human Genetic Variation," 68–74.
3. Jocelyn Kaiser," A Plan to Capture Human Diversity in 1000 Genomes," *Science* 319 (2008): 395.
4. Genevieve L. Wojcik et al., "Genetic Analyses of Diverse Populations Improves Discovery for Complex Traits," *Nature* (2019): 514–18, 514.

5. Jevin M. Waters, Mohamed T. Hassanein, Loïc Le Marchand, Lynne R. Wilkens, Gertraud Maskarinec, Kristine R. Monroe, Laurence N. Kolonel, David Altshuler, Brian E. Henderson, and Christopher A. Haiman, "Consistent Association of Types 2 Diabetes Risk Variants Found in Europeans in Diverse Racial and Ethnic Groups," *PLoS Genetics* 6, no. 8 (2010): e1001078.
6. Genevieve L. Wojcik et al., "Genetic Analyses of Diverse Populations Improves Discovery for Complex Traits," *Nature* (2019): 514–18, 514.
7. Waters et al., "Consistent Association of Type 2 Diabetes Risk Variants."
8. Samuel Pattillo Smith, Sahar Shahamatdar, Wei Cheng, Selena Zhang, Joseph Paik, Misa Graff, Christopher Haiman, T. C. Matise, Kari E. North, Ulrike Peters, Eimear Kenny, Chris Gignoux, Genevieve Wojcik, Lorin Crawford, and Sohini Ramachandran, "Enrichment Analyses Identify Shared Associations for 25 Quantitative Traits in Over 600,000 Individuals from Seven Diverse Ancestries," *American Journal of Human Genetics* 109, no. 5 (2022): 871–84.
9. National Institutes of Health, "Epidemiologic Investigation of Putative Causal Genetic Variants—Study Investigators (U01): Request For Applications (RFA) Number: RFA-HG-07-014," 2007, https://grants.nih.gov/grants/guide/rfa-files/RFA-HG-07-014.html.
10. National Institutes of Health, "Population Architecture Using Genomics and Epidemiology (PAGE), Phase II—Study Investigators (U01): Request For Applications (RFA) Number: RFA-HG-12-010," 2013, https://grants.nih.gov/grants/guide/rfa-files/RFA-HG-12-010.html.
11. Nicholas Wade, "A Decade Later, Genetic Map Yields Few New Cures," *New York Times*, June 13, 2010, https://www.nytimes.com/2010/06/13/health/research/13genome.html.
12. Brendan Maher, "Personal Genomes: The Case of the Missing Heritability," *Nature News* 456, no. 7218 (2008): 18–21, 19.
13. Maher, "Personal Genomes," 19. As another article noted in 2017, "GWAS have many limitations, such as their inability to fully explain the genetic/familial risk of common diseases; the inability to assess rare genetic variants; the small effect sizes of most associations; the difficulty in figuring out true causal associations; and the poor ability of findings to predict disease risk." Juliet Preston, "Are Genome-Wide Association Studies Fundamentally Fawed?," *MedCity News*, June 16, 2017. https://medcitynews.com/2017/06/genome-wide-association-studies-flawed/.
14. Muin Khoury, "Public Health Impact of Genome-Wide Association Studies: Glass Half Full or Half Empty?," *Genomics and Health Impact*, August 1, 2013, https://blogs.cdc.gov/genomics/2013/08/01/public-health-impact/.
15. Maher, "Personal Genomes," 18–21, 20.

16. Evan A. Boyle, Yang I. Li, and Jonathan K. Pritchard, "An Expanded View of Complex Traits: From Polygenic to Omnigenic," *Cell* 169, no. 7 (2017): 1177–86.
17. International HapMap 3 Consortium, "Integrating Common and Rare Genetic Variation in Diverse Human Populations," *Nature* 467, no. 7311 (2010): 52.
18. Jon McClellan and Mary-Claire King, "Genetic Heterogeneity in Human Disease," *Cell* 141, no. 2 (2010): 210–17, 210, 213.
19. Giorgio Sirugo, Scott M. Williams, and Sarah A. Tishkoff, "The Missing Diversity in Human Genetic Studies," *Cell* 177, no. 1 (2019): 26–31, 28.
20. C. Choi, "Prescriptions Go Personal," *The Scientist*, July 5, 2005, accessed June 20, 2005, https://go.gale.com/ps/i.do?id=GALE%7CA133836790&sid=googleScholar&v=2.1&it=r&linkaccess=abs&issn=08903670&p=AONE&sw=w&userGroupName=mlin_oweb&aty=ip.
21. Francis Collins, *The Language of Life: DNA and the Revolution in Personalized Medicine* (New York: Profile Books, 2010), 231.
22. Maher, "Personal Genomes," 18–21, 19.
23. Jonathan Kahn, *Race in a Bottle: The Story of BiDil and Racialized Medicine in a Post-Genomic Age* (New York: Columbia University Press, 2012), 157–93.
24. Ben Harder, "The Race to Prescribe," *Science News*, April 12, 2005. https://www.sciencenews.org/article/race-prescribe.
25. Lea Harty, Keith Johnson, and Aidan Power, "Race and Ethnicity in the Era of Emerging Pharmacogenomics," *Journal of Clinical Pharmacology* 46 (2006): 405–7.
26. Collins, *The Language of Life*, 163.
27. Kadija Ferryman, Mikaela Pitcan, and Concerns Primer, "What is Precision Medicine: Contemporary Issues and Concerns Primer," *Data & Society* (2018): 5, https://datasociety.net/library/what-is-precision-medicine/.
28. National Research Council, Committee on A Framework for Developing a New Taxonomy of Disease, *Toward Precision Medicine: Building a Knowledge Network for Biomedical Research and a New Taxonomy of Disease* (Washington, D.C.: National Academies Press, 2011), https://www.ncbi.nlm.nih.gov/books/n/nap13284/pdf/.
29. Alla Katsnelson, "Momentum Grows to Make 'Personalized' Medicine More 'Precise,'" *Nature Medicine* (2013): 249.
30. Katsnelson, "Momentum Grows," 249.
31. National Research Council, *Toward Precision Medicine*, 124, 125.
32. Eric T. Juengst, Michelle L. McGowan, Jennifer R. Fishman, and Richard A. Settersten Jr., "From 'Personalized' to 'Precision' Medicine: The Ethical and Social Implications of Rhetorical Reform in Genomic Medicine." *Hastings Center Report* 46, no. 5 (2016): 21–33, 23.

33. Including scientists, translational researchers, commercial and nonprofit developers, research funders, clinician-researchers, clinicians in private practice, health professional educators, medical journal editors, and health insurers. Juengst et al., "From 'Personalized' to 'Precision' Medicine," 21–33, 22, 28.
34. Muin Khoury, "The Shift From Personalized Medicine to Precision Medicine and Precision Public Health: Words Matter!" Posted April 21, 2016, https://blogs.cdc.gov/genomics/2016/04/21/shift/.
35. Juengst et al., "From 'Personalized' to 'Precision' Medicine," 28, 29.
36. The White House, Office of the Press Secretary, "Remarks by the President in State of the Union Address," January 20, 2015, https://obamawhitehouse.archives.gov/the-press-office/2015/01/20/remarks-president-state-union-address-January-20-2015; The White House, "Fact Sheet: President Obama's Precision Medicine Initiative," January 30, 2015, https://www.whitehouse.gov/the-press-office/2015/01/30/fact-sheet-president-obama-s-precision-medicine-initiative.
37. Kathy Hudson, Rick Lifton, and B. Patrick-Lake, "The Precision Medicine Initiative Cohort Program-Building a Research Foundation for 21st Century Medicine," Precision Medicine Initiative (PMI) Working Group Report to the Advisory Committee to the Director (2015): v., https://acd.od.nih.gov/documents/reports/DRAFT-PMI-WG-Report-9-11-2015-508.pdf.
38. Hudson et al, "The Precision Medicine Initiative Cohort Program," v.
39. "All of Us: Program Overview," https://web.archive.org/web/20190314204454/https://www.joinallofus.org/en/program-overview.
40. For discussion of some of these issues, see, e.g., Pamela L. Sankar and Lisa S. Parker, "The Precision Medicine Initiative's All of Us Research Program: An Agenda for Research on Its Ethical, Legal, and Social Issues," *Genetics in Medicine* 19, no. 7 (2017): 743–50; Stephanie A. Kraft, Mildred K. Cho, Katherine Gillespie, Meghan Halley, Nina Varsava, Kelly E. Ormond, Harold S. Luft, Benjamin S. Wilfond, and Sandra Soo-Jin Lee, "Beyond Consent: Building Trusting Relationships with Diverse Populations in Precision Medicine Research," *American Journal of Bioethics* 18, no. 4 (2018): 3–20; and Maya Sabatello and Paul S. Appelbaum, "The Precision Medicine Nation," *Hastings Center Report* 47, no. 4 (2017): 19–29.
41. See, e.g., the *All of Us* program's "Overview," which states "we're asking one million people to lead the way to provide the types of information that can help us create *individualized* prevention, treatment, and care for all of us. . . . The *All of Us* Research Program is part of the Precision Medicine Initiative. Precision medicine is health care that is based on you as an *individual*." *All of Us: Program Overview*, https://web.archive.org/web/20190314204454/https://www.joinallofus.org/en/program-overview (emphasis added).

42. The brochures can be downloaded at *"All of Us* Sharable Resources: All of Us Research Program," National Institutes of Health, https://www.joinallofus.org/en/community/community-resources.
43. Lucia A. Hindorff, Vence L. Bonham, Lawrence C. Brody, Margaret E. C. Ginoza, Carolyn M. Hutter, Teri A. Manolio, and Eric D. Green, "Prioritizing Diversity in Human Genomics Research," *Nature Reviews Genetics* 19, no. 3 (2018): 175–85, 175.
44. See, e.g., Courtney R. Lyles, Mitchell R. Lunn, Juno Obedin-Maliver, and Kirsten Bibbins-Domingo, "The New Era of Precision Population Health: Insights for the All of Us Research Program and Beyond," *Journal of Translational Medicine* 16, no. 1 (2018): 211–214, 212; and Leading Edge, "Addressing Diversity and Inclusion in Human Genetics Research," *Cell* (2018): 175–177, 176.
45. Lori Knowles, Westerly Luth, and Tania Bubela, "Paving the Road to Personalized Medicine: Recommendations on Regulatory, Intellectual Property and Reimbursement Challenges," *Journal of Law and the Biosciences* 4, no. 3 (2017): 453–506, 455, 465.
46. Food and Drug Administration, *Paving the Way for Personalized Medicine: FDA's Role in a New Era of Medical Product Development* (Washington, D.C.: U.S. Department of Health and Human Services, 2013), 13. https://www.agemed.org/Portals/0/PDF/FDA%20-%20Paving%20the%20Way%20for%20Personalized%20medicine.pdf.
47. Troy Duster, "Social Diversity in Humans: Implications and Hidden Consequences for Biological Research," *Cold Spring Harbor Perspectives in Biology* 6, no. 5 (2014): a008482, 9.
48. Hindorff et al., "Prioritizing Diversity," 175–85, 175.
49. Including scientists, translational researchers, commercial and nonprofit developers, research funders, clinician-researchers, clinicians in private practice, health professional educators, medical journal editors, and health insurers. Juengst et al., "From 'Personalized' to 'Precision' Medicine," 21–33, 28.
50. Hindorff et al., "Prioritizing Diversity," 175–85, 175.
51. Russ B. Altman, Snehit Prabhu, Arend Sidow, Justin M. Zook, Rachel Goldfeder, David Litwack, Euan Ashley, George Asimenos, Carlos D. Bustamante, Katherine Donigan, Kathleen M Giacomini, Elaine Johansen, Natalia Khuri, Eunice Lee, Xueying Sharon Liang, Marc Salit, Omar Serang, Zivana Tezak, Dennis P. Wall, Elizabeth Mansfield, and Taha Kass-Hout, "A Research Roadmap for Next-Generation Sequencing Informatics," *Science Translational Medicine* 8, no. 335 (2016): 1–5, 3.
52. Hudson et al., "The Precision Medicine Initiative," 23.
53. See, e.g., Lyles et al., "New Era of Precision," 212; and Edge, "Addressing Diversity," 176.

54. Edward Ramos, Shawneequa L. Callier, and Charles N. Rotimi, "Why Personalized Medicine Will Fail If We Stay the Course," *Personalized Medicine* 9, no. 8 (2012): 839–47.
55. A. Panofsky and C. Bliss, "Ambiguity and Scientific Authority: Population Classification in Genomic Science," *American Sociological Review*, 82, no. 1(2017): 59–87, 59.
56. Maya Sabatello, Shawneequa Callier, Nanibaa' A. Garrison, and Elizabeth G. Cohn, "Trust, Precision Medicine Research, and Equitable Participation of Underserved Populations," *American Journal of Bioethics* 18 (2018): 34–36.
57. Katherine Weatherford Darling, Sara L. Ackerman, Robert H. Hiatt, Sandra Soo-Jin Lee, and Janet K Shim, "Enacting the Molecular Imperative: How Gene-Environment Interaction Research Links Bodies and Environments in the Post-Genomic Age," *Social Science & Medicine* 155 (2016): 51–60, 51.
58. Muin J. Khoury, Michael F. Iademarco, and William T. Riley, "Precision Public Health for the Era of Precision Medicine," *American Journal of Preventive Medicine* 50, no. 3 (2016): 398–401, 399.
59. White House Office of the Press Secretary, "Remarks by the President, Prime Minister Tony Blair of England (via satellite), Dr. Francis Collins, Director of the National Human Genome Research Institute, and Dr. Craig Venter, President and Chief Scientific Officer, Celera Genomics Corporation, on the Completion of the First Survey of the Entire Human Genome Project," June 26, 2000, accessed February 20, 2019, https://www.genome.gov/10001356/june-2000-white-house-event/.
60. See, e.g., Lundy Braun, "Race, Ethnicity, and Health: Can Genetics Explain Disparities?," *Perspectives in Biology and Medicine* 45, no. 2 (2002): 159–74; William W. Dressler, Kathryn S. Oths, and Clarence C. Gravlee, "Race and Ethnicity in Public Health Research: Models to Explain Health Disparities," *Annual Review of Anthropology* 34 (2005): 231–52; Chandra L. Ford and Nina T. Harawa, "A New Conceptualization of Ethnicity for Social Epidemiologic and Health Equity Research," *Social Science & Medicine* 71, no. 2 (2010): 251–58; Timothy Caulfield, Stephanie M. Fullerton, Sarah E. Ali-Khan, Laura Arbour, Esteban G. Burchard, Richard S. Cooper, Billie-Jo Hardy, Simrat Harry, Robyn Hyde-Lay, Jonathan Kahn, Rick Kittles, Barbara A Koenig, Sandra S. J. Lee, Michael Malinowski, Vardit Ravitsky, Pamela Sankar, Stephen W. Scherer, Béatrice Séguin, Darren Shickle, Guilherme Suarez-Kurtz, and Abdallah S. Daar, "Race and Ancestry in Biomedical Research: Exploring the Challenges," *Genome Medicine* 1, no. 1 (2009): 8; Kahn, *Race in a Bottle*, 200–19.
61. Edward O. Wilson, *The Diversity of Life* (New York: Norton, 1999 [1992]). Perhaps even more on point is Jonathan Marks's 1995 book, *Human Biodiversity: Genes,*

Race, and History (New York: Routledge, 2017 [1995]); see *Bakke v. Regents of the University of California*, 438 U.S. 265, 313–314 (1978).

8. GENETIC ENTANGLEMENTS OF SOCIO-LEGAL DIVERSITY IN THE 2010s

1. Lars Larson Show, "Ralph Taylor: Should You Be Able to Use DNA to Get Minority Status for State Contracts?," SoundCloud, September 18, 2018, https://soundcloud.com/thelarslarsonshow/ralph-taylor-should-you-be-able-to-use-dna-to-get-minority-status-for-state-contracts.
2. *Orion Ins. Grp. v. Wash. State Office of Minority & Women's Bus. Enters.*, No. 16–5582 RJB, 2017 WL 3387344, at 8–9 (W.D. Wash. Aug. 7, 2017); see also Christine Willmsen, "For Years He Identified as White. Now He's Using a DNA Test to Claim Minority Status for His Business," *News Tribune*, September 23, 2018.
3. *Orion Ins. Grp. v. Wash. State Office of Minority & Women's Bus. Enters.*, No. 16–5582 RJB, 2017 WL 3387344, at 6–7.
4. *Orion Ins. Grp. v. Wash. State Office of Minority & Women's Bus. Enters.*, No. 16–5582 RJB, 2017 WL 3387344, at 7.
5. 49 C.F.R. § 26 App. E (2023).
6. Lars Larson Show, "Ralph Taylor."
7. *Orion Ins. Grp. v. Wash. State Office of Minority & Women's Bus. Enters.*, No. 16–5582 RJB, 2017 WL 3387344, at 11.
8. *Orion Ins. Grp. v. Wash. State Office of Minority & Women's Bus. Enters.*, No. 16–5582 RJB, 2017 WL 3387344, at 3 (Transcript of Record); see also *Orion Ins. Grp v. Wash. State Office of Minority & Women's Bus*, at Amended Complaint.
9. *Orion Ins. Grp. v. Wash. State Office of Minority & Women's Bus. Enters.*, No. 16–5582 RJB, 2017 WL 3387344, at 14.
10. 49 C.F.R. § 26.63(a) (2023).
11. *Orion Ins. Grp. v. Wash. State Office of Minority & Women's Bus. Enters.*, No. 16–5582 RJB, 2017 WL 3387344, at 25.
12. Adjudicating racial identity, of course, exists in diverse legal realms beyond affirmative action and has deep historical roots in American legal practice. See, e.g., Ian Haney López, *White by Law: The Legal Construction of Race* (New York: New York University Press, 1995); Ariela Gross, *What Blood Won't Tell: A History of Race on Trial in America* (Cambridge, MA: Harvard University Press, 2009); The infamous 1896 separate but equal case of *Plessy v. Ferguson* hinged upon a tram conductor's characterization of Homer Plessy as Black. Jonathan Kahn, "Controlling Identity: Plessy, Privacy, and Racial Defamation," *DePaul Law Review* 54 (2004): 755.

13. Luther Wright Jr., "Who's Black, Who's White, and Who Cares: Reconceptualizing the United States Definition of Race and Racial Classifications," *Vanderbilt Law Review* 48 (1995): 513, 515–16.
14. Tseming Yang, "Choice and Fraud in Racial Identification: The Dilemma of Policing Race in Affirmative Action, the Census, and a Color-Blind Society," *Michigan Journal of Race & Law* 11 (2006): 367, 368.
15. Melissa Korn and Jennifer Levitz, "Students Were Advised to Claim to Be Minorities in College-Admissions Scandal," *Wall Street Journal*, May 19, 2019, https://www.wsj.com/articles/students-were-advised-to-claim-to-be-minorities-in-college-admissions-scandal-11558171800.
16. Paul Pringle and Adam Elmahrek, "House Majority Leader Kevin McCarthy's Family Benefited from U.S. Program for Minorities Based on Disputed Ancestry," *Los Angeles Times*, October. 14, 2018, https://www.latimes.com/local/california/la-na-pol-mccarthy-contracts-20181014-story.html.
17. Paul Pringle and Adam Elmahrek, "Minority Contractors Claiming to be 'Native American' to Undergo Nationwide Review," *Los Angeles Times*, September 18, 2019, https://www.latimes.com/world-nation/story/2019-09-17/minority-contractors-native-american-review.
18. *Orion Ins. Grp. v. Washington State Office of Minority & Women's Bus. Enters.*, No. 16–5582 RJB, 2017 WL 3387344, at 7–8 (W.D. Wash. Aug. 7, 2017); Pringle and Elmahrek, "House Majority Leader."
19. *Orion Ins. Grp. v. Washington State*, 2017 WL 3387344, 10, 27.
20. *Orion Ins. Grp. v. Washington State*, 2017 WL 3387344, 26.
21. Kim TallBear, *Native American DNA: Tribal, Belonging and the False Promise of Genetic Science* (Minneapolis: University of Minnesota Press, 2013), 86–88, 86.
22. *Orion Ins. Grp. v. Washington State*, 2017 WL 3387344, 35.
23. U.S. Department of Transportation, "Disadvantaged Business Enterprise (DBE) Program," https://www.transportation.gov/civil-rights/disadvantaged-business-enterprise.
24. *Orion Ins. Grp. v. Washington State*, 2017 WL 3387344, 35–36.
25. Douglas L. T. Rohde, Steve Olson, and Joseph T. Chang, "Modelling the Recent Common Ancestry of all Living Humans," *Nature* 431 (2004): 562.
26. *Orion Ins. Grp. v. Washington State*, 2017 WL 3387344, 6, 10, 11.
27. Lars Larson Show, "Ralph Taylor"; Caroline Modarressy-Tehrani, "A DNA Test Revealed This Man Is 4% Black. Now He Wants to Abolish Affirmative Action," *Huffpost*, September 19, 2019, https://www.huffpost.com/entry/dna-test-affirmative-action_n_5d824762e4b0957256afa986.
28. *Orion Ins. Grp. v. Washington State*, 56–57.
29. *Orion Ins. Grp. v. Washington State*, 2017 WL 3387344, 56–57.

30. Transcript of Oral Argument, *Orion Ins. Grp. v. Washington State.*, 2017 WL 3387344.
31. *Orion Ins. Grp. v. Wash. State Office of Minority & Women's Bus. Enters.*, 754 Fed. Appx 556, 558 (9th Cir. 2018) (noting that "[t]his disposition is not appropriate for publication and is not precedent except as provided by Ninth Circuit Rule 36–3.").
32. Modarressy-Tehrani, "A DNA Test."
33. *Orion Ins. Grp. v. Washington State*, 2017 WL 3387344, 29, 42–43.
34. *Orion Ins. Grp. v. Washington State*, 2017 WL 3387344, 43 (noting that "[i]t is again not clear whether Plaintiffs intend to assert this claim against the State Defendants, and whether they intend to make the claim via the APA or 42 U.S.C. § 1983.").
35. Reply Brief for Appellants at 17, *Orion Ins. Grp. v. Wash. State Office of Minority & Women's Bus. Enters.*, No. 17–35749, slip op. (9th Cir. Dec. 19, 2018).
36. Reply Brief for Appellants, 17, 18.
37. See, e.g., David A. Hollinger, "Amalgamation and Hypodescent: The Question of Ethnoracial Mixture in the History of the United States," *American Historical Review* 108 (2003): 1363.
38. *Orion Ins. Grp. v. Wash. State Office of Minority & Women's Bus. Enters.*, No. 16–5582 RJB, 2017 WL 3387344, at *46.
39. Mary Ziegler, "What Is Race?: The New Constitutional Politics of Affirmative Action," 50 *Connecticut Law Review* 279 (2018).
40. 572 U.S. 291 (2014); 136 S. Ct. 2198 (2016).
41. 572 U.S. 291 (2014).
42. Ziegler, "What Is Race?," 282 (citing *Schuette*, 572 U.S. at 307).
43. Ziegler, "What is Race?," (citing Fisher v. Univ. of Tex.,136 S. Ct. 2198, 2230 (2016).
44. Ziegler, "What is Race?," 283.
45. See, e.g., Jenny Reardon, *Race to the Finish: Identity and Governance in an Age of Genomics* (Princeton, NJ: Princeton University Press, 2009), 17–49; Dorothy Roberts, *Fatal Invention: How Science, Politics, and Big Business Re-Create Race in the Twenty-First Century* (2011), 3–26.
46. Press Release, The White House Office of the Press Secretary, "Remarks by the President, Prime Minister Tony Blair of England (via satellite), Dr. Francis Collins, Director of the National Human Genome Research Institute, and Dr. Craig Venter, President and Chief Scientific Officer, Celera Genomics Corporation, on the Completion of the First Survey of the Entire Human Genome Project," June 26, 2000, https://clintonwhitehouse5.archives.gov/WH/New/html/genome-20000626.html.
47. Jonathan Kahn, *Race in a Bottle: The Story of BiDil and Racialized Medicine in a Post-Genomic Age* (New York: Columbia University Press, 2012), 193–224; Roberts, *Fatal Invention*, 287–308.

48. Roberts, *Fatal Invention*, 293.
49. Sally Satel, "I am a Racially Profiling Doctor," *New York Times*, May 5, 2002, https://www.nytimes.com/2002/05/05/magazine/i-am-a-racially-profiling-doctor.html.
50. Kahn, *Race in a Bottle*, 75–101, 216–17; Roberts, *Fatal Invention*, 181–86.
51. Kahn, *Race in a* Bottle, 157–92; For a full discussion of the case of BiDil, see Kahn, *Race in a Bottle*.
52. Roberts, *Fatal Invention*, 292.
53. Roberts, *Fatal Invention*, 293.
54. Mary Ziegler, "What is Race?," 283.
55. *Fisher v. Univ. of Tex. II*, 645 F. Supp. 2d 587 (W.D. Texas 2009).
56. *Fisher v. Univ. of Tex. II*, 645 F. Supp. 2d, 590.
57. *Grutter v. Bollinger*, 539 U.S. 306 (2003); *Fisher v. Univ. of Tex. II*, 645 F. Supp. 2d, 590, 612–13.
58. *Fisher v. Univ. of Tex.*, 631 F. 3d 213 (5th Cir. 2011); *Fisher v. Univ. of Tex.*, 136 S. Ct. 2198 (2016).
59. *Fisher v. Univ. of Tex.*, 631 F. 3d, 264 n. 22.
60. *Fisher v. Univ. of Tex.*, 631 F. 3d, 247.
61. *Fisher v. Univ. of Tex.*, 631 F. 3d, 247. Notably, none of the briefs or arguments before the court mentioned the issue of genetics or the biological basis of race, indicating that this argument was introduced *sua sponte* by Garza.
62. Mary Ziegler, "What is Race?," 279–338.
63. *Fisher v. Univ. of Tex.*, 631 F. 3d, 264 n. 22; Larry Alexander and Maimon Schwarzschild, "Grutter or Otherwise: Racial Preferences and Higher Education," *Constitutional Commentary* 21, no. 1 (2004): 3, 21. Both Alexander and Schwarzschild are listed as "contributors" in events sponsored by the conservative Federalist Society. Federalist Society, "Prof. Lawrence Alexander," accessed September 9, 2024, https://fedsoc.org/contributors/lawrence-alexander; Federalist Society, "Prof. Maimon Schwarzschild," https://fedsoc.org/contributors/maimon-schwarzschild. The previous year, Yale law professor, Peter Schuck anticipated this argument in his critique of affirmative action in his book, *Diversity in America*, wherein he stated: "Scientists have long discredited the notion of race that underlies affirmative action policy, and the latest DNA research provides further evidence, were any needed, of its artificiality and incoherence." Peter H. Schuck, *Diversity in America: Keeping Government at a Safe Distance* (New York: Routledge, 2003), 144.
64. Joseph L. Graves Jr., *The Emperor's New Clothes: Biological Theories of Race at the Millennium* (New Brunswick, NJ: Rutgers University Press, 2001), 2, 3.
65. Brief for American Social Science Researchers as Amici Curiae Supporting Respondents, *Fisher v. Univ. of Tex.*, 631 F. 3d 213 (5th Cir. 2011); Brief for

American Social Science Researchers as Amici Curiae Supporting Respondents, *Fisher v. Univ. of Tex*, 136 S. Ct. 2198 (2016).

66. Joseph L. Graves Jr., "Why the Nonexistence of Biological Races Does Not Mean the Nonexistence of Racism," *American Behavioral Scientist* 59 (2015): 1474.

67. Graves, "Nonexistence of Racism," 1481.

68. See, generally, *Fisher v. Univ. of Tex.*, 570 U.S. 297 (2013); *Fisher v. Univ. of Tex.*, 758 F.3d 633 (2014); *Fisher v. Univ. of Tex.*, 136 S.Ct. 2198 (2016); Brief for Judicial Watch, Inc. and Allied Educational Foundation as Amici Curiae Supporting Petitioner, *Fisher v. Univ. of Tex.*, 136 S.Ct. 2198 (2016); Brief for Judicial Watch, Inc. and Allied Educational Foundation as Amici Curiae Supporting Petitioner, *Fisher v. Univ. Tex.*, 570 U.S. 297 (2013); Brief for Judicial Watch, Inc. and Allied Educational Foundation as Amici Curiae Supporting Petitioner, *Fisher v. Univ. of Tex.*, 136 S.Ct. 2198 (2016); Brief for the American Center for Law and Justice in Support of Rehearing En Banc, *Fisher v. Univ. of Tex.*, 136 S.Ct. 2198 (2016).

69. Judicial Watch, "About," accessed September 8, 2024. https://www.judicialwatch.org/about/#mission.

70. Alex Leary, "Meet the Conservative Group That's Driving Clinton's Email Scandal," *Tampa Bay Times*, October 7, 2016, https://www.miamiherald.com/news/politics-government/election/article106738447.html. For a fuller exploration of the influence of conservative money in American politics, see Jane Mayer, *Dark Money: The Hidden History of the Billionaires Behind the Rise of the Radical Right.* (New York: Doubleday, 2016).

71. *Fisher v. Univ. of Tex.*, 570 U.S. 297, 299 (2013).

72. See, generally, *Adarand Constructors v. Pena*, 515 U.S. 200 (1995); Brief for Judicial Watch, Inc. and Allied Educational Foundation as Amici Curiae Supporting Petitioner, *Fisher v. Univ. Tex.*, 570 U.S. 297, at 3 (2013).

73. *Grutter v. Bollinger*, 539 U.S. 306 (2003).

74. Brief for Judicial Watch, Inc. and Allied Educational Foundation as Amici Curiae Supporting Petitioner, *Fisher v. Univ. Tex.*, 570 U.S. 297, at 3–4 (2013).

75. See, e.g., Press Release, The White House Office of the Press Secretary, "Remarks by the President"; Edward R. B. McCabe, "2009 Presidential Address: Beyond Darwin? Evolution, Coevolution, and the American Society of Human Genetics," *American Journal of Human Genetics* 86 (2010): 311–15.

76. Brief for Judicial Watch, Inc. and Allied Educational Foundation as Amici Curiae Supporting Petitioner, *Fisher v. Univ. Tex.*, 570 U.S. 297, at 5–6, 8 (2013).

77. Brief for Judicial Watch, Inc., 22–23 (citing *McMillan v. City of New York*, 253 F.R.D. 247, 249 (E.D.N.Y. 2008) (Weinstein, J.).

78. Brief for Judicial Watch, Inc., 6.

79. Brief for Judicial Watch, Inc. and Allied Educational Foundation as Amici Curiae Supporting Petitioner, at 9, *Fisher v. Univ. of Tex.*, 136 S.Ct. 2198 (2016).
80. Annie Linskey, "Elizabeth Warren Releases Results of DNA Test," *Boston Globe*, October 15, 2018, https://www.bostonglobe.com/news/politics/2018/10/15/warren-addresses-native-american-issue/YEUaGzsefBogPBe2AbmSVO/story.html.
81. Brief for the American Center for Law and Justice as Amici Curiae Supporting Petitioner, *Fisher v. Univ. of Tex.*, 136 S.Ct. 2198 (2016); Brief for the American Center for Law and Justice as Amici Curiae Supporting Rehearing En Banc, *Fisher v. University of Texas at Austin*, 2014 WL 4058032; Brief for the American Center for Law and Justice as Amici Curiae Supporting Rehearing En Banc, at 1, *Fisher v. University of Texas at Austin*, 2014 WL 4058032.
82. Elizabeth Williamson, "In Jay Sekulow, Trump Taps Longtime Loyalist for Impeachment Defense," *New York Times*, January 17, 2020, https://www.nytimes.com/2020/01/17/us/politics/jay-sekulow-trump-impeachment.html.
83. Jay Alan Sekulow and Walter M. Weber, "Fisher v. University of Texas at Austin: The Incoherence and Unseemliness of State Racial Classification," *University of Miami Business Law Review* 24 (2016): 91.
84. Brief for the American Center for Law and Justice as Amici Curiae Supporting Rehearing En Banc, at 2, *Fisher v. University of Texas at Austin*, 2014 WL 4058032.
85. Brief for the American Center for Law and Justice as Amici Curiae, 5 (quoting *St. Francis College v. Al-Khazraji*, 481 U.S. 604, 610 n.4 (1987)).
86. Jonathan Kahn, *Race on the Brain: What Implicit Bias Gets Wrong About the Struggle for Racial Justice* (New York: Columbia University Press, 2017), 97–110.
87. Jerry Kang and Mahzarin R. Banaji, "Fair Measures: A Behavioral Realist Revision of Affirmative Action," *California Law Review* 94 (2006): 1063, 1065.
88. "If the facts are against you, argue the law. If the law is against you, argue the facts. If the law and the facts are against you, pound the table and yell like hell." Attributed to the poet, Carl Sandberg. See Joseph L. Smith, "Law, Fact, and the Threat of Reversal from Above," *Politics Research Quarterly* 42 (2014): 226 (quoting Carl Sandberg).
89. Margaret Thatcher Foundation, "Interview for Women's Own," https://www.margaretthatcher.org/document/106689.
90. Federalist Society, "Ward Connerly," https://fedsoc.org/contributors/ward-connerly.
91. Mary C. Waters, "Counting and Classifying by Race: The American Debate," *Tocqueville Review* 29 (2008): 1, 8, 9.

92. Michael K. Brown, Martin Carnoy, Elliott Currie, Troy Duster, David B. Oppenheimer, Marjorie M. Shultz, and Davis Wellman, *Whitewashing Race: The Myth of a Color-Blind Society* (Berkeley: University of California Press, 2003), 55, 85; Richa Amar, "Unequal Protection and the Racial Privacy Initiative," *UCLA Law Review* 52 (2004): 1279.
93. Ward Connerly and Mike Gonzales, "It's Time the Census Bureau Stops Dividing America," *Washington Post*, January 3, 2018, https://www.washingtonpost.com/opinions/its-time-the-census-bureau-stops-dividing-america/2018/01/03/a914a176-f0af-11e7-97bf-bba379b809ab_story.html.
94. Connerly and Gonzales, "It's Time the Census Bureau."
95. Sydney Trent, "Race Isn't Real, Science Says. Advocates Want the Census to Reflect That," *Washington Post*, October 16, 2023, https://www.washingtonpost.com/dc-md-va/2023/10/16/census-race-eliminate-race-box/.
96. Judicial Watch, "Comment in Response to Proposals from the Federal Interagency Working Group for Revision of the Standards for Maintaining, Collecting, and Presenting Federal Data on Race and Ethnicity, 82 Fed. Red. 12242 (March 2017), OMB 2016–0008 and OMB-2017-0003 ('Notice and Comments')." April 30, 2017. https://www.judicialwatch.org/wp-content/uploads/2017/06/Federal-Data-on-Race-and-Ethnicity-letter.pdf.
97. Brief for Judicial Watch as Amici Curiae Supporting Petitioners, *U.S. Department of Commerce v. New York*, 139 S. Ct. 2551 (2019).
98. Adriel I. Cepeda Derieux, Jonathan S. Topaz, and Dale E. Ho, " 'Contrived' : The Voting Rights Act Pretext for the Trump Administration's Failed Attempt to Add a Citizenship Question to the 2020 Census," *Yale Law & Policy Review* 38 (2020): 322.
99. For a brief discussion of the myriad ways in which the collection of racial data can impact federal programs and rights, see Kahn, *Race in a Bottle*, 27–29.
100. Jennifer Hamilton, "The Case of the Genetic Ancestor," in *Genetics and the Unsettled Past: The Collision of DNA, Race, and History*, ed. Keith Wailoo, Alondra Nelson, and Catherine Lee (New Brunswick, NJ: Rutgers University Press, 2012), 266–78.
101. TallBear, *Native American DNA*, 55–56; Will Chavez, "Cherokee Nation Responds to Senator Warren's DNA Test," *Cherokee Phoenix*, October 16, 2018, https://www.cherokeephoenix.org/Article/index/62699.
102. See, e.g., TallBear, *Native American DNA*; Krystal Tsosie and Matthew Anderson, "Two Native American Geneticists Interpret Elizabeth Warren's DNA Test," *The Conversation*, October 22, 2018, https://theconversation.com/two-native-american-geneticists-interpret-elizabeth-warrens-dna-test-105274; Chavez, "Cherokee Nation."

103. TallBear, *Native American DNA*, 83–84.
104. Brendan Koerner, "Blood Feud," *Wired*, September 1, 2005, https://www.wired.com/2005/09/seminoles/.
105. Koerner, "Blood Feud."
106. Hina Walajahi, David R. Wilson, and Sara Chandros Hull, "Constructing Identities: The Implications of DTC Ancestry Testing for Tribal Communities," *Genetics in Medicine* 21 (2019): 1744, 1746.
107. TallBear, *Native American DNA*, 55, 58–59.
108. Walajahi et al., "Constructing Identities," 1747.
109. *Brackeen v. Zinke*, 338 F. Supp. 3d 514, 525 (N.D. Tex. 2018), reversed, *Brackeen v. Bernhardt*, 937 F.3d 406 (5th Cir. 2019).
110. *Brackeen v. Bernhardt*, 937 F.3d 406, 416 (citations omitted).
111. *Brackeen v. Bernhardt*, 937 F.3d 406, 417.
112. *Brackeen v. Bernhardt*, 937 F.3d 406, 417 (citing 25 U.S.C. § 1915(a)).
113. *Brackeen v. Bernhardt*, 937 F.3d 406, 417 (citing 25 U.S.C. § 1903(4)).
114. *Brackeen v. Bernhardt*, 338 F. Supp. 3d 514, 523.
115. *Morton v. Mancari*, 417 U.S. 535 (1974).
116. *Brackeen v. Bernhardt*, 937 F.3d 406, 523.
117. *Adarand Constructors v. Pena*, 515 U.S. 200 (1995); *Brackeen v. Bernhardt*, 937 F.3d 406, 523.
118. Motion for Leave to File and Brief Amicus Curiae for the Pacific Legal Foundation Supporting Petitioners, *S. S. v. Colorado River Indian Tribes*, 138 S.Ct. 380 (2017); Brief for the Pacific Legal Foundation as Amici Curiae Supporting Petitioners, *Renteria v. Superior Court of California*, Tulare County, 138 S.Ct. 986 (2018).
119. Motion for Leave to File and Brief Amicus Curiae for the Pacific Legal Foundation Supporting Petitioners, at 9–10, *S. S. v. Colorado River Indian Tribes*, 138 S.Ct. 380 (2017); Brief for Pacific Legal Foundation as Amici Curiae Supporting Petitioners, at 11, *Renteria v. Superior Court of California*, Tulare County, 138 S.Ct. 986 (2018).
120. See, generally, *Hawaii v. Office of Hawaiian Affairs*, 556 U.S. 163 (2009); Brief for the Pacific Legal Foundation, the Cato Institute, and the Center for Equal Opportunity as Amici Curiae Supporting Petitioners, at 18–19, *Hawaii v. Office of Hawaiian Affairs*, 556 U.S. 163 (2009).
121. *Hawaii v. Office of Hawaiian Affairs*, 556 U.S. 163, 166 (2009) (For a fuller discussion of this history); see Troy J. H. Andrade, "(Re)Righting History: Deconstructing the Court's Narrative of Hawai'i's Past," *University of Hawaii Law Review* 39 (2017): 631.
122. Brief for the Pacific Legal Foundation, 18–19, 20, 21.
123. *Brackeen v. Bernhardt*, 937 F.3d 406, 416.

124. Brackeen v. Haaland, 994 F.3d 249 (2021).
125. *Haaland v. Brackeen*, 143 S.Ct. 1609 (2023).
126. Brief Amici Curiae, Goldwater Institute, Cato Institute, Texas Public Policy Foundation, and Families Affected by Icwa in Support of Brackeen, et al., and State of Texas. 2022 WL 2020308. June 1, 2022, 12.
127. Brief For Petitioner the State of Texas. 2022 WL 1785628. May 26, 2022, 1; Brief For Petitioner the State of Texas, 20.
128. 25 U.S.C. § 1903(4)(a).
129. Abi Fain and Mary Kathryn Nagle, "Close to Zero: The Reliance on Minimum Blood Quantum Requirements to Eliminate Tribal Citizenship in the Allotment Acts and the Post-Adoptive Couple Challenges to the Constitutionality of ICWA," *Mitchell Hamline Law Review* 43 (2017): 801, 873.
130. *Brackeen v. Bernhardt*, 937 F.3d 406, 428–429 (5th Cir. 2019).
131. *Adoptive Couple v. Baby Girl*, 570 U.S. 637, 641 (2013).
132. Fain and Nagle, "Close to Zero," 803 (citing Transcript of Oral Argument at 42–43, *Adoptive Couple v. Baby Girl*, 570 U.S. 637 (2013)).
133. Roxanna Asgarian, "How a White Evangelical Family Could Dismantle Adoption Protections for Native Children," *Vox*, February 20, 2020, https://www.vox.com/identities/2020/2/20/21131387/indian-child-welfare-act-court-case-foster-care.
134. Rebecca Nagle, "The Story of Baby O—and the Case That Could Gut Native Sovereignty," *The Nation*, November 9, 2022, https://www.thenation.com/article/society/icwa-supreme-court-libretti-custody-case/.
135. Stephanie Pappas, "What Does Elizabeth Warren's 'Native' Ancestry Mean?," *Live Science*, October 16, 2018, https://www.livescience.com/63848-elizabeth-warren-native-american-ancestry-explained.html; @KimTallBear, Twitter post, October 15, 2018, https://twitter.com/KimTallBear/status/1052017467021651969/photo/1; Tsosie and Anderson, "Two Native American Geneticists"; Chavez, "Cherokee Nation."
136. *Haaland v. Brackeen*, 143 S.Ct. 1609, 1648 (2023).
137. Richard Herrnstein and Charles Murray, *The Bell Curve: Intelligence and Class Structure in American Life* (New York: Free Press, 1994, 447–509; Steven Fraser, *The Bell Curve Wars: Race, Intelligence, and the Future of America* (New York: Basic Books, 1995); Christine Ma and Michael Schapira, *An Analysis of Richard J. Herrnstein and Charles Murray's* The Bell Curve*: Intelligence and Class Structure in American life* (London: Taylor & Francis, 2017), 25–34; Matthew Yglesias, "The Bell Curve is About Policy. And It's Wrong," *Vox*, April 10, 2018, https://www.vox.com/2018/4/10/17182692/bell-curve-charles-murray-policy-wrong; Ryan Fortson, "Affirmative Action, The Bell Curve, and Law School Admissions," *Seattle University Law Review* 24 (2001): 1087.

138. George M. Fredrickson, *Racism: A Short History* (Princeton, NJ: Princeton University Press, 2015), 49–96; Charles Murray and Richard J. Herrnstein, "Race, Genes and I.Q.—An Apologia," *New Republic*, October 31, 1994, https://newrepublic.com/article/120887/race-genes-and-iq-new-republics-bell-curve-excerpt; *Grutter v. Bollinger*, 539 U.S. 306 (2003).
139. Press Release, The White House Office of the Press Secretary, "Remarks by the President."
140. Alexander and Schwarzschild, "Grutter or Otherwise," 3, 6.
141. Modarressy-Tehrani, "A DNA Test."
142. Lars Larson Show, "Ralph Taylor"; Modarressy-Tehrani, "A DNA Test"; VDARE, "Flight From White: Businessman Tries To Prove He's Non-White For Minority-Owned Business Money," accessed September 8, 2024, https://web.archive.org/web/20190130170933/https://vdare.com/posts/flight-from-white-businessman-tries-to-prove-he-s-non-white-for-minority-owned-business-money; Gregory Hood, "Norwegian Girl Disappointed to Discover She's 'So White,'" *American Renaissance*, April 24, 2020, https://www.amren.com/commentary/2020/04/norwegian-girl-crushed-to-discover-shes-so-white/.
143. See, e.g., Sal Restivo and Jennifer Croissant, "Social Constructionism in Science and Technology Studies," in *Handbook of Constructionist Research*, ed. Jaber F. Gubrium and James Holstein (New York: Guilford Press, 2008), 213, 222; Ann Pettifor, *Just Money: How Society Can Break the Despotic Power of Finance* (London: Sed Books, 2014). As Sociologist Phillip Cohen put it:

> That race is a "social construction" does not imply that it does not exist. We need to dispel that confusion for two reasons. First, at the risk of stating the obvious, things that are socially constructed are still constructed-they exist socially. The vast, historically persistent, life-and-death consequences of race in human societies cannot be ignored or dismissed as figments of our collective imagination. Race was not the cause of Africans being stolen from their homes and sold into slavery in the Americas; it was a result of that process.

Philip Cohen, "How Troubling Is Our Inheritance? A Review of Genetics and Race in the Social Sciences," *Annals of the American Academy of Political and Social Science* 661 (2015): 65, 71.

9. DIVERSITY AND THE FRAMES OF REPRESENTATION

1. NHGRI, *Roundtable on Inclusion and Engagement of Underrepresented Populations in Genomics: Executive Summary*, September 16, 2015, 1, 2. https://www.genome.gov/Pages/About/NACHGR/February2016AgendaDocuments/2015_09_16_Roundtable_Report_final.pdf.

2. NHGRI, *Roundtable on Inclusion*, 1.
3. Food and Drug Administration, *Collection, Analysis, and Availability of Demographic Subgroup Data for FDA -Approved Medical Products*, August 2013, https://web.archive.org/web/20170404220257/https://www.fda.gov/downloads/RegulatoryInformation/LawsEnforcedbyFDA/SignificantAmendmentstotheFDCAct/FDASIA/UCM365544.pdf,
4. Food and Drug Administration, *Collection*, 7–8.
5. Food and Drug Administration, *FDA Action to Enhance the Collection and Availability of Demographic Subgroup Data*, August 2014, 23, https://www.fda.gov/media/89307/download.
6. Food and Drug Administration. *FDA Action*, 8.
7. Rose Marie Robertson, MD, FAHA, Chief Science and Medical Officer, American Heart Association, *Testimony Before the FDA Public Hearing on the Action Plan for the Collection, Analysis, and Public Availability of Demographic Subgroup Data*, April 1, 2014, Re: Docket No. FDA-2013-N-0745, Comments on FDASIA Section 907 Report: "Collection, Analysis, and Availability of Demographic Subgroup Data for FDA-Approved Medical Products"; Nancy A. Brown, Chief Executive Officer American Heart Association; Phyllis Greenberger President and CEO Society for Women's Health Research; Lisa M. Tate Chief Executive Officer WomenHeart: The National Coalition for Women with Heart Disease; Cynthia A. Pearson, Executive Director National Women's Health Network. *Comments on FDASIA Section 907 Report: "Collection, Analysis, and Availability of Demographic Subgroup Data for FDA-Approved Medical Products,"* November 20, 2013.
8. Brown et al., *Comments on FDASIA Section 907 Report*, Re: Docket No. FDA-2013-N-0745, November 20, 2013, 2.
9. Biotechnology Industry Organization, *Comments*, Submitted by Andrew W. Womack, PhD, Director, Science and Regulatory Affairs Biotechnology Industry Organization (BIO). Re: Docket No. FDA–2013–N–0745 May 16, 2014, 2.
10. Boehringer Ingelheim Pharmaceuticals Inc. *Comments*. Submitted by Joanne Palmisano, MD, FACP, Vice President-Regulatory Affairs Boehringer Ingelheim Pharmaceuticals, Inc. Re: Docket No. FDA–2013–N–0745, November. 20, 2013, 3.
11. Boehringer, *Comments*, 4.
12. Jonathan Kahn, "Harmonizing Race: Competing Regulatory Paradigms of Racial Categorization in International Drug Development," *Santa Clara Journal of International Law* 5 (2007): 34.
13. Robert M. Califf, "2016: The Year of Diversity in Clinical Trials," *FDA Voice Blog*, January 27, 2016, https://web.archive.org/web/20160128085201/http://blogs

.fda.gov/fdavoice/index.php/2016/01/2016-the-year-of-diversity-in-clinical-trials/.
14. Alice B. Popejoy and Stephanie M. Fullerton, "Genomics Is Failing on Diversity," *Nature News* 538, no. 7624 (2016): 161–64, 161 By September 2024, the article had 1,741 citations on Google Scholar, https://scholar.google.com/scholar?hl=en&as_sdt=0%2C22&q=%22Genomics+is+failing+on+diversity.%22+&btnG=.
15. Popejoy and Fullerton, "Genomics Is Failing on Diversity," 161, 164.
16. Popejoy and Fullerton, "Genomics Is Failing on Diversity," 161–164.
17. Amy R. Bentley, Shawneequa Callier, and Charles N. Rotimi, "Diversity and Inclusion in Genomic Research: Why the Uneven Progress?," *Journal of Community Genetics* 8 (2017): 255–66, 255
18. Bentley et al., "Diversity and Inclusion," 255.
19. Bentley et al., "Diversity and Inclusion," 255–266, 258, 259.
20. American Society of Human Genetics, *Press Release*, April 10, 2019, "Program to Increase, Network, Mentor, and Retain Diverse Early-Career Researchers," https://www.ashg.org/publications-news/press-releases/201904-hgsi/.
21. American Society of Human Genetics, "Program to Increase."
22. This dynamic was particularly evident in the call to enlist "diverse" researchers to help enroll more "diverse" subjects into trials for COVID-19 vaccines in 2020.
23. American Society of Human Genetics, "Program to Increase."
24. Derrick A. Bell Jr., "*Brown v. Board of Education* and the Interest-Convergence Dilemma," *Harvard Law Review* (1980): 518–33. The paradigmatic example of this was the Supreme Court's decision in *Brown v. Board of Education*. An interest-convergence theory reading of the case focuses on how dismantling Jim Crow in the early 1950s was of critical importance to, among others, the Eisenhower administration in its efforts to position the United States as a role model for nations emerging from colonialism in the aftermath of World War II and the growing Cold War competition with the Soviet Union.
25. Democratic Party Platform, 2008, https://www.presidency.ucsb.edu/documents/1960-democratic-party-platform.
26. Prabarna Ganguly, *Press Release*, National Human Genome Research Institute, "Putting Diversity Front and Center," June 19, 2019, https://www.genome.gov/news/news-release/Putting-diversity-front-and-center.
27. Genevieve L. Wojcik et al., "Genetic Analyses of Diverse Populations Improve Discovery for Complex Traits," *Nature* 570, no. 7762 (2019): 514–18.
28. Wojcik et al. "Genetic Analyses," 514.
29. Deepti Gurdasani, Tommy Carstensen, Fasil Tekola-Ayele, Luca Pagani, Ioanna Tachmazidou, Konstantinos Hatzikotoulas, Savita Karthikeyan et al., "The

African Genome Variation Project Shapes Medical Genetics in Africa," *Nature* 517, no. 7534 (2015): 327–32.
30. Gurdasani et al., "The African Genome Variation Project," 327.
31. Nicholas Wade, *A Troublesome Inheritance: Genes, Race, and Human History* (New York: Penguin, 2015).
32. Agustín Fuentes, "*A Troublesome Inheritance*: Nicholas Wade's Botched Interpretation of Human Genetics, History, and Evolution," *Human Biology* 86, no. 3 (2014): 215–220, 215; citing Wade at 123.
33. Philip N. Cohen, "How Troubling Is Our Inheritance? A Review of Genetics and Race in the Social Sciences," *ANNALS of the American Academy of Political and Social Science* 661, no. 1 (2015): 65–84, 74.
34. Graham Coop, Michael Eisen, Rasmus Nielsen, Molly Przworsky, and Noah Rosenberg, "Letters: A Troublesome Inheritance," *New York Times*, August 8, 2014, https://www.nytimes.com/2014/08/10/books/review/letters-a-troublesome-inheritance.html.
35. Charles Murray, "Book review: *A Troublesome Inheritance* by Nicholas Wade," *Wall Street Journal*, May 2, 2014, http://online.wsj.com/news/articles/SB10001424052702303380004579521482247869874; Jared Taylor, "Nicholas Wade Takes on the Regime," *American Renaissance*, March 2, 2014, www.amren.com/features/2014/03/attack-on-the-regime/.
36. Jennifer Raff, "Nicholas Wade and Race: Building a Scientific Façade," *Human Biology* 86, no. 3 (2014): 227–232, 230.
37. Sonja Soo, "Scientists Release a New Human 'Pangenome' Reference,'" *NHGRI*, May 10, 2023, https://www.genome.gov/news/news-release/scientists-release-a-new-human-pangenome-reference.
38. Soo, "'Pangenome' Reference."
39. Ting Wang et al., "The Human Pangenome Project: a Global Resource to Map Genomic Diversity," *Nature* 604, no. 7906 (2022): 437–46, 438.
40. Wang et al., "Human Pangenome Project," 440.
41. Elie Dolgin, "Scientists Unveil a More Diverse Human Genome," *New York Times*, May 10, 2023, https://www.nytimes.com/2023/05/10/science/pangenome-human-dna-genetics.html?searchResultPosition=1.
42. Wen-Wei Liao et al., "A Draft Human Pangenome Reference," *Nature* 617 (2023): 312–24, 313.
43. Dolgin, "Diverse Human Genome."
44. National Academies of Sciences, Engineering, and Medicine, *Using Population Descriptors in Genetics and Genomics Research: A New Framework for an Evolving Field* (Washington, D.C.: National Academies Press, 2023), 2, 7, 9, https://doi.org/10.17226/26902.

45. Max Kozlov, "Ambitious Survey of Human Diversity Yields Millions of Undiscovered Genetic Variants," *Nature* (2024), https://www.nature.com/articles/d41586-024-00502-0.
46. The All of Us Research Program Genomics Investigators, "Genomic Data in the All of Us Research Program," *Nature* 627 (2024): 340–46. https://doi.org/10.1038/s41586-023-06957-x; Max Kozlov, "'All of Us' Genetics Chart Stirs Unease Over Controversial Depiction of Race," *Nature*, February 23, 2024, https://www.nature.com/articles/d41586-024-00568-w#ref-CR1.
47. Kozlov, "'All of Us.'"
48. The All of Us Research Program Genomics Investigators, "Genomic Data," 342.
49. Jonathan Pritchard, Twitter post, February 19, 2024, https://twitter.com/jkpritch/status/1759769445759893832.
50. Max Kozlov, "'All of Us.'"
51. Steven Epstein, *Inclusion: The Politics of Difference in Medical Research* (Chicago: University of Chicago Press, 2008), 58–59.
52. David Plotke, "Representation is Democracy," *Constellations* 4, no. 1 (1997): 19–34, 19.
53. Epstein, *Inclusion*, 89–90.
54. Sandra Soo-Jin Lee, Stephanie M. Fullerton, Caitlin E McMahon, Michael Bentz, Aliya Saperstein, Melanie Jeske, Emily Vasquez, Nicole Foti, Larissa Saco, and Janet K Shim, "Focus: Bioethics: Targeting Representation: Interpreting Calls for Diversity in Precision Medicine Research," *Yale Journal of Biology and Medicine* 95, no. 3 (2022): 317.
55. Hannah Pitkin, *The Concept of Representation* (Berkeley: University of California Press, 1967), 11.
56. Pitkin, *Concept of Representation*, 88.
57. Jane Mansbridge, "Clarifying the Concept of Representation," *American Political Science Review* 105, no. 3 (2011): 621–30.
58. Lee et al., "Focus," 317.

10. POLITICAL VALENCES OF CONTEMPORARY GENETIC DIVERSITY

1. David Rotman, "DNA Databases are Too White: This Man Aims to Fix That," *MIT Technology Review*, October 15, 2018, https://www.technologyreview.com/2018/10/15/139472/dna-databases-are-too-white-this-man-aims-to-fix-that/.
2. David Reich, *Who We Are and How We Got Here: Ancient DNA and the New Science of the Human Past* (New York: Vintage Books, 2018), xix, 171.

3. David Reich, "How Genetics Is Changing Our Understanding of 'Race,'" *New York Times*, March 23, 2018, https://www.nytimes.com/2018/03/23/opinion/sunday/genetics-race.html.
4. Jonathan Kahn et al., "How Not to Talk About Race and Genetics," *BuzzFeed*, March 30, 2018, https://www.buzzfeednews.com/article/bfopinion/race-genetics-david-reich.
5. Reich, *Who We Are and How We Got Here*, xix.
6. Maggie Astor, Christina Caron, and Daniel Victor, "A Guide to the Charlottesville Aftermath," *New York Times*, August 12, 2017, https://www.nytimes.com/2017/08/13/us/charlottesville-virginia-overview.html; Glenn Kessler, "The 'Very Fine People' at Charlottesville: Who Were They?," *Washington Post*, May 8, 2020, https://www.washingtonpost.com/politics/2020/05/08/very-fine-people-charlottesville-who-were-they-2/.
7. Southern Poverty Law Center, *The Year in Hate and Extremism, 2019*, March 18, 2020, https://www.splcenter.org/news/2020/03/18/year-hate-and-extremism-2019.
8. Elspeth Reeve, "Alt-Right Trolls Are Getting 23andMe Genetic Tests to 'Prove' Their Whiteness," *Vice*, October 8, 2016, https://www.vice.com/en/article/vbygqm/alt-right-trolls-are-getting-23andme-genetic-tests-to-prove-their-whiteness.
9. Sarah Zhang, "Will the Alt-Right Promote a New Kind of Racist Genetics," *The Atlantic*, December 29, 2016, https://www.theatlantic.com/science/archive/2016/12/genetics-race-ancestry-tests/510962/.
10. Michael Price, "'It's a Toxic Place.' How the Online World of White Nationalists Distorts Population Genetics," *Science*, May 22, 2018, https://www.science.org/content/article/it-s-toxic-place-how-online-world-white-nationalists-distorts-population-genetics.
11. Amanda Hess, "The Racial Spectacle of DNA Test Result Videos," *New York Times*, May 6, 2018, https://www.nytimes.com/2018/05/06/arts/dna-testing-race.html.
12. Courtney Subramanian and Jordan Culver, "Donald Trump Sidesteps Call to Condemn White Supremacists—and the Proud Boys Were 'Extremely Excited' About It," *USA Today*, September 29, 2020, https://www.usatoday.com/story/news/politics/elections/2020/09/29/trump-debate-white-supremacists-stand-back-stand-by/3583339001/.
13. Hess, "Racial Spectacle."
14. Aaron Panofsky and Joan Donovan, "Genetic Ancestry Testing Among White Nationalists: From Identity Repair to Citizen Science," *Social Studies of Science* 49, no. 5 (2019): 653–81, 653.

15. Ari Feldman, "Human Biodiversity: The Pseudoscientific Racism of the Alt-Right," *Forward*, August 5, 2016, https://forward.com/opinion/346533/human-biodiversity-the-pseudoscientific-racism-of-the-alt-right/.
16. Jonathan M. Marks, *Human Biodiversity: Genes, Race, and History* (New York: DeGruyter, 1995), and "I Coined the Phrase 'Human Biodiversity.' Racists Stole It," *Anthropomics*, December 28, 2019, http://anthropomics2.blogspot.com/2019/12/i-coined-phrase-human-biodiversity.html.
17. Aaron Panofsky, Kushan Dasgupta, and Nicole Iturriaga, "How White Nationalists Mobilize Genetics: From Genetic Ancestry and Human Biodiversity to Counterscience and Metapolitics," *American Journal of Physical Anthropology* 175, no. 2 (2021): 387–98.
18. Quoted in Panofsky et al., "How White Nationalists Mobilize Genetics," 393.
19. Panofsky et al., "How White Nationalists Mobilize Genetics," 393.
20. Jedidiah Carlson and Kelley Harris, "Quantifying and Contextualizing the Impact of bioRxiv Preprints Through Automated Social Media Audience Segmentation," *PLoS Biology* 18, no. 9 (2020): e3000860.
21. Charles Murray, *Human Diversity: The Biology of Gender, Race, and Class* (New York: Twelve, 2020).
22. Philip Ball, "The Gene Delusion," *New Statesman*, June 10, 2020, https://www.newstatesman.com/long-reads/2020/06/class-race-genetics-science-human-diversity-charles-murray-review.
23. Murray, *Human Diversity*, 178.
24. K. Paige Harden, "Genetic Determinism, Essentialism, and Reductionism: Semantic Clarity for Contested Science," *Nature Reviews Genetics* 24, no. 3 (2023): 197–204, 198, n.5, https://www.nature.com/articles/s41576-022-00537-x.
25. Deborah A. Bolnick, "Individual Ancestry Inference and the Reification of Race as a Biological Phenomenon," in Barbara A. Koenig, Sandra Soo-Jin Lee, and Sarah S. Richardson, eds., *Revisiting Race in a Genomic Age* (New Brunswick, NJ: Rutgers University Press, 2008), 70–85, 77; see also Rob DeSalle and Ian Tattersall, *Troublesome Science: The Misuse of Genetics and Genomics in Understanding Race* (New York: Columbia University Press, 2018), 143–48.
26. Ball, "The Gene Delusion."
27. Murray, *Human Diversity*, 216–17, 216.
28. Angela Saini, *Superior: The Return of Race Science* (Boston: Beacon Press, 2019), 63–65, and "Why Race Science Is on the Rise Again," *The Guardian*, May 18, 2019, https://www.theguardian.com/books/2019/may/18/race-science-on-the-rise-angela-saini; Emil O. W. Kirkegaard, "Book Review Article: Human Biodiversity for Beginners: A Review of Charles Murray's Human Diversity," *Mankind Quarterly* 60, no. 3 (2020).

29. Amy Harmon, "Why White Supremacists Are Chugging Milk (and Why Geneticists Are Alarmed)," *New York Times*, October 17, 2018, https://www.nytimes.com/2018/10/17/us/white-supremacists-science-dna.html.
30. John Jackson, "Racial Science and Strawman Arguments," March 30, 2018, https://altrightorigins.com/2018/03/30/racial-science-strawman/.
31. Harmon, "White Supremacists."
32. Tom Lehrer, "Wernher von Braun," *That Was the Year That Was* (1965).
33. Megan Molteni, "Buffalo Shooting Ignites a Debate Over the Role of Genetics Researchers in White Supremacist Ideology," *STAT News*, May 23, 2022, https://www.statnews.com/2022/05/23/buffalo-shooting-ignites-debate-genetics-researchers-in-white-supremacist-ideology/.
34. Robbee Wedow, Daphne O. Martschenko, and Sam Trejo, "Scientists Must Consider the Risk of Racist Misappropriation of Research," *Scientific American* (2022). https://www.scientificamerican.com/article/scientists-must-consider-the-risk-of-racist-misappropriation-of-research/.
35. Michael Barbaro, "The Racist Theory Behind So Many Mass Shootings," *New York Times*, May 16, 2022, https://www.nytimes.com/2022/05/16/podcasts/the-daily/buffalo-shooting-replacement-theory.html.
36. Megan Molteni, "Buffalo Shooting."
37. Monika Pronczuk and Koba Ryckewaert, "A Racist Researcher, Exposed by a Mass Shooting," *New York Times*, June 9, 2022, https://www.nytimes.com/2022/06/09/world/europe/michael-woodley-buffalo-shooting.html; Gene E. Robinson, Christina M. Grozinger, and Charles W. Whitfield, "Sociogenomics: Social Life in Molecular Terms," *Nature Reviews Genetics* 6 (2005): 257–70, https://doi.org/10.1038/nrg1575; Aaron Panofsky, *Misbehaving Science: Controversy and the Development of Behavior Genetics* (Chicago: University of Chicago Press, 2014).
38. David B. Braudt, "Sociogenomics in the 21st Century: An Introduction to the History and Potential of Genetically Informed Social Science," *Sociology Compass* 12, no. 10 (2018): e12626.
39. Nathaniel Comfort, "Sociogenomics Is Opening a New Door to Eugenics," *MIT Technology Review* (2018), https://www.technologyreview.com/2018/10/23/139420/sociogenomics-is-opening-a-new-door-to-eugenics/.
40. Price, "It's a Toxic Place."
41. Wedow et al., "Risk of Racist Misappropriation of Research."
42. Mahzarin R. Nanaji and Anthony G. Greenwald, *Blind Spot: Hidden Biases of Good People* (New York: Delacorte, 2013), 176.
43. *Shelby County v. Holder*, 133 S.Ct. 2612 (2013), 2625.
44. Charles Murray, *Human Diversity*, 238–39.

45. Phil Han, "Donald Trump: I have the Genes for Success," *CNN*, February 11, 2010, https://web.archive.org/web/20240408052207/https://www.cnn.com/2010/SHOWBIZ/02/11/donald.trump.marriage.apprentice/index.html.
46. Adam Cohen, "Op-Ed: Eugenics is Making a Comeback. Resist, Before History Repeats Itself," *Los Angeles Times*, October 14, 2020, https://www.latimes.com/opinion/story/2020-10-14/trump-eugenics-politics-history; Jonathan Kahn, Marcy Darnovsky, and Jonathan Marks, "Trump's 'Racehorse Theory' and Why It Matters," *Biopolitical Times*, October 5, 2020, https://www.geneticsandsociety.org/biopolitical-times/trumps-racehorse-theory-and-why-it-matters.
47. Kahn et al., "Trump's Racehorse Theory."
48. Michael Gold, "Trump's Long Fascination With Genes and Bloodlines Gets New Scrutiny," *New York Times*, December 23, 2023, https://www.nytimes.com/2023/12/22/us/politics/trump-blood-comments.html.
49. Harmon, "White Supremacists."
50. The forbidden words were "vulnerable," "entitlement," "diversity," "transgender," "fetus," "evidence-based," and "science-based." Lena Sun and Juliet Eilperin, "CDC Gets List of Forbidden Words: Fetus, Transgender, Diversity," *Washington Post*, December 15, 2017, https://www.washingtonpost.com/national/health-science/cdc-gets-list-of-forbidden-words-fetus-transgender-diversity/2017/12/15/f503837a-e1cf-11e7-89e8-edec16379010_story.html.
51. United States, Executive Office of President Donald Trump, *Executive Order 13950. On Combating Race and Sex Stereotyping*, September 22, 2020, https://trumpwhitehouse.archives.gov/presidential-actions/executive-order-combating-race-sex-stereotyping/.
52. Fabiola Cineas, "Critical Race Theory, and Trump's War on It, Explained," *Vox*, September 24, 2020, https://www.vox.com/2020/9/24/21451220/critical-race-theory-diversity-training-trump.
53. Jessica Guynn, "Trump Executive Order on Diversity Training Roils Corporate America," *USA Today*, September 27, 2020, https://www.usatoday.com/story/money/2020/09/25/trump-executive-order-diversity-training-race-gender/3537241001/.
54. *Santa Cruz Lesbian and Gay Community Center v. Donald Trump*, Order Granting in Part Motion for Nationwide Preliminary Injunction. United States District Court, Northern District of California, Case No. 5:20-cv-07741-BLF, December 22, 2020.
55. United States, Executive Office of President Joseph Biden, *Executive Order 13985: Advancing Racial Equity and Support for Underserved Communities Through the Federal Government*, January 20, 2021, https://www.federalregister

.gov/documents/2021/01/25/2021-01753/advancing-racial-equity-and-support-for-underserved-communities-through-the-federal-government.

56. "'America First 2.0': Vivek Ramaswamy's Presidential Campaign Explained," *Yahoo News Video and Transcript*, August 11, 2023, https://uk.news.yahoo.com/america-first-2-0-vivek-130035358.html.

57. White House Office of the Press Secretary, "Remarks by the President, Prime Minister Tony Blair of England (via satellite), Dr. Francis Collins, Director of the National Human Genome Research Institute, and Dr. Craig Venter, President and Chief Scientific Officer, Celera Genomics Corporation, on the Completion of the First Survey of the Entire Human Genome Project," June 26, 2000, https://clintonwhitehouse5.archives.gov/WH/New/html/genome-20000626.html; Brief for the American Center for Law and Justice as Amici Curiae Supporting Petitioner, *Fisher v. Univ. of Tex.*, 136 S.Ct. 2198 (2016); Brief for the American Center for Law and Justice as Amici Curiae Supporting Rehearing En Banc, *Fisher v. University of Texas at Austin*, 2014 WL 4058032; Brief for the American Center for Law and Justice as Amici Curiae Supporting Rehearing en Banc, at 1, *Fisher v. University of Texas at Austin*, 2014 WL 4058032.

58. Parents Involved in *Community Schools v. Seattle School District No. 1*, 551 U.S. 701, 748 (2007); *SFFA v Harvard*, 600 U.S. 181, 206. (2023); *Schuette v. BAMN*, 134 S. Ct. 1623, 1638 (2014).

59. Reva Siegel, "The Supreme Court, 2012 Term Foreword: Equality Divided," *Harvard Law Review* 127, no. 1 (2013): 1–94, 42.

60. Eddie Glaude, "DEI Is Not the Monster Here," *Time*, December 13, 2023, https://time.com/6458406/dei-campus-speech/.

61. *Students for Fair Admissions, Inc. v. Harvard*, 980 F.3d 157 (2020); Brief for the Pacific Legal Foundation as Amici Curiae Supporting Appellants, at 12, *Students for Fair Admissions, Inc. v. President and Fellows of Harvard University*, 2020 WL 1469644 (1st Cir. 2020) (emphasis added).

62. Brief for the Pacific Legal Foundation as Amici Curiae Supporting Appellants, 13.

63. *SFFA. v. Harvard University et al.*, 143 S.Ct. 2141 (2023).

64. *SFFA v. Harvard University et al.*, 143 S.Ct. 2141 (2023), 2167–68.

65. Larry Alexander and Maimon Schwarzschild, "Grutter or Otherwise: Racial Preferences and Higher Education," *Constitutional Commentary* 21 (2004): 3, 6.

66. Ward Connerly and Mike Gonzalez. "It's Time the Census Bureau Stops Dividing America," *Washington Post*, January 3, 2018, https://www.washingtonpost.com/opinions/its-time-the-census-bureau-stops-dividing-america/2018/01/03/a914a176-f0af-11e7-97bf-bba379b809ab_story.html.

67. Michael Barone, "Genetics Is Undercutting the Case for Racial Quotas," *National Review*, April 6, 2018, https://www.nationalreview.com/2018/04/genetic-science-does-not-support-racial-quotas/.
68. *SFFA v. Harvard University et al.*, 143 S.Ct. 2141, 2210 (2023) (Gorsuch, J., concurring).
69. *SFFA v. Harvard University et al.*, 143 S.Ct. 2141, 2166 (2023).
70. *SFFA v. Harvard University et al.*, 143 S.Ct. 2141, 2166, 2171 (2023).
71. *SFFA v. Harvard University et al.*, 143 S.Ct. 2141, 2160 (2023).
72. See *Brown v. Board of Education of Topeka*, 347 U.S. 483, 494 (1954).
73. William Rehnquist, "A Random Thought on the Segregation Cases," https://www.govinfo.gov/content/pkg/GPO-CHRG-REHNQUIST/pdf/GPO-CHRG-REHNQUIST-4-16-6.pdf.
74. Chronicle Staff, "DEI Legislation Tracker," *Chronicle of Higher Education*, accessed August 15, 2023, https://www.chronicle.com/article/here-are-the-states-where-lawmakers-are-seeking-to-ban-colleges-dei-efforts.
75. Alia Wong, "DEI Came to Colleges with a Bang. Now, These Red States Are on a Mission to Snuff It Out," *USA Today*, March 23, 2023, https://www.usatoday.com/story/news/education/2023/03/23/dei-diversity-in-colleges-targeted-by-conservative-red-states/11515522002/.
76. Heather McGhee, *The Sum of Us: What Racism Costs Everyone and How We Can Prosper Together* (New York: One World, 2022).
77. Benjamin Wallace-Wells, "How a Conservative Activist Invented the Conflict Over Critical Race Theory," *New Yorker*, June 18, 2021, https://www.newyorker.com/news/annals-of-inquiry/how-a-conservative-activist-invented-the-conflict-over-critical-race-theory.
78. See, e.g., Lauren B. Edelman, Sally Riggs Fuller, and Iona Mara-Drita, "Diversity Rhetoric and the Managerialization of Law," *American Journal of Sociology* 106, no. 6 (2001): 1589–1641; Brent K. Nakamura and Lauren B. Edelman, "Bakke at 40: How Diversity Matters in the Employment Context," *UC Davis Law Review* 52 (2018): 2627; and Frank Dobbin, *Inventing Equal Opportunity* (Princeton, NJ: Princeton University Press, 2009).
79. Wallace-Wells, "Critical Race Theory."
80. Jason Wilson, "Activist Who Led Puster of Harvard President Linked to 'Scientific Racism' Journal," *The Guardian*, January 31, 2024, https://www.theguardian.com/world/2024/jan/31/rightwing-activist-christopher-rufo-ties-scientific-racism-journal?CMP=Share_iOSApp_Other.
81. R. Roosevelt Thomas Jr., "From Affirmative Action to Affirming Diversity," *Harvard Business Review* 68, no. 2 (1990): 107–17, 112, https://hbr.org/1990/03/from-affirmative-action-to-affirming-diversity%202.

82. Thomas, "From Affirmative Action," 112; Christopher Rufo, "D.E.I. Programs Are Getting in the Way of Liberal Education," *New York Times*, July 27, 2023, https://www.nytimes.com/2023/07/27/opinion/christopher-rufo-diversity-desantis-florida-university.html.
83. *Richmond v. Croson*, 488 U.S. 469 (1989).
84. Ellen Berrey, *The Enigma of Diversity: The Language of Race and the Limits of Racial Justice* (Chicago: University of Chicago Press, 2015), 251–52.
85. Robert Post, "Affirmative Action and Higher Education: The View from Somewhere," *Yale Law & Police Review* 23 (2005): 25–32, 26.
86. Quinetta M. Roberson, "Disentangling the Meanings of Diversity and Inclusion in Organizations," *Group & Organization Management* 31, no. 2 (2006): 212–36, 212.
87. Society for Diversity, "Evolution of DEI," https://web.archive.org/web/20230720141012/https://www.societyfordiversity.org/evolution-of-dei.
88. Institute for Diversity Certification, "Why is DEIA important?," https://www.diversitycertification.org/deia-matters/what-is-deia/deia-dictionary.
89. See, e.g., Lauryn Burnett and Herman Aguinis, "How to Prevent and Minimize DEI Backfire," *Business Horizons* (2023); Rosalind M. Chow, L. Taylor Phillips, Brian S. Lowery, and Miguel M. Unzueta, "Fighting Backlash to Racial Equity Efforts," *MIT Sloan Management Review* 62, no. 4 (2021): 25–31.
90. Frank Dobbin and Alexandra Kalev, "Why Doesn't Diversity Training Work? The Challenge for Industry and Academia," *Anthropology Now* 10, no. 2 (2018): 48–55.
91. Elizabeth Levy Paluck, Roni Porat, Chelsey S. Clark, and Donald P. Green, "Prejudice Reduction: Progress and Challenges," *Annual Review of Psychology* 72 (2021): 533–60.
92. Nicole Hannah Jones, ed., "The 1619 Project," *New York Times*, August 14, 2019, https://www.nytimes.com/interactive/2019/08/14/magazine/1619-america-slavery.html.
93. Jake Silverstein, "The 1619 Project and the Long Battle Over US History," *New York Times Magazine* (2021), https://www.nytimes.com/2021/11/09/magazine/1619-project-us-history.html.
94. *The 1619 Project*, "Educational Materials," https://1619education.org/curricular-resources.
95. Silverstein, "1619 Project."
96. Brooke Migdon, "What is DeSantis's 'Stop WOKE Act?,'" *The Hill*, August 19, 2022, https://thehill.com/changing-america/respect/diversity-inclusion/3608241-what-is-desantiss-stop-woke-act/.
97. See, e.g., Sylvia Yanagisako and Carol Delaney, eds., *Naturalizing Power: Essays in Feminist Cultural Analysis* (New York: Routledge, 1994).

98. Elizabeth Harris and Alexandra Alter, "Book Ban Efforts Spread Across the U.S.," *New York Times*, June 22, 2023, https://www.nytimes.com/2022/01/30/books/book-ban-us-schools.html.
99. *SFFA v Harvard University et al.*, 143 S.Ct. 2141, 2200–01 (2023) (Thomas, J., concurring).
100. *SFFA v Harvard University et al.*, 143 S.Ct. 2141, 2200–02 (2023) (Thomas, J., concurring). Succinctly stating the heart of his colorblind jurisprudence, he intoned:

> The solution to our Nation's racial problems thus cannot come from policies grounded in affirmative action or some other conception of equity. Racialism simply cannot be undone by different or more racialism. Instead, the solution announced in the second founding is incorporated in our Constitution: that we are all equal, and should be treated equally before the law without regard to our race.

101. Richard C. Lewontin, "Race and Intelligence," *Bulletin of the Atomic Scientists* 26, no. 3 (1970): 2–8, 8.

EPILOGUE: DIVERSITY'S PANDEMIC DISTRACTIONS

1. See, e.g., Priscilla Wald, *Contagious: Cultures, Carriers, and the Outbreak Narrative* (Durham, NC: Duke University Press, 2008); Susan Craddock, "City of Plagues: Disease, Poverty, and Deviance in San Francisco," *Journal of the History of Medicine and Allied Sciences* 56 (2001): 302.
2. See Keith Wailoo, "Spectacles of Difference: The Racial Scripting of Epidemic Disparities," *Bulletin of the History of Medicine* 94 (2020): 602, 607–8.
3. Tyler T. Reny and Matt A. Barreto, "Xenophobia in the Time of Pandemic: Othering, Anti-Asian Attitudes, and COVID-19," *Politics, Groups, and Identities* 10, no. 2 (2022): 209–32.
4. See, e.g., Weiyi Cai, Audra D. S. Burch, and Jugal K. Patel, "Swelling Anti-Asian Violence: Who Is Being Attacked Where," *New York Times*, April 3, 2021, https://www.nytimes.com/interactive/2021/04/03/us/anti-asian-attacks.html.
5. The literature on this is voluminous. For a brief overview, see Jonathan Kahn et al., "How Not to Talk About Race and Genetics," *BuzzFeed News*, March 30, 2018, https://www.buzzfeednews.com/article/bfopinion/race-genetics-david-reich.
6. As sociologists, Michael Omi and Howard Winant have argued that race is perhaps best understood as "an unstable and 'decentered' complex of social meanings constantly being transformed by political struggle." Michael Omi and

Howard Winant, *Racial Formation in the United States: From the 1960s to the 1990s* 2nd edition (New York: Routledge, 1994), 55.

7. Alan Goodman, "Race Is Real, But It's Not Genetic," *Sapiens*, March 13, 2020, https://www.sapiens.org/biology/is-race-real/.
8. Nancy Krieger, "If 'Race' Is the Answer, What Is the Question?—On 'Race,' Racism, and Health: A Social Epidemiologist's Perspective," *Is Race "Real?,"* June 7, 2006, https://forums.ssrc.org/race-and-genomics/author/nkrieger/.
9. *Grutter v. Bollinger*, 539 U.S. 306, 343 (2003).
10. Derrick Bell, "Diversity's Distractions," *Columbia Law Review* 103 (2003):1622.
11. Reny and Barreto, "Xenophobia in the Time of a Pandemic," 213.
12. Lisa Betty, "'Black Death': Race and Representations of the Ebola Epidemic and COVID-19 Pandemic," *Medium*, May 30, 2020, https://lbetty1.medium.com/black-death-race-and-representations-of-the-ebola-epidemic-and-covid-19-pandemic-479195d590 [https://perma.cc/XJ89-D4KF].
13. Wailoo, "Spectacles of Difference," 604, 605, 606.
14. National Conference of State Legislatures, "President Trump Declares State of Emergency for COVID-19," March 25, 2020, https://www.ncsl.org/ncsl-in-dc/publications-and-resources/president-trump-declares-state-of-emergency-for-covid-19.aspx.
15. The COVID Tracking Project, "About the Racial Data Tracker," https://covidtracking.com/race/about [https://perma.cc/K97D-ZYJP] (last visited May 21, 2021).
16. Catherine Powell, "Color of Covid: The Racial Justice Paradox of Our New Stay-at-Home Economy," *CNN*, April 18, 2020, https://www.cnn.com/2020/04/10/opinions/covid-19-people-of-color-labor-market-disparities-powell/index.html.
17. Marc A. Garcia, Patricia A Homan, Catherine García, and Tyson H Brown, "The Color of COVID-19: Structural Racism and the Pandemic's Disproportionate Impact on Older Racial and Ethnic Minorities," *Gerontological Society of America* (2020): 1, 5.
18. Lisa Rubin-Miller, Christopher Alban, Samantha Artiga, and Sean Sullivan, "COVID-19 Racial Disparities in Testing, Infection, Hospitalization, and Death: Analysis of Epic Patient Data," September 16, 2020, https://www.kff.org/coronavirus-covid-19/issue-brief/covid-19-racial-disparities-testing-infection-hospitalization-death-analysis-epic-patient-data/.
19. Charles DiMaggio, Michael Klein, Cherisse Berry, and Spiros Frangos, "Black/African American Communities Are at Highest Risk of COVID-19: Spatial Modeling of New York City ZIP Code-Level Testing Results," *Annals Epidemiology* 51 (2020): 7, 9.
20. Clarence Gravlee, "Racism, Not Genetics, Explains Why Black Americans Are Dying of COVID-19," *Scientific American*, June 7, 2020, https://blogs

.scientificamerican.com/voices/racism-not-genetics-explains-why-black-americans-are-dying-of-covid-19/.
21. David E Winickoff and Osagie K. Obasogie. "Race-Specific Drugs: Regulatory Trends and Public Policy," *Trends in Pharmacological Sciences* 29, no. 6 (2008): 277–79.
22. "False Claim: African Skin Resists the Coronavirus," Reuters, March 10, 2020, https://www.reuters.com/article/uk-factcheck-coronavirus-ethnicity/false-claim-african-skin-resists-the-coronavirus-idUSKBN20X27G.
23. Dahleen Glanton, "Column: Let's Stop the Spread—of the Myth Black People Are Immune to the Coronavirus," *Chicago Tribune*, March 18, 2020, https://www.chicagotribune.com/columns/dahleen-glanton/ct-dahleen-glanton-coronavirus-black-immunity-myth-idris-elba-20200319-5auoqjzrmbcsphitbhpocth3qa-story.html.
24. Hugo Zeberg and Svante Pääbo, "The Major Genetic Risk Factor for Severe COVID-19 is Inherited From Neanderthals," *Nature* 587 (2020): 610–12.
25. Janie F. Shelton, Anjali J. Shastri, Chelsea Ye, Catherine H. Weldon, Teresa Filshtein-Somnez, Daniella Coker, Antony Symons, Jorge Esparza-Gordillo, The 23andMe COVID-19 Team, Stella Aslibekyan, and Adam Auton, "Trans-Ethnic Analysis Reveals Genetic and Non-Genetic Associations with COVID-19 Susceptibility and Severity," *medRxiv*, September 7, 2020, https://www.medrxiv.org/content/10.1101/2020.09.04.20188318v1 [https://perma.cc/QM8Z-Y3J7].
26. Jon Entine and Patrick Whittle, *"What's 'Race' Got to Do With It? Sub-Saharan Africa Emerges as Coronavirus 'Cold Spot', Offering Clues to Develop COVID-19 Vaccines,"* Genetic Literacy Project, March 31, 2020, https://geneticliteracyproject.org/2020/03/31/whats-race-got-to-do-with-it-most-of-sub-saharan-africa-emerges-as-coronavirus-cold-spot-which-may-offer-clues-to-finding-covid-19-vaccine/.
27. Erola Pairo-Castineira et al., "Genetic Mechanisms of Critical Illness in COVID-19," *Nature* 591 (2021): 92, 96.
28. Hugo Zeberg and Svante Pääbo, "A Genomic Region Associated with Protection Against Severe COVID-19 is Inherited from Neandertals," *Proceedings of the National Academy of Sciences* (2021): 1, 1–2, 4.
29. Nathaniel Sharping, "How Much Neanderthal DNA Do Humans Have?," *Discover Magazine*. April 28, 2020, https://www.discovermagazine.com/planet-earth/how-much-neanderthal-dna-do-humans-have.
30. Anne Soy, "Coronavirus in Africa: Five Reasons Why Covid-19 Has Been Less Deadly Than Elsewhere," *BBC*, October 8, 2020, https://www.bbc.com/news/world-africa-54418613.
31. Anna Jones, "How Did New Zealand Become Covid-19 Free?," *BBC*, July 10, 2020, https://www.bbc.com/news/world-asia-53274085.

32. Austin Nguyen, Julianne K. David, Sean K. Maden, Mary A. Wood, Benjamin R. Weeder, Abhinav Nellore, and Reid F. Thompson, "Human Leukocyte Antigen Susceptibility Map for Severe Acute Respiratory Syndrome Coronavirus 2," *J. Virology* (2020): 1; Monojit Debnath, Moinak Banerjee, and Michael Berk, "Genetic Gateways to COVID-19 Infection: Implications for Risk, Severity, and Outcomes," *FASEB J.* 34 (2020): 8787; David J. Langton, Stephen C Bourke, Benedicte A Lie, Gabrielle Reiff, Shonali Natu, Rebecca Darlay, John Burn, and Carlos Echevarria, "The Influence of HLA Genotype on Susceptibility to, and Severity of, COVID-19 Infection," *medRxiv*, January 4, 2021, https://www.medrxiv.org/content/10.1101/2020.12.31.20249081v1.full; Debnath et al., "Genetic Gateways," 8787; Yanan Cao, Lin Li, Zhimin Feng, Shengqing Wan, Peide Huang, Xiaohui Sun, Fang Wen, Xuanlin Huang, Guang Ning, and Weiqing Wang, "Comparative Genetic Analysis of the Novel Coronavirus (2019-nCoV/SARS-CoV-2) Receptor ACE2 in Different Populations," *Cell Discovery* (2020): 1; see Supinda Bunyavanich, Chantal Grant, and Alfin Vicencio, "Racial/Ethnic Variation in Nasal Gene Expression of Transmembrane Serine Protease 2 (TMPRSS2)," *JAMA* 324 (2020):1567; see, generally, J. M. Carethers, "Insights into Disparities Observed with COVID-19," *Journal of Internal Medicine* 289 (2021), https://pubmed.ncbi.nlm.nih.gov/33164230/; John R. Giudicessi, Dan M. Roden, Arthur A. M. Wilde, and Michael J. Ackerman, "Genetic Susceptibility for COVID-19-Associated Sudden Cardiac Death in African Americans," *Heart Rhythm* 17 (2020): 1487; See, generally, Rahul Chaudhary, Kevin P. Bliden, Rolf P. Kreutz, Young-Hoon Jeong, Udaya S. Tantry, Jerrold H. Levy, and Paul A. Gurbel, "Race-Related Disparities in COVID-19 Thrombotic Outcomes: Beyond Social and Economic Explanations," *EClinical Medicine* 29 (2020): 1.
33. Nguyen et al., "Human Leukocyte," 1
34. Debnath et al, "Genetic Gateways," 8791.
35. Nguyen et al, "Human Leukocyte," 6–7, 7.
36. K. Cao, A. M Mo.ormann, K. E. Lyke, C. Masaberg, O. P. Sumba, O. K. Doumbo, D. Koech, A. Lancaster, M. Nelson, D. Meyer, R. Single, R. J. Hartzman, C. V. Plowe, J. Kazura, D. L. Mann, M. B. Sztein, G. Thomson, and M. A. Fernández-Viña, "Differentiation Between African Populations Is Evidenced by the Diversity of Alleles and Haplotypes of HLA Class I Loci," *Tissue Antigens* 63 (2004): 293–94.
37. Cao et al., "Comparative Genetic Analysis," 11.
38. Debnath et al, "Genetic Gateways," 8787.
39. Nicole Phillips, In-Woo Park, Janie R Robinson, and Harlan P Jones, "The Perfect Storm: COVID-19 Health Disparities in US Blacks," *Journal of Racial and Ethnic Health Disparities* 8 (2021): 1153–1160, 1153
40. For an extended discussion of this phenomenon, see Jonathan Kahn, *Race in a Bottle: The Story of BiDil and Racialized Medicine in a Post-Genomic Age* (New York: Columbia University Press, 2012), 157–92.

41. Philips et al., "The Perfect Storm," 1157.
42. Jacqui Wise, "Covid-19: Known Risk Factors Fail to Explain the Increased Risk of Death Among People from Ethnic Minorities," *BMJ: British Medical Journal (Online)* 369 (2020): 1873.
43. Wise, "Covid-19," 1873.
44. U. K. Cabinet Office, Race Disparity Unit, *Quarterly Report on Progress to Address Covid-19 Health Inequalities*, October 2020, 53
45. "Socioeconomic Factors Drive COVID Risks for Minorities—UK Government Report," Reuters, October 21, 2020, https://www.reuters.com/article/us-health-coronavirus-britain-ethnicity/socioeconomic-factors-drive-covid-risks-for-minorities-uk-government-report-idUSKBN2763CF.
46. Geetanjali Saini, Monica H Swahn, and Ritu Aneja, "Disentangling the Coronavirus Disease 2019 Health Disparities in African Americans: Biological, Environmental, and Social Factors," *Open Forum Infectious Diseases* (2021): 1–11, 8 (emphasis added).
47. Lourdes Ortiz-Fernández and Amr H. Sawalha, "Genetic Variability in the Expression of the SARS-CoV-2 Host Cell Entry Factors Across Populations," *Genes & Immunity* 21 (2020): 269, 271.
48. Supinda Bunyavanich, Chantal Grant, and Alfin Vicencio, "Racial/Ethnic Variation in Nasal Gene Expression of Transmembrane Serine Protease 2 (TMPRSS2)," *JAMA* 324 (2020): 1568.
49. Merlin Chowkwanyun and Adolph L. Reed Jr., "Racial Health Disparities and Covid-19—Caution and Context," *New England Journal of Medicine* 383 (2020): 201, 202.
50. Giudicessi et al., "Genetic Susceptibility," 1487, 1490, 1491.
51. Chaudhary et al., "Race-Related Disparities," 20.
52. Jay S. Kaufman, Lena Dolman, Dinela Rushani, and Richard S. Cooper, "The Contribution of Genomic Research to Explaining Racial Disparities in Cardiovascular Disease: A Systematic Review," *American Journal of Epidemiology* 187 (2015): 464–72, 464.
53. Apoorva Mandavilli, "Medical Journals Blind to Racism as Health Crisis, Critics Say," *New York Times*, June 2, 2021, https://www.nytimes.com/2021/06/02/health/jama-racism-bauchner.html.
54. Gravlee, "Racism, Not Genetics."
55. Clyde W. Yancy, "COVID-19 and African Americans," *JAMA* 323 (2020): 1891–92.
56. Ruqaiijah Yearby and Seema Mohapatra, "Law, Structural Racism, and the COVID-19 Pandemic," *Journal of Law and the Biosciences* 7 (2020): 2, 4; Harriet A. Washington, "How Environmental Racism is Fueling the Coronavirus Pandemic," *Nature*, May 19, 2020, https://www.nature.com/articles/d41586-020-01453-y.

57. Center for Disease Control and Prevention, *Health Equity Considerations and Racial and Ethnic Minority Groups,* 2021, https://stacks.cdc.gov/view/cdc/103876; Monia Webb Hooper, Anna Maria Nápoles, and Eliseo J. Pérez-Stable, "COVID-19 and Racial/Ethnic Disparities," *JAMA* (2020): 2466; Eddie Burkhalter et al., "Incarcerated and Infected: How the Virus Tore Through the U.S. Prison System," *New York Times,* April 10, 2021, https://www.nytimes.com/interactive/2021/04/10/us/covid-prison-outbreak.html?
58. Yearby and Mohapatra, "Law, Structural Racism."
59. Marilyn D. Thomas, Eli K. Michaels, Sean Darling-Hammond, Thu T. Nguyen, M. Maria Glymour, and Eric Vittinghoff, "Whites' County-Level Racial Bias, COVID-19 Rates, and Racial Inequities in the United States," *International Journal of Environmental Research and Public Health* (2020), https://www.ncbi.nlm.nih.gov/pmc/articles/PMC7700363/pdf/ijerph-17-08695.pdf.
60. National Academies Sciences Engineering Medicine, "Framework for Equitable Allocation of COVID-19 Vaccine" (2020), https://nap.nationalacademies.org/catalog/25917/framework-for-equitable-allocation-of-covid-19-vaccine.
61. U. K. Cabinet Office, "Quarterly Report"; National Academies Sciences Engineering and Medicine, "National Academies Release Draft Framework for Equitable Allocation of a COVID-19 Vaccine, Seek Public Comment" (2020), https://www.nationalacademies.org/news/2020/09/national-academies-release-draft-framework-for-equitable-allocation-of-a-covid-19-vaccine-seek-public-comment.
62. Gina Kolata, "Nothing to Do with Genes': Racial Gaps in Pandemic Stem from Social Inequities, Studies Find," *New York Times,* December 9, 2020, https://www.nytimes.com/2020/12/09/world/nothing-to-do-with-genes-racial-gaps-in-pandemic-stem-from-social-inequities-studies-find.html.
63. Gbenga Ogedegbe, Joseph Ravenell, Samrachana Adhikari, Mark Butler, Tiffany Cook, Fritz Francois, Eduardo Iturrate, Girardin Jean-Louis, Simon A Jones, Deborah Onakomaiya, Christopher M Petrilli, Claudia Pulgarin, Seann Regan, Harmony Reynolds, Azizi Seixas, Frank Michael Volpicelli, and Leora Idit Horwitz, "Assessment of Racial/Ethnic Disparities in Hospitalization and Mortality in Patients with COVID-19 in New York City," *JAMA Network Open* (2020): 1, 2.
64. Kolata, "Nothing to Do with Genes."
65. Eboni G. Price-Haywood, Jeffrey Burton, Daniel Fort, and Leonardo Seoane, "Hospitalization and Mortality Among Black Patients and White Patients with Covid-19," *New England Journal of Medicine* 382 (2020): 2534.
66. Baligh R. Yehia, Angela Winegar, Richard Fogel, Mohamad Fakih, Allison Ottenbacher, Christine Jesser, Angelo Bufalino, Ren-Huai Huang, and Joseph

Cacchione, "Association of Race with Mortality Among Patients Hospitalized With Coronavirus Disease 2019 (COVID-19) at 92 US Hospitals," *JAMA Network Open* (2020): 1, 6.
67. Chaudhary et al., "Race-Related Disparities."
68. U.S. Food and Drug Administration, *FDA Briefing Document Pfizer-BioNTech COVID-19*, 7, December 10, 2020, https://www.fda.gov/media/144246/download; Tal Zaks, Chief Medical Officer at ModernaTX, Presentation to FDA, Emergency Use Authorization (EUA) Application for mRNA-1273, December 17, 2020.
69. Jon Zelner, Rob Trangucci, Ramya Naraharisetti, Alex Cao, Ryan Malosh, Kelly Broen, Nina Masters, and Paul Delamater, "Racial Disparities in Coronavirus Disease 2019 (COVID-19) Mortality Are Driven by Unequal Infection Risks," *Criminal Infectious Disease* 72 (2021): e88; Gina Kolata, "Nothing to Do with Genes."
70. Harriet A. Washington, *Medical Apartheid: The Dark History of Medical Experimentation on Black American from Colonial Times to the Present* (New York: Doubleday, 2006).
71. Darshali A. Vyas, Leo G. Einstein, and David S. Jones, "Hidden in Plain Sight—Reconsidering the Use of Race Correction in Clinical Algorithms," *New England Journal of Medicine* 383 (2020): 874, 879.
72. Vyas et al., "Hidden in Plain Sight," 879.
73. Michael W. Sjoding, Robert P. Dickson, Theodore J. Iwashyna, Steven E. Gay, and Thomas S. Valley, "Racial Bias in Pulse Oximetry Measurement," *New England Journal of Medicine* 383 (2020): 2477.
74. Sjoding et al., "Racial Bias," 2478.
75. Letter from Elizabeth Warren, Ron Wyden, and Cory A. Booker, Senators, to Dr. Janet Woodcock, Acting Comm'r, U.S. Food & Drug Admin, January 25, 2021. https://www.warren.senate.gov/imo/media/doc/2020.01.25%20Letter%20to%20FDA%20re%20Bias%20in%20Pulse%20Oximetry%20Measurements.pdf.
76. U.S. Food and Drug Administration, *Safety Communication on Pulse Oximeter Accuracy and Limitation* (2021).
77. U.S. Food and Drug Administration, *Pulse Oximeters*, https://www.fda.gov/medical-devices/products-and-medical-procedures/pulse-oximeters#discussionpaper.
78. U.S. Food and Drug Administration, *Anesthesiology and Respiratory Therapy Devices Panel of the Medical Devices Advisory Committee Meeting Announcement*, February 2, 2024, https://www.fda.gov/advisory-committees/advisory-committee-calendar/february-2-2024-anesthesiology-and-respiratory-therapy-devices-panel-medical-devices-advisory#event-materials.

79. Meredith A. Anderson, Atul Malhotra, and Amy L Non, "Could Routine Race-Adjustment of Spirometers Exacerbate Racial Disparities in COVID-19 Recovery?," *Lancet Respiratory Medicine* 9 (2021): 124.
80. Lundy Braun, *Breathing Race into the Machine: The Surprising Career of the Spirometer from Plantation of Genetics* (Minneapolis: University of Minnesota Press, 2021): 204–5.
81. Anderson, "Could Routine Race-Adjustment," 124.
82. Jon Levine, "RFK Jr. Says COVID May Have Been 'Ethnically Targeted' to Spare Jews," *New York Post*, July 23, 2023, https://nypost.com/2023/07/15/rfk-jr-says-covid-was-ethnically-targeted-to-spare-jews/.
83. Ryan King, "RFK Jr. Defends His 'Ethnically Targeted' COVID-19 Comments," *New York Post*, July 16, 2023, https://nypost.com/2023/07/16/rfk-jr-defends-his-ethnically-targeted-covid-19-comments/.
84. Yuan Hou, Junfei Zhao, William Martin, Asha Kallianpur, Mina K. Chung, Lara Jehi, Nima Sharifi, Serpil Erzurum, Charis Eng, and Feixiong Cheng, "New Insights Into Genetic Susceptibility of COVID-19: An ACE2 and TMPRSS2 Polymorphism Analysis," *BMC Medicine* 18, no. 1 (2020): 1–8, 4.
85. *United States v. Johnson*, No. 17-CR-162, 2020 U.S. Dist. WL 2770266, at 1, 5 (E.D. Wis. May 28, 2020).
86. *United States v. Alexander*, No. CV 19–32, 2020 U.S. Dist. WL 2507778, at 4 (D.N.J. May 15, 2020).
87. *United States v. Billings*, No.19-CR-00099-REB, U.S. Dist. WL 4705285, at *3 n. 5 (D. Colo. Aug. 13, 2020) (citing Gravlee, "Racism, Not Genetics.").
88. Yearby and Mohapatra, "Law, Structural Racism," 36.
89. Steven Epstein, *Inclusion: The Politics of Difference in Medical Research* (Chicago: University of Chicago Press, 2008), 1–2, 82.
90. Otis W. Brawley, "Response to 'Inclusion of Women and Minorities in Clinical Trials and the NIH Revitalization Act of 1993—The Perspective of NIH Clinical Trialists,' " *Controlled Clinical Trials* 16 (1995): 293.
91. Food and Drug Administration Modernization Act of 1997, Pub. L. No. 105–115, 111 Stat. 2296 (1997) (codified as amended in scattered sections of 21 U.S.C.); Guidance for Industry on Population Pharmacokinetics, 64 Fed. Reg. 6663 (Feb. 10, 1999); Guidance for Industry on the Collection of Race and Ethnicity Data in Clinical Trials, 70 Fed. Reg. 54,946 (Sept. 19, 2005); U.S. Food and Drug Administration, *Collection of Race and Ethnicity Data in Clinical Trials: Guidance for industry and Food and Drug Administration Staff* (2016).
92. Letter from Robert Menendez et al., to David A. Ricks, Chief Exec. Officer, Eli Lily & Co., April 20, 2020, https://www.warner.senate.gov/public/_cache/files/9/b/9b8e1c06-0174-462f-900a-3d9261cb97e3/7028435ABFB2FDD2F7C780508CAA307D.drug-trial-diversity-eli-lilly-.pdf.

93. Mary Chris Jaklevic, "Researchers Strive to Recruit Hard-Hit Minorities Into COVID-19 Vaccine Trials," *JAMA* 324 (2020): 826, 827.
94. Elizabeth Cohen, "Moderna Increases Minority Numbers in Its Vaccine Trial, But Still Not Meeting Fauci's Goal," *CNN*, August 29, 2020, https://www.cnn.com/2020/08/29/health/moderna-coronavirus-vaccine-minorities-goal/index.html.
95. U.S. Food and Drug Administration, *Development and Licensure of Vaccines to Prevent COVID-19: Guidance for Industry* (2020).
96. "Enhancing the Diversity of Clinical Trial Populations—Eligibility, Criteria, Enrollment Practices, and Industry," 85 Fed. Reg. 71,654 (November 10, 2020).
97. "Why Is Diversity So Important In Vaccine Trials?" *Henry Ford Health* (blog), November 24, 2020, https://www.henryford.com/blog/2020/11/diversity-in-vaccine-trials; Samantha Artiga, Jennifer Kates, Josh Michaud, and Latoya Hill, "Racial Diversity Within COVID-19 Vaccine Clinical Trials: Key Questions and Answers," Kaiser Family Foundation, January. 26, 2021, https://www.kff.org/racial-equity-and-health-policy/issue-brief/racial-diversity-within-covid-19-vaccine-clinical-trials-key-questions-and-answers/.
98. "COVID-19 Treatments Must Work for Communities of Color," National Medical Association, June 11, 2020, https://www.nmanet.org/news/512553/COVID-19-Treatments-Must-Work-for-Communities-of-Color.htm.
99. Walter M. Kimbrough and C. Reynold Verret, "A Message from the Presidents of Dillard and Xavier," *Dillard University* (September 2, 2020), https://www.dillard.edu/du-news/a-message-from-the-presidents-of-dillard-and-xavier/.
100. Wayne A. I. Frederick, David M. Carlisle, Valerie Montgomery Rice, and James Hildreth, "Joint Statement on the Integrity of Vaccine Trials and the Inclusion of Black, Indigenous and People of Color," *Howard University Newsroom*, September 17, 2020, https://newsroom.howard.edu/newsroom/article/13236/joint-statement-integrity-vacc.
101. Renee Mahaffey Harris, "More People of Color Needed In COVID-19 Vaccine Trials," Interview by Lulu Garcia-Navarro, *NPR*, August 23, 2020, 7:49 a.m., https://www.npr.org/2020/08/23/905181731/more-people-of-color-needed-in-covid-19-vaccine-trials.
102. Adam Feuerstein, Damian Garde, and Rebecca Robbins, "Covid-19 Clinical Trials Are Failing to Enroll Diverse Populations, Despite Awareness Efforts," *STAT*, August 14, 2020, https://www.statnews.com/2020/08/14/covid-19-clinical-trials-are-are-failing-to-enroll-diverse-populations-despite-awareness-efforts/. [https://perma.cc/4PZ8-FRVV].
103. Curtis Bunn, "A COVID-19 Vaccine Will Work Only if Trials Include Black Participants, Experts Say," *NBC News*, June 7, 2020, https://www.nbcnews.com

/news/nbcblk/covid-19-vaccine-will-only-work-if-trials-include-black-n1228371.

104. Brian Gormley, "Researchers Look for Ways to Make Drug Trials More Diverse," *Wall Street Journal*, March 28, 2021, https://www.wsj.com/articles/researchers-look-for-ways-to-make-drug-trials-more-diverse-11616885108; Wayne A.I. Frederick et al., "We Need to Recruit More Black Americans in Vaccine Trials," *New York Times*, September 11, 2020, https://www.nytimes.com/2020/09/11/opinion/vaccine-testing-black-americans.html; Karen Bass, "People of Color Are Disproportionately Affected By Covid-19. Yet They Are Underrepresented in Vaccine Trials," *Washington Post*, September 1, 2020, https://www.washingtonpost.com/opinions/karen-bass-covid-trials-people-of-color-underrepresented-/2020/09/01/9f5c8502-ebae-11ea-ab4e-581edb849379_story.html.

105. Quita Highsmith and Jamie Freedman, "A Call For More Inclusive COVID-19 Research: Tackling Disparities During a Pandemic," *Genentech*, May 22, 2020, https://www.gene.com/stories/a-call-for-more-inclusive-covid-19-research.

106. Macaya Douoguih, "Pathway to a Vaccine: Efforts to Develop a Safe, Effective, and Accessible COVID-19 Vaccine," Virtual Hearing before House Energy and Commerce Subcommittee on Oversight and Investigation, July 21, 2020, https://www.congress.gov/event/116th-congress/house-event/110926.

107. *PhRMA*, "Principles on Conduct of Clinical Trials Communication of Clinical Trial Results," October. 14, 2020, https://www.phrma.org/-/media/Project/PhRMA/PhRMA-Org/PhRMA-Org/PDF/P-R/PhRMAPrinciples-of-Clinical-Trials-FINAL.pdf.

108. *U.S. Food and Drug Administration Center for Biologics Evaluation and Research 162nd Vaccines and Related Biological Products Advisory Committee Meeting*, December 10, 2020, https://www.fda.gov/media/144859/download (transcript of the statement of William Gruber, Pfizer Senior Vice President of Vaccine Clinical Research & Development available at 211), 352

109. Julie Steenhuysen, "Moderna Vaccine Trial Contractors Fail to Enroll Enough Minorities, Prompting Slowdown—Sources," Reuters, October 6, 2020, https://www.reuters.com/article/idUSL1N2GW1F0.

110. Scrip (@PharmaScrip), Twitter (now X) post, April 17, 2021, https://twitter.com/PharmaScrip/status/1383329928867762180

111. *U.S. Food and Drug Administration Center for Biologics Evaluation and Research 162nd Vaccines and Related Biological Products Advisory Committee Meeting.*

112. Letter from Mark Del Monte, Chief Exec. Officer, Am. Acad. of Pediatrics, to Hana El Sahly, Chair, U.S. Food & Drug Admin.: Vaccines & Biological Products Advisory Comm. & Prabhakara Atreya, Director, U.S. Food & Drug

Admin: Div. of Sci. Advisors & Consultants, October 15, 2020. Docket #FDA-2020-N-1898; *U.S. Food and Drug Administration Center for Biologics Evaluation and Research 162nd Vaccines and Related Biological Products Advisory Committee Meeting*; Cynthia A. Pearson, Executive Director, National Women's Health Network, to U.S. Food & Drug Administration. Vaccines and Related Biological Products Advisory Committee, December 17, 2020, https://www.regulations.gov/comment/FDA-2020-N-2242-0379.

113. *U.S. Food and Drug Administration, Center for Biologics Evaluation and Research 162nd Vaccines and Related Biological Products Advisory Committee Meeting*, 211.

114. Robert Menendez et al., Senators, to David A. Ricks, Chief Exec. Officer, Eli Lily & Co., April 20, 2020 (on file with the U.S. Senate), citing Leah Cairns, "Diversity in Clinical Trials," *Johns Hopkins Science Policy Group*, September 6, 2017, https://www.jhscipolgroup.org/blog-1/2017/9/6/diversity-in-clinical-trials.

115. Artiga et al., "Racial Diversity."

116. Bunn, "A COVID-19 Vaccine."

117. Feuerstein et al, "Covid-19 Clinical Trials."

118. Laura E. Flores, Walter R. Frontera, Michele P. Andrasik, Carlos del Rio, Antonio Mondríguez-González, Stephanie A. Price, Elizabeth M. Krantz, Steven A. Pergam, and Julie K. Silver, "Assessment of the Inclusion of Racial/Ethnic Minority, Female, and Older Individuals in Vaccine Clinical Trials," *JAMA Network Open* 4 (2021): 3.

119. Kristin Toussaint, "COVID-19 Vaccine Trials Are Being Undermined by a Lack of Diversity," *FAST CO.*, September 28, 2020, https://www.fastcompany.com/90555722/why-covid-19-vaccine-trials-still-need-more-diversity.

120. World Health Organization, *Design of Vaccine Efficacy Trials to Be Used During Public Health Emergencies—Points of Considerations and Key Principles*, https://www.who.int/docs/default-source/blue-print/working-group-for-vaccine-evaluation-(4th-consultation)/ap1-guidelines-online-consultation.pdf.

121. Natalie E. Dean, Pierre-Stéphane Gsell, Ron Brookmeyer, Victor De Gruttola, Christl A. Donnelly, M. Elizabeth Halloran, Momodou Jasseh, Martha Nason, Ximena Riveros, Conall H. Watson, Ana Maria Henao-Restrepo, and Ira M. Longini, "Design of Vaccine Efficacy Trials During Public Health Emergencies," *Science Translational Medicine* 11 (2019): 499.

122. Petra Zimmermann and Nigel Curtis, "Factors That Influence the Immune Response to Vaccination," *Clinical Microbiology Reviews* 32 (2019): 1, 5.

123. Zimmermann and Curtis, "Factors," 37. It also cites three studies by one researcher who uses broader racial groups to identify variable immune

responses to certain vaccines but without identifying any underlying genetic mechanism to explain the difference or finding any difference in safety or efficacy.

124. Three of the articles cited by the study, all with the same lead author, do purport to find racial difference in immune response to certain vaccines, but these findings are both limited and equivocal. As one of these studies states: "Ethnicity and race-specific data on infectious disease susceptibility and clinical course, and/or differences in immune responses to pathogens and vaccines is limited in the literature, and the underlying mechanisms for the reported observations are still unknown," and goes on to note that "the observed statistically significant effects (cytokine response differences) in our study are relatively small and there is no known correlate of protection for vaccinia-specific cell-mediated immunity." Iana H. Haralambieva, Inna G. Ovsyannikova, Richard B. Kennedy, Beth R. Larrabee, V. Shane Pankratz, and Gregory A. Poland, "Race and Sex-Based Differences in Cytokine Immune Response to Smallpox Vaccine in Healthy Individuals," *Human Immunology* 74 (2013): 1, 4.
125. Zimmermann and Curtis, "Factors," 3.
126. U.S. Food and Drug Administration, Vaccines and Related Biological Products Advisory Committee Meeting, "Pfizer BioNTech COVID-19 Vaccine Briefing Document," (2020), https://www.fda.gov/media/144245/download.
127. U.S. Food and Drug Administration, Vaccines and Related Biological Products Advisory Committee Meeting, "Moderna COVID-19 Vaccine Briefing Document" (2020), https://www.fda.gov/media/144434/download [https://perma.cc/A249-CXLK].
128. U.S. Food and Drug Administration, Vaccines and Related Biological Products Advisory Committee Meeting, "Janssen Ad 26. COV2.S Vaccine for the Prevention of COVID-19 Briefing Document," (2021), https://www.fda.gov/media/146217/download.
129. Aravinda Chakravarti et al., "Using Population Descriptors in Genetics and Genomics Research: A New Framework for an Evolving Field," *National Academies of Sciences, Engineering, and Medicine* (Washington, D.C., The National Academies Press, 2023): 25. https://doi.org/10.17226/26902.
130. Robert Menendez et al., "Diversity in Clinical Trials."
131. Gökhan S. Hotamisligil, "Inflammation, Metaflammation and Immunometabolic Disorders," *Nature* 542 (2017): 177.
132. S. U. Yasuda, L. Zhang, and S-M. Huang, "The Role of Ethnicity in Variability in Response to Drugs: Focus on Clinical Pharmacology Studies," *Clinical Pharmacology and Therapeutics* 84 (2008): 1–3.
133. A classic example of this is the widely prescribed anticoagulant drug warfarin, which is commonly prescribed to patients who are at risk of developing blood

clots, such as persons with atrial fibrillation, recurrent strokes, deep venous thrombosis, pulmonary embolism, or those who have received heart valve replacements. It is difficult to calibrate the right dose for an individual patient because warfarin has a narrow therapeutic window of efficacy and a wide range of interindividual variability in response; that is, people tend to metabolize the drug at different rates. If you are fast metabolizer and do not get a high enough dose, you risk developing blood clots. If you are a slow metabolizer and get too high a dose, you risk having a dangerous hemorrhage. Many studies over the years have noted correlations between certain broad ethnic or racial groups and the likelihood of being a rapid or slow metabolizer. Drug labels have noted this for years. In the past decade specific genetic variations have been identified that greatly affect warfarin metabolization and dosing algorithms have been developed to replace the cruder use of racial categories with genetic testing. Nonetheless, the use of race as a crude proxy in drug dosing for warfarin (and some other drugs) persists. See Kahn, *Race in a Bottle*, 157–92.

134. Cairns, "Diversity in Clinical Trials."
135. Meghan Coakley, Emmanuel Olutayo Fadiran, L. Jo Parrish, Rachel A. Griffith, Eleanor Weiss, and Christine Carter, "Dialogues on Diversifying Clinical Trials: Successful Strategies for Engaging Women and Minorities in Clinical Trials," *Journal of Women's Health* (2012): 713, 714–15 (emphasis added).
136. Coakley et al., "Dialogues on Diversifying Clinical Trials," 714 (citing Y. Zopf, C. Rabe, A. Neubert et al., "Women Encounter ADRs More Often Than Do Men," *European Journal of Clinical Pharmacology* 64 (2008): 999).
137. Y. Zopf, C. Rabe, A. Neubert et al., "Women Encounter ADRs More Often." (emphasis added).
138. Janet Heinrich, Director Health Care—Public Health Issues, U.S. General Accounting Office, to Senators Tom Harkin, Olympia J. Snowe, and Barbara A. Mikulski and Representative Henry A. Waxman, January 19, 2001, https://www.gao.gov/assets/gao-01-286r.pdf.
139. Angela Ballantyne and Agomoni Ganguli-Mitra, "To What Extent Are Calls for Greater Minority Representation in COVID Vaccine Research Ethically Justified?," *American Journal of Bioethics* 21 (2021): 99.
140. Zimmermann and Curtis, "Factors," 3.
141. Revisions to the Standards for the Classification of Federal Data on Race and Ethnicity, 62 Fed. Reg. 58,782 (October 30, 1997).
142. See, generally, Kahn, *Race in a Bottle*, 25–47; Epstein, *Inclusion*, 74–93.
143. See, e.g., Esteban González Burchard, Elad Ziv, Natasha Coyle, Scarlett Lin Gomez, Hua Tang, Andrew J. Karter, Joanna L. Mountain, Eliseo J. Pérez-Stable, Dean Sheppard, and Neil Risch, "The Importance of Race and Ethnic Background in Biomedical Research and Clinical Practice," *New England Journal of Medicine*

348 (2003): 1170; Luisa N. Borrell, Jennifer R. Elhawary, Elena Fuentes-Afflick, Jonathan Witonsky, Nirav Bhakta, Alan H.B. Wu, Kirsten Bibbins-Domingo, José R. Rodríguez-Santana, Michael A. Lenoir, James R. Gavin III, Rick A. Kittles, Noah A. Zaitlen, David S. Wilkes, Neil R. Powe, Elad Ziv, and Esteban G. Burchard, "Race and Genetic Ancestry in Medicine—A Time for Reckoning with Racism," *New England Journal of Medicine* 384 (2021): 474.

144. Nguyen et al., "Human Leukocyte"; Debnath et al., "Genetic Gateways," 8787; David J. Langton, Stephen C. Bourke, Benedicte A. Lie, Gabrielle Reiff, Shonali Natu, Rebecca Darlay, John Burn, and Carlos Echevarria, "The Influence of HLA Genotype on Susceptibility to, and Severity of, COVID-19 Infection," *HLA*, January 4, 2021, https://www.medrxiv.org/content/10.1101/2020.12.31.20249081v1.full.

145. "Genes Associated with Left-Handedness Linked with Shape of the Brain's Language Regions," Oxford University, September 5, 2019, https://www.ox.ac.uk/news/2019-09-05-genes-associated-left-handedness-linked-shape-brains-language-regions.

146. Niall McCarthy, "The Countries with the Most Left-Handed People," *Statista*, February 4, 2020, https://www.statista.com/chart/20708/rate-of-left-handedness-in-selected-countries/.

147. "Enhancing the Diversity of Clinical Trial Populations," 85 Fed. Reg. 71 (November 10, 2020)," 654; U.S. Food and Drug Administration, *Enhancing the Diversity of Clinical Trial Populations—Eligibility Criteria Enrollment Practices, and Trial Designs Guidance for Industry*, 2020, https://www.fda.gov/media/127712/download.

148. U.S. Food and Drug Administration, *Development and Licensure of Vaccines to Prevent COVID-19: Guidance for Industry*, 2020

149. Jaklevic, "Researchers Strive," 827.

150. Food and Drug Administration, Comment from Infectious Diseases Society of America, December 4, 2020, https://www.regulations.gov/comment/FDA-2020-N-1898-0105.

151. "Why is Diversity Important," *Henry Ford Health*.

152. Brennen Jensen, "Boosting Diversity in COVID-19 Vaccine Clinical Trials," *Hopkins Bloomberg Public Health Magazine*, November 9, 2020, https://magazine.publichealth.jhu.edu/2020/boosting-diversity-covid-19-vaccine-clinical-trials.

153. See, generally, Catherine Bliss, *Race Decoded: The Genomic Fight for Social Justice* (Stanford, CA: Stanford University Press, 2020); Ann Morning, *The Nature of Race: How Scientists Think and Teach About Human Difference* (Berkeley: University of California Press, 2011); Darshali A. Vyas, Leo G. Eisenstein, and David S. Jones, "Hidden in Plain Sight—Reconsidering the Use of

Race Correction in Clinical Algorithms," *New England Journal of Medicine* 383, no. 9 (2020): 874–82.
154. Harris, "More People of Color Needed In COVID-19 Vaccine Trials."
155. *PhRMA*, "Principles on Conduct of Clinical Trials Communication of Clinical Trial Results."
156. Jaklevic, "Researchers Strive."
157. Jaklevic, "Researchers Strive."
158. Elizabeth Cohen, "Moderna Increases Minority Numbers."
159. Leon McDougle, "NMA COVID-19 Task Force on Vaccines and Therapeutics, Advisory Statement on Federal Drug Administration's Emergency Use Authorization Approval for Pfizer and Moderna Vaccine," *National Medical Association*, December 21, 2020, https://www.nmanet.org/news/544970/NMA-COVID-19-Task-Force-on-Vaccines-and-Therapeutics.htm.
160. U.S. Food and Drug Administration, *Step 3: Clinical Research*, https://www.fda.gov/patients/drug-development-process/step-3-clinical-research.
161. Paul Griffin, "Explaining Vaccine Clinical Trial Phases," *Medical Xpress*, August 27, 2020, https://medicalxpress.com/news/2020-08-vaccine-clinical-trial-phases.html.
162. John T. Schiller, Xavier Castellsagué, and Suzanne M. Garland, "A Review of Clinical Trials of Human Papillomavirus Prophylactic Vaccines," *Vaccine* (2012): 1, 3.
163. Steven H. Weinberg, Amy T. Butchart, and Matthew M. Davis, "Size of Clinical Trials and Introductory Prices of Prophylactic Vaccine Series," *Human Vaccines & Immunotherapeutics* 8 (2012): 1, 2.
164. Admittedly, combining numbers from two different vaccine trials is problematic for numerous reasons. Nonetheless, the vaccines were based on similar technologies and had similar safety and efficacy profiles. Combining the numbers here is not meant to make a specifical biomedical point but rather as a heuristic device for thinking about how numbers are being conceptualized in relation to "diversity" and "representation" in these trials.
165. See Anne L. Taylor, Susan Ziesche, Clyde Yancy, Peter Carson, Ralph D'Agostino Jr, Keith Ferdinand, Malcolm Taylor, Kirkwood Adams, Michael Sabolinski, Manuel Worcel, and Jay N. Cohn, "Combination of Isosorbide Dinitrate and Hydralazine in Blacks with Heart Failure," *New England Journal of Medicine* 351 (2004): 2049–50.
166. See, e.g., Artiga et al., "Racial Diversity"; Zoe Christen Jones, "Fauci Urges Black Community to Be Confident in COVID-19 Vaccine: 'The Time is Now to Put Skepticism Aside,' " *CBS News*, December 8, 2020, https://www.cbsnews.com/news/fauci-black-community-covid-19-vaccine/; Shadim Hussain, "We

Need 'Horizontal' Trust to Overcome Vaccine Skepticism," *Wired*, November 21, 2020, https://www.wired.com/story/we-need-horizontal-trust-to-overcome-vaccine-skepticism.
167. Noni E. MacDonald and SAGE Working Group on Vaccine Hesitancy, "Vaccine Hesitancy: Definition, Scope and Determinants," *Vaccine* 33 (2015): 4161, 4163.
168. Lauren Bunch, "A Tale of Two Crises: Addressing Covid-19 Vaccine Hesitancy as Promoting Racial Justice," *HFC Forum* 33 (2021): 143, 147, 244.
169. Perry Bacon Jr., "Why a Big Bloc of Americans is Wary of the COVID-19 Vaccine—Even as Experts Hope to See Widespread Immunization," *Five Thirty-Eight*, December 11, 2020, https://fivethirtyeight.com/features/many-black-americans-republicans-women-arent-sure-about-taking-a-covid-19-vaccine/; Cary Funk and Alec Tyson, "Growing Share of Americans Say They Plan to Get a COVID-19 Vaccine—Or Already Have," *Pew Research Center*, March 5, 2021, https://www.pewresearch.org/science/2021/03/05/growing-share-of-americans-say-they-plan-to-get-a-covid-19-vaccine-or-already-have/; "One in Five Still Shun Vaccine," Monmouth University Poll, April 14, 2021, https://www.monmouth.edu/polling-institute/reports/monmouthpoll_US_041421/.
170. Emily K. Brunson and Monica Schoch-Spana, "A Social and Behavioral Research Agenda to Facilitate COVID-19 Vaccine Uptake in the United States," *Health Security* 18 (2020): 338, 340–41.
171. Angus Thomson, Karis Robinson, and Gaëlle Vallée-Tourangeau, "The 5As: A Practical Taxonomy for the Determinants of Vaccine Uptake," *Vaccine* 34 (2016): 1018.
172. Caitlin Jarrett, Rose Wilson, Maureen O'Leary, Elisabeth Eckersberger, Heidi J. Larson, "Strategies for Addressing Vaccine Hesitancy—A Systematic Review," *Vaccine* 33 (2015): 4180, 4186.
173. Artiga et al., "Racial Diversity."
174. Flores et al.., "Assessment of the Inclusion," 3, 6 (emphasis added).
175. Ballantyne and Ganguli-Mitra, "Vaccine Research Ethically Justified," 101.
176. Allison T. Walker, Philip J Smith, and Maureen Kolasa, "Reduction of Racial/Ethnic Disparities in Vaccination Coverage, 1995–2011," *Morbidity & Mortality Weekly Report* 7 (2014): 7.
177. Ralph V. Katz, S. Steven Kegeles, Nancy R. Kressin, B. Lee Green, Min Qi Wang, Sherman A. James, Stefanie Luise Russell, and Cristina Claudio, "The Tuskegee Legacy Project: Willingness of Minorities to Participate in Biomedical Research," *Journal of Health Care for the Poor and Underserved* 17 (2006): 698, 703; Ralph V. Katz, B. Lee Green, Nancy R. Kressin, Cristina Claudio, Min Qi Wang, and Stefanie L. Russell, "Willingness of Minorities to Participate in

Biomedical Studies: Confirmatory Findings from a Follow-up Study Using the Tuskegee Legacy Project Questionnaire," *Journal of National Medicine Association* (2007): 1052–53.

178. Blake Farmer, "As COVID-19 Vaccine Trials Move at Warp Speed, Recruiting Black Volunteers Takes Time," *NPR*, September 11, 2020, https://www.npr.org/sections/health-shots/2020/09/11/911885577/as-covid-19-vaccine-trials-move-at-warp-speed-recruiting-black-volunteers-takes.

179. Reuben C. Warren, Lachlan Forrow, David Augustin Hodge Sr., and Robert D. Truog, "Trustworthiness Before Trust—Covid-19 Vaccine Trials and the Black Community," *New England Journal of Medicine* (2020): e121(1), e121(1)-(2).

180. Farmer, "As COVID-19 Vaccine Trials Move."

181. Wayne A. I. Frederick, David M. Carlisle, Valerie Montgomery Rice, and James Hildreth, "Joint Statement on the Integrity of Vaccine Trials and the Inclusion of Black, Indigenous and People of Color," *Morehouse School of Medicine*, September 17, 2020.

182. Cary Funk and Alec Tyson, "Intent to Get a COVID-19 Vaccine Rises to 60% as Confidence in Research and Development Process Increases," *Pew Research Center*, December 3, 2020, https://www.pewresearch.org/science/2020/12/03/intent-to-get-a-covid-19-vaccine-rises-to-60-as-confidence-in-research-and-development-process-increases/.

183. Funk and Tyson, "Intent to Get a COVID-19 Vaccine"; see also Alec Tyson and Giancarlo Pasquini, "U.S. Public Now Divided Over Whether to Get COVID-19 Vaccine," *Pew Research Center*, September 12, 2020, https://www.pewresearch.org/science/2020/09/17/u-s-public-now-divided-over-whether-to-get-covid-19-vaccine/.

184. "Growing Interest in Swine Flu, Many See Press Overstating Its Danger," *Pew Research Center*, October 15, 2009, https://www.pewresearch.org/politics/2009/10/15/growing-interest-in-swine-flu-many-see-press-overstating-its-danger/.

185. Matthew A. Baum, "Red State, Blue State, Flu State: Media Self-Selection and Partisan Gaps in Swine Flu Vaccinations," *Journal of Health Politics, Policy and Law* 36 (2011): 1021, 1022.

186. Matt Nisbet, "Partisan Pandemics," *Skeptical Inquirer* 40 (2016): 21, 22.

187. Mark R. Joslyn and Steven M. Sylvester, "The Determinants and Consequences of Accurate Beliefs About Childhood Vaccinations," *American Politics Research* 47 (2019): 628, 633; Bert Baumgaertner, Juliet E Carlisle, and Florian Justwan, "The Influence of Political Ideology and Trust on Willingness to Vaccinate," *PloS One* 13, 1 (2018): 8.

188. Arthur Allen, "How the Anti-Vaccine Movement Crept into the GOP Mainstream," *Politico*, May 27, 2019, https://www.politico.com/story/2019/05/27/anti-vaccine-republican-mainstream-1344955.

189. Bunch, "A Tale of Two Crises," 244.
190. Funk and Tyson, "Growing Share of Americans."
191. "Coronavirus: Vaccination: Registered Voters," Civiqs (last visited Mar. 28, 2021) (reporting 49,573 responses as of March 28, 2021). Kaiser Family Foundation Tracking polls sound similar trends. *KFF Health Tracking Poll/KFF COVID-19 Vaccine Monitor*, Kaiser Fam. Found. (2021), https://files.kff.org/attachment/Topline-KFF-COVID-19-Vaccine-Monitor-KFF-Health-Tracking-Poll-February-2021.pdf.
192. Monmouth University Polling Institute. "One in Five Still Shun Vaccine."
193. Liz Hamel, Lunna Lopes, Grace Sparks, Ashley Kirzinger, Audrey Kearney, Mellisha Stokes, and Mollyann Brodie, "KFF COVID-19 Vaccine Monitor: January 2022," *KFF*, January 28, 2022, https://www.kff.org/coronavirus-covid-19/poll-finding/kff-covid-19-vaccine-monitor-january-2022/.
194. See, e.g., Hunt Allcott, Levi Boxell, Jacob Conway, Matthew Gentzkow, Michael Thaler, David Yang, "Polarization and Public Health: Partisan Differences in Social Distancing During the Coronavirus Pandemic," *Journal of Public Economics* 191(2020): 1, 3; see also Anton Gollwitzer, Cameron Martel, William J. Brady, Philip Pärnamets, Isaac G. Freedman, Eric D. Knowles, and Jay J. Van Bavel, "Partisan Differences in Physical Distancing Are Linked to Health Outcomes During the COVID-19 Pandemic," *Nature Human Behaviour* 4 (2020): 1186, 1186–91; J. Clinton, J. Cohen, J. Lapinski, M. Trussler, "Partisan Pandemic: How Partisanship and Public Health Concerns Affect Individuals' Social Mobility During COVID-19," *ScienceAdvances* 7 (2021): 1, 1–2.
195. Epstein, *Inclusion*, 42–44, 187–95.
196. See, e.g., Bunn, "A COVID-19 Vaccine"; Allan M. Brandt, "Racism and Research: The Case of the Tuskegee Syphilis Study," *Hastings Center Report* 8 (1978): 21.
197. Debbie Elliott, "In Tuskegee, Painful History Shadows Efforts to Vaccinate African Americans," *NPR*, February 16, 2021, https://www.npr.org/2021/02/16/967011614/in-tuskegee-painful-history-shadows-efforts-to-vaccinate-african-americans.
198. Katz et al., "The Tuskegee Legacy Project," 707.
199. April Dembosky, "Stop Blaming Tuskegee, Critics Say. It's Not an 'Excuse' for Current Medical Racism," *Kaiser Health News*, March 25, 2021, https://khn.org/news/article/stop-blaming-tuskegee-critics-say-its-not-an-excuse-for-current-medical-racism.
200. Lauren D. Nephew, "Systemic Racism and Overcoming My COVID-19 Vaccine Hesitancy," *EClinicalMedicine* 32 (2021): 100713.
201. Nephew, "Systemic Racism," 100713.

202. John Eligon, "Black Doctor Dies of Covid-19 After Complaining of Racist Treatment," *New York Times,* December 23, 2020, https://www.nytimes.com/2020/12/23/us/susan-moore-black-doctor-indiana.html
203. Dembosky, "Stop Blaming Tuskegee."
204. Susan M. Reverby, "Racism, Disease, and Vaccine Refusal: People of Color Are Dying for Access to COVID-19 Vaccines," *PLoS biology* 19 (2021): e3001167.
205. Rep. Nanette Diaz Barragán to Francis Collins, Director, National Institutes of Health and Anthony S. Fauci, Director, National Institutes of Allergy and Infectious Disease, August 25, 2020, https://barragan.house.gov/wp-content/uploads/2020/08/Vaccine-Trial-Locations-Barragan.pdf.
206. Rep. Sean Casten et al. to Stephen Hahn, Commissioner, Food and Drug Administration, and Alex M. Azar II, Secretary, Department of Health and Human Services, June 16, 2020, https://luria.house.gov/sites/luria.house.gov/files/wysiwyg_uploaded/6.16.20_Letter%20to%20Commissioner%20Hahn%20and%20Secretary%20Azar%20on%20COVID-19%20Vaccine%20Inclusion.pdf.
207. Paul Slovic, "Trust, Emotion, Sex, Politics, and Science: Surveying the Risk Assessment Battlefield," *Risk Analysis: An Official Publication of the Society for Risk Analysis* 19 (1999): 689.
208. See, e.g., "Make the Pledge to Share Your Intellectual Property in the Fight against COVID-19," The Open Covid Pledge. https://opencovidpledge.org/.
209. See, e.g., Geetanjali Saini, Monica H Swahn, and Ritu Aneja, "Disentangling the Coronavirus Disease 2019 Health Disparities in African Americans: Biological, Environmental, and Social Factors," *Open Forum Infectious Diseases* (2021): 1, 9.; Frederick et al., "Integrity of Vaccine Trials."
210. Thomson et al., "The 5As,"1018.
211. Keeanga-Yamahtta Taylor, *Race for Profit: How Banks and the Real Estate Industry Undermined Black Homeownership* (Chapel Hill: University of North Carolina Press, 2019).
212. See, e.g., Sheryl Gay Stolberg, Thomas Kaplan, and Rebecca Robbins, "Pressure Mounts to Lift Patent Protections on Coronavirus Vaccines," *New York Times,* May 3, 2021, https://www.nytimes.com/2021/05/03/us/politics/biden-coronavirus-vaccine-patents.html [https://perma.cc/PW62-LQJK]; Walden Bello, "The West Has Been Hoarding More Than Vaccines," *New York Times,* May 3, 2021, https://www.nytimes.com/2021/05/03/opinion/covid-biden-wto-vaccine.html.
213. *Open Covid Pledge,* "About Us," https://opencovidpledge.org/about/ [https://perma.cc/VJ9A-3JAL] (last visited Feb. 6, 2022).
214. Council for Trade-Related Aspects of Intellectual Property Rights, "Waiver from Certain Provisions of the Trips Agreement for the Prevention,

Containment and Treatment of Covid-19," WTO Doc. IP/C/W/669 (October 2, 2020).

215. "Warren, Colleagues Call on Pfizer, Moderna, and Johnson & Johnson to Expand Access to COVID-19 Vaccines Across the Globe," Elizabeth Warren, April 28, 2021, https://www.warren.senate.gov/oversight/letters/warren-collea gues-call-on-pfizer-moderna-and-johnson-and-johnson-to-expand-access-to -covid-19-vaccines-across-the-globe [https://perma.cc/6999-SN99].

216. Rebecca Robbins and Peter S. Goodman, "Pfizer Reaps Hundreds of Millions in Profits from Covid Vaccine," *New York Times*, May 4, 2021, https://www .nytimes.com/2021/05/04/business/pfizer-covid-vaccine-profits.html.

217. Ed Silverman, "U.S. Will Back Proposal to Waive Intellectual Property Rights and Boost Covid-19 Vaccine Production," *STAT*, May 5, 2021, https://www .statnews.com/pharmalot/2021/05/05/biden-covid19-vaccine-patent-rights/; Thomas Kaplan, Sheryl Gay Stolberg, and Rebecca Robbins, "Taking 'Extraordinary Measures,' Biden Backs Suspending Patents on Vaccines," *New York Times*, May 5, 2021, https://www.nytimes.com/2021/05/05/us/politics/covid -vaccine-patent-biden.html [https://perma.cc/8U66-ELLK].

218. Damian Garde, Helen Branswell, and Matthew Herper, "Waiver of Patent Rights on Covid-19 Vaccines, in Near Term, May Be More Symbolic Than Substantive," *STAT*, May 6, 2021, https://www.statnews.com/2021/05/06/waiver-of -patent-rights-on-covid-19-vaccines-in-near-term-may-be-more-symbolic -than-substantive/.

219. Melody Schreiber, "There's Something Missing From Biden's Move to Free the Covid Vaccines," *New Republic*, May 6, 2021, https://newrepublic.com/article /162320/theres-something-missing-bidens-move-free-covid-vaccines.

220. Charles R. Lawrence III, "Two Views of the River: A Critique of the Liberal Defense of Affirmative Action," *Columbia Law Review* 101 (2001): 928, 930, 931.

221. Lawrence, "Two Views of the River," 931.

222. Lawrence, "Two Views of the River," 931.

223. Bell, "Diversity's Distractions," 1622.

224. See, e.g. Jonathan Kahn, "Genes, Race, and Population: Avoiding a Collision of Categories," *American Journal of Public Health* 96 (2006): 1965; Timothy Caulfield, Stephanie M. Fullerton, Sarah E Ali-Khan, Laura Arbour, Esteban G. Burchard, Richard S. Cooper, Billie-Jo Hardy, Simrat Harry, Robyn Hyde-Lay, Jonathan Kahn, Rick Kittles, Barbara A. Koenig, Sandra Sj Lee, Michael Malinowski, Vardit Ravitsky, Pamela Sankar, Stephen W. Scherer, Béatrice Séguin, Darren Shickle, Guilherme Suarez-Kurtz, and Abdallah S. Daar, "Race and Ancestry in Biomedical Research: Exploring the Challenges," *Genome Medicine* (2009): 8.1–8.2; Joan H. Fujimura and Ramya Rajagopalan, "Different Differences: The Use

of 'Genetic Ancestry' Versus Race in Biomedical Human Genetic Research," *Social Studies of Science* 41 (2011): 5, 5–6; Reanne Frank, "What to Make of It? The (Re)emergence of a Biological Conceptualization of Race in Health Disparities Research," *Social Science & Medicine* 64 (1982): 1977; Osagie K. Obasogie, "The Return of Biological Race: Regulating Innovations in Race and Genetics Through Administrative Agency Race Impact Assessments," *Southern California Interdisciplinary Law Journal* 22 (2012): 1, 5–6; Osagie K. Obasogie, "Beyond Best Practices: Strict Scrutiny as a Regulatory Model for Race-Specific Medicines," *Journal of Law, Medicine & Ethics* 36 (2008): 491; Michael Yudell, Dorothy Roberts, Rob DeSalle, and Sarah Tishkoff, "NIH Must Confront the Use of Race in Science," *Science* 396 (2020): 1313.

Index

Abbott, Greg, 254
A.D. v. Washburn (2015), 206
Adarand v. Pena, 55–57, 80
Adoptive Couple v. Baby Girl (2013), 206
adverse drug reactions (ADRs), 293
affirmative action: civil rights and, 96–98; diversity machines and, 113–14, 117–21, 125–28, 131–33; diversity rhetoric and, 91–93, 96–99, 106; emergence of diversity and, 24–29; *Fisher v. University of Texas*, 185, 188–92; genetic politics of, 185–97; genomic diversity and, 221–22; *Grutter v. Bollinger*, 7, 93, 95, 98–108, 111, 116–17, 127–28, 134, 188–89, 192–93, 208, 221–22, 251, 256–57, 260, 264, 313–14; Herrnstein on, 41–42; introduction to, 24, 91–93; law and political diversity, 93–96; liberal defense of, 313; McGowen on, 45–46; naturalness tropes of diversity, 40–42; politics of genetic diversity and, 250–54; representation and, 58, 117, 221–22; *Schuette v. Coalition to Defend Affirmative Action*, 185; sociolegal diversity and, 208–10; R. Roosevelt Thomas on, 39–41
African American Heart Failure trial (A-HeFT), 136–37
African genetic ancestry, 225–27
African Genome Variation Project (AGVP), 224–26
Alexander, Larry, 111, 201
Alito, Samuel, 206–7
Al-Khazraji, Majid Ghaidan, 33–34, 59
All of Us program, 167–68, 172, 215, 227, 230, 261
Alper, Joseph, 48
alt-right, 4, 53, 240–44, 259–61
American Academy of Pediatrics, 287
American Anthropological Association (AAA), 49, 193, 199–200
American Association of Law Schools, 107
American Center for Law and Justice (ACLJ), 195, 248

American Civil Liberties Union, 195
American Civil Rights Institute, 197
American Heart Association, 213
American Institute for Managing Diversity, 39
American Renaissance, 244
American Society of Human Genetics (ASHG), 219–20, 223, 244
ancestral diversity, 211–13
ancestry. *See* genetic ancestry
AncestrybyDNA, 177, 179
Ancestry Informative Markers (AIMs), 114–17, 127, 211, 218, 244
angiotensin-converting enzyme-2 (ACE2) receptor, 269–73
"Apportionment of Human Diversity, The" (Lewontin), 9, 19–21, 35
Association of Black Cardiologists, 186
Azar, Alex, 310

Baldwin, James, 249
Ball, Philip, 243–44
Ballantyne, Angela, 294, 302–3
Banaji, Mahzarin, 246
Barragán, Nanette Diaz, 310
Barrett, Amy Coney, 207
Bassett, Mary, 274
BC Biolibrary project, 150
Beck, Glenn, 305
Beckwith, Jonathan, 23, 48
Bell, Derrick, 103–4, 264, 313–15
Bell Curve, The (Herrnstein, Murray), 7, 10, 23, 48–51, 94, 107–12, 208–9
Berrey, Ellen, 101
Biden, Joe, 248, 313
BiDil drug, 8, 134–38, 155, 162–66, 186–88, 213, 220–21, 234, 300
Big Pharma, 213–15

biodiversity (biological diversity): cultural competence model, 42–43; emergence of, 16, 19–24; genetic variation and, 19–24; institutionalization of, 53–55; introduction to, 3–4, 20; in medical practice, 43; politics of genetic diversity and, 261; regression and, 225–26; in scientific investigations, 43–48; sociolegal diversity and, 189, 190–91
bioethics/bioethicists, 7, 45, 46, 60, 149, 294, 308
biological child terminology, 203–6
biological understandings of race, 2, 8–10, 108, 263, 299
biomedical research diversity: BiDil drug, 8, 134–38, 155, 162–66, 186–88, 213, 220–21, 234, 300; DEI and, 1, 4, 16, 219–21, 246, 262; geneticizing race in, 129–38; genomic medicine, 140–42, 154–57, 162–64, 173, 186–87, 211, 228; health disparities, 217–19; human subject/volunteers as barrier to, 155–57; introduction to, 14; personalized medicine, 1, 62, 154, 162–66, 171–75; precision medicine, 4, 14, 163–76, 223, 232; representation and, 15–16, 53–55, 137–38
Biotechnology Industry Organization (BIO), 214
Black American ancestry, 14–15, 176–84
Blackburn, Robert, 282–83
Black Lives Matter movement, 16
Blind Spot (Banaji, Greenwald), 246
Bliss, Catherine, 43
Bloch, Ofra, 100–101
blood group analysis, 19, 24, 58–59, 71, 268–70

Boehringer Ingelheim Pharmaceuticals Inc., 214
Bok, Derek, 98, 104
Bolnick, Deborah, 126
Booker, Cory, 280
Borno, Hala, 289
Bowen, William, 98, 104
Brackeen, Chad, 203
Brackeen, Jennifer, 203
Brakeen v. Bernhard, 207
Braun, Lundy, 281
Brawley, Otis, 54, 284
Bridges, Khiara, 35–36, 59, 80, 195
broadcast diversity, 56
Brown, Michael, 88
Brown v. Board of Education, 253
Budowle, Bruce, 74–75
Bunch, Lauren, 305
Bush, George H. W., 189
Bustamante, Carlos, 237, 246
Butler, John, 82–83

Carlson, Jedidiah, 240, 242, 245
Carsen, Peter, 136
Cassidy, Bill, 266–67
Cavalli-Sforza, Luca, 45, 48, 50, 60
Celera Genomics, 3
census categories, 2, 28–29, 46, 53, 78, 198–99, 214, 252, 294–96
Centers for Disease Control (CDC), 32, 247, 285, 303
Chakraborty, Ranajit, 70, 72–73, 79
Chandler, Paulette, 290
Chen, Anthony, 26
Cherokee Freedmen, 201–2
civil rights, 28–29, 42, 50, 63, 95–99, 102, 132–33, 188, 200, 285, 315
Civil Rights Act (1866), 33–34
Civil Rights Act (1964), 25, 26

Civil Rights Commission, 32
civil rights movement, 26, 132
Civiqs, 306
Clinton, Bill, 3, 12, 85, 91, 138, 248
Clinton, Hillary, 192
Closing the Health Care Gap Act (2004), 132
Cobbs, Price, 39
Code of Federal Regulations (CFR), 177
Cohen, Philip, 226
Cohn, Jay, 134–37
Cole, Simon, 88
Coleman, Mary Sue, 99
Collins, Francis, 3, 64–65, 140–41, 145–46, 153–56, 285
color blindness approach, 25, 97–98, 102, 125–28, 185–87, 190, 197–98, 251, 254, 260
Comfort, Nathaniel, 245
Committee on Human Genome Diversity of the National Research Council, 60
"Common Destiny: Blacks and American Society, A" (National Research Council), 41
Common Disease/Common Variant (CD/CV) hypothesis, 124–25, 139, 158, 174
Concept of Representation, The (Pitkin), 232
Condit, Celeste, 94, 130
Congressional Black Caucus, 137, 137, 186
Connerly, Ward, 197–98, 254, 259
continental ancestry, 70, 115, 171–72, 223, 316
Convention on Biological Diversity, 44
Copeland, Lennie, 39
Cotton, Tom, 258

COVID-19 pandemic: diagnostic algorithms for race, 278–82; diversity in vaccine trials, 294–307; empirical concerns for diversity, 289–94; genetic correlations to racial disparities, 275–78; introduction to, 17–18, 262–65; law and diversity, 282–83; liberal responses to, 313–14; political rationales for diversity in vaccine trials, 310–13; racialization of disease, 265–75; racialization of trust, 307–10; racialization of vaccine trials, 283–88; rationales for diversity in vaccine trials, 301–7; scientific rationales for vaccine trials, 288–301
COVID Racial Data Tracker, 266
Cox, Archibald, 9, 24–27
critical race theory, 103, 248, 255, 258
Cross, Terry, 42–43
cultural competence model, 42–43
cultural diversity, 30–31, 42, 228, 242
cultural essentialism, 99

Dannemeyer, William, 55
Darwin, Charles, 19–20
Dawes Roll, 201–2
Decker, Amy, 82–83
deCODE Genetics, 146
DeFunis, Marco, 24–28
DeFunis v. Odegaard, 9, 24–28
Democratic Party, 29–30, 44, 59, 93, 221, 292, 304–7, 312–13
Department of Energy (DOE), 43
Department of Health and Human Services (DHHS), 32, 133
DeSantis, Ron, 4, 255, 258
desegregation, 28, 30, 32, 132, 249, 254. *See also* segregation

Directive 15 classifications, 28–29, 63–64, 70, 84
direct-to-consumer (DTC) genetic ancestry companies, 114–15, 117, 212, 241, 268
Disadvantaged Business Enterprise (DBE), 177, 208
discrimination model of racial inequality, 41–42
disentanglement of race and biology, 3, 12, 15, 92–93, 108, 112
distributional model of racial inequality, 41
Diversity, Equity, and Inclusion (DEI), 1, 4, 16, 219–21, 246, 248, 250, 254–58, 262
Diversity (Wood), 6, 37, 103
Diversity Challenged (Orfield, Kurlaender), 99
Diversity in America (Schuck), 6, 102
diversity in education, 4–9, 30, 103
Diversity Machine, The (Lynch), 95
diversity machines: affirmative action and, 113–14, 117–21, 125–28, 131–33; geneticizing race, 129–38; International HapMap Project, 121–26, 130; introduction to, 13, 113–17; STRUCTURE program, 12, 13, 126–30
diversity management, 10, 37–40, 42–43, 47, 52, 55–60, 65, 95–101, 104, 220, 224, 234, 241, 249, 256–59
diversity mandates, 53, 121, 213–15
Diversity of Life, The (Wilson), 44, 174
diversity rhetoric, 38, 44, 51, 91–93, 96–99, 106
diversity training, 247–49, 254, 257, 259
DNA Mystique, The (Nelkin, Lindee), 51
DNA Polymorphism Discovery Resource (PDR), 125

DNA testing, 68, 79, 119–21, 176–85, 202, 240–41
Dobbin, Frank, 255
Donnelly, Peter, 126
Donovan, Joan, 241
Durbin, Anna, 297
Duster, Troy, 87, 114, 168–69

Edelman, Lauren, 255
educational attainment, 27, 245
education diversity, 4–13, 16, 30, 103
Elba, Idris, 267
electronic health records (EHR), 166
Eliot, Charles, 26
Ellison, George, 133
Emperor's New Clothes, The (Graves), 190
employment discrimination, 37–38, 132, 275–76
Endangered Species Act, 44
Enigma of Diversity, The (Berrey), 101
Epstein, Steven, 44, 54–55, 143, 231, 283–84
Equal Protection Clause (Fourteenth Amendment), 25, 38, 92
Ethical, Legal, and Social Implications (ELSI) of the Human Genome Project, 60, 64
"Ethnic-Affiliation Estimation by Use of Population-Specific DNA Markers" (Shriver), 114
European populations, 47, 71, 123, 224–25
Evaluation of Forensic DNA Evidence, The (NRCII), 75–81
experiential diversity, 99

Falsey, Ann, 296–97
Farmer-Paellmann, Deadria, 118–21, 183

Federal Bureau of Investigation (FBI), 68, 74
Feldman, Ari, 241
Feldman, Marcus, 108, 127–28
Fisher, Abigail, 185, 188–92
Fisher v. University of Texas, 185, 188–92
Floyd, George, 248, 258
Food and Drug Administration (FDA), 11, 137–38, 168, 213–15, 284–85, 293, 296–97
Food and Drug Administration Modernization Act (FDAMA), 11, 62, 284
Food and Drug Administration Safety and Innovation Act (FDASIA), 213
forensic diversity: CODIS loci, 82, 85, 87; *The Evaluation of Forensic DNA Evidence* (NRCII), 75–81; inertial power of race, 86–90; introduction to, 11; nonracial general population base, 84–86; overview of, 67–81; race *vs.* technology, 82–83; random match probability, 68, 75–77, 84–86, 89; restriction fragment length polymorphism, 68, 81; Short Tandem Repeats, 81–83; social *vs.* genetic race, 83–84; variable number of tandem repeats, 68–70, 75, 81–82
Fortun, Mike, 146
Fourteenth Amendment, 25, 38, 102, 188
Framingham Heart Study, 142
Fullerton, Stephanie Malia, 215

Ganguli-Mitra, Agomoni, 294, 302–3
Garza, Emilio, 188–94, 209, 255
Gay, Claudine, 255
gene-environment interactions (GEI), 142, 144–45, 172–74, 275

Index 409

genetic ancestry: African, 225–27; Black American, 14–15, 176–84; continental ancestry and, 70, 115, 171–72, 223, 316; direct-to-consumer (DTC) companies, 114–15, 117; DNA testing, 68, 79, 119–21, 176–85, 202, 240–41; genetic standing and, 120; International HapMap Project, 121–26; law and diversity, 117–21; minority status claim, 14–15, 176–84, 188, 202, 208–9; Native American, 118, 179–84, 189, 194, 198, 201–7, 237–38; racial diversity and, 170–72; slavery reparations and, 13, 118–21, 183, 201; sociolegal diversity and, 194–97; test/testing, 15, 117–21, 176–85, 188, 194, 201–2, 207–8; tracing of, 114–19; variation of, 171

genetic databases, 4, 6, 13, 114, 124–25, 139, 161, 167, 216–20, 223–27, 233–35, 247, 261

Genetic Information Nondiscrimination Act (GINA), 14, 152–55

geneticizing race in biomedical research, 129–38

genetic race, 83–84

Genetics and Public Policy Center (GPPC), 13–14, 142–44, 147–55, 217

genetic standing, 120

genetic variation: ancestry tests/tracing and, 117, 176, 229; biodiversity (biological diversity) and, 19–24; biomedicine and, 186, 215–16; blood group analysis and, 19, 24, 59, 71, 268–70; COVID-19 pandemic and, 268–70, 273; GWAS and, 160, 224; HapMap Project and, 121–24, 158, 161; HGDP and, 44–47, 52, 54, 60–64, 89, 261; human subject/volunteer enrollments, 139, 141–44; mapping of, 10; NIH research on, 169–71; NRC report on, 79–82, 163; politics and, 216; population databases, 84–86; race and biological *vs.*, 2, 5–10; rare variants, 14; regression and, 225; representation and, 16, 232–34; Request for Applications and, 159; SNPs (single nucleotide polymorphisms), 83, 122–25, 243–44; social behaviors and, 245; socially identified racial groupings and, 71–73, 94; STRUCTURE program and, 128–29, 211, 261; UMAP algorithm and, 230

gene x environment interactions (GxE), 167, 170, 172

Genome Wide Association Studies (GWAS), 61, 139–40, 215, 222, 224, 243

genomic diversity, 169–71, 221–25, 228

genomic enterprises, 14, 139–41, 154, 168, 221

genomic medicine, 140–42, 154–57, 162–64, 173, 186–87, 211, 228

Gingrich, Newt, 55, 59, 95, 186

Glaude, Eddie, 249–50

Goldberg, Jonah, 118

Gonzalez, Mike, 198

Goodman, Alan, 3–4, 132, 263

Gore, Al, 93

Gorsuch, Neil, 207, 252–54

Gould, Stephen J., 23, 34

Gratz v. Bollinger (2003), 100

Graves, Joseph, 3–4

Graves, Joseph, Jr., 190–91

Gravlee, Lance, 266–67, 275, 282–83

Great Society programs, 21

Greely, Hank, 45, 47

Green, Eric, 227–28

Greenwald, Anthony, 246

Griggs, Lewis, 39, 107

Gruber, William, 288
Grutter, Barbara, 91
Grutter v. Bollinger (1994), 7, 93, 95, 98–108, 111, 116–17, 127–28, 134, 188–89, 192–93, 208, 221–22, 251, 256–57, 260, 264, 313–14
Gulati, Mitu, 101

Hahn, Stephen, 310
Hamilton, Jennifer, 200–201
HapMap Project. *See* International HapMap Project
Harden, K. Paige, 243
Hardiman, Rachel, 303
Harmon, Amy, 118, 121
Harris, Renee Mahaffey, 286, 298
Harry, Debra, 46
Hartl, Daniel, 70–71, 78–80
Head Start, 21
health disparities, 217–19
Heckler, Margaret, 30–32, 44
Hedgecoe, Adam, 62
Heller, Jean, 19–21
Henry Ford Health System, 285
Herrnstein, Richard, 9, 11, 21–23; on affirmative action, 41–42; *The Bell Curve*, 7, 10, 23, 48–51, 94, 107–12, 208–9; discrimination model of racial inequality, 41–42; distributional model of racial inequality, 41
Historically Black Colleges and Universities (HBCUs), 30–31, 285, 303–4
Hopwood v. Texas (1996), 10–11, 58–60, 125
Hoyt, Carlos, 199
Hudson, Kathy, 143, 154–55, 166
Hudson Institute, 37
human biodiversity, 16, 53, 241–44, 255
Human Biodiversity (Marks), 52, 241

Human Diversity (Murray), 23, 127, 242–43
Human Genetic Cell Repository at the Coriell Institute, 20
Human Genome Diversity Project (HGDP): affirmative action and, 92; Ethical, Legal, and Social Implications (ELSI) of, 60, 64; evaluation of, 60–64; forensic diversity and, 72–73, 77, 89; genetic variation and, 44–47, 52, 54, 60–64, 89, 261; geographic sampling, 225; political fallout from, 122, 138
Human Genome Project (HGP): affirmative action and, 92; biological diversity and, 43–48; emergence of, 3, 10, 33, 92; human subject/volunteer enrollments and, 140–42, 153; Judicial Watch and, 191–93; original promise of, 14; relation to health outcomes, 132–33
Human Leukocyte Antigen (HLA) system, 269–70, 295
Human Pangenome Reference Consortium, 227–29
human subject/volunteer enrollments: approaches to, 140–52; as barrier to genomic medicine, 155–57; Genetic Information Nondiscrimination Act and, 14, 152–55; genetic variation, 139, 141–44; Human Genome Project (HGP) and, 140–42, 153; introduction to, 139–40; large-scale population studies, 139, 142–43, 147–48, 155; law and, 152–55

incarceration rates, 88–89, 120, 276
Inclusion (Epstein), 284
Indian Child Welfare Act (ICWA), 15, 202–7

indigenous peoples: biodiversity in science, 45, 48; biological child terminology, 203–6; Indian Child Welfare Act, 202–7; racializing indigeneity, 200–207

Indigenous Peoples Council on Biocolonialism, 46

inertial power of race, 86–90

"Inference of Population Structure Using Multilocus Genotype Data" (Pritchard, Stephens, Donnelly), 126

Institute of Medicine (IOM), 133–34

International HapMap Consortium (IHC), 124

International HapMap Project, 121–26, 130, 158, 161, 228

Interstate Blood Bank, Inc., 83

ion channel genetic variants, 269

Jackson Heart Study, 142
Jeffreys, Alec, 67–68
Jensen, Arthur, 9, 21–23
Johnson & Johnson, 287
Jones, Ian Reese, 133
Judicial Watch, 191–93, 196–200, 207, 209
Juengst, Eric, 46

Kaiser Family Foundation, 285, 289, 302, 306
Kastelic, Sarah, 207
Kaye, David, 75
Kennedy, Anthony, 249
Kennedy, Robert F., Jr., 281–82
Khoury, Muin, 165, 172–73
Kidd, Kenneth, 70, 72–73, 79, 89
Kilgore, Paul, 297
Kittles, Rick, 119, 201–2
Kolata, Gina, 277

Krieger, Nancy, 4, 263
Kurlaender, Michael, 99

Lander, Eric, 61, 74–75
large-scale population studies (LPSs), 139, 142–43, 147–48, 155
law and diversity: affirmative action and, 93–96; COVID-19 pandemic and, 282–83; diversity management and, 55–60; emergence of, 24–28; human subject/volunteer enrollments, 152–55; introduction to, 14–15; Proposition 209 (California), 57–60
Lawrence, Charles, III, 103, 104–5, 313–14
Lee, Catherine, 61
Lee, Sandra, 86, 125
Lehrer, Tom, 244
Levinson, Sanford, 6, 95
Lewontin, Richard, 9, 19–24, 70–71, 76, 78–80, 261
LGBTQ+ persons, 285
Lincoln, Karen, 309
Lind, Michael, 50
Lindee, Susan, 50–51
Litvin, Deborah, 44
Livingstone, Frank, 20
Lury, Celia, 107
Lynch, Frederick, 95

major histocompatibility complex (MHC) class 1 genes, 269–70
Malone, Paul, 178–79
Malone, Philip, 178–79
Marks, Jonathan, 16, 46, 52, 241
Marshall, Thurgood, 56
Martin-Arnold, Edwina, 178
McGovern, George, 30
McGowen, Miranda Oshige, 45–46

Merikangas, Kathleen, 61
Metro Broadcasting, Inc. v. FCC, 55–56
Michalewicz, Rachel, 185, 188–91
Miller, Steven, 242
Minority Business Enterprise (MBE), 15, 38, 177
Minority Health and Health Disparities Research and Education Act (2000), 132–33
minority status claim, 14–15, 176–84, 188, 202, 208–9
model-based clustering method, 126
Moderna, 287–88, 300
Moldawer, Alan, 118
molecularized race, 173–74
Montague, Ashley, 34
Moore, Susan, 309
Morton v. Mancari (1974), 203, 207
Murray, Charles, 11, 23, 127, 226, 242–44; *The Bell Curve*, 7, 10, 23, 48–51, 94, 107–12, 208–9

NAACP, 137, 182, 186
Nagle, Rebecca, 207
National Academies of Science, Engineering, and Medicine (NASEM), 229, 276–77
National Academies of Science (NAS), 69, 74, 163
National Center on Minority Health and Health Disparities, 133
National Commission on the Future of DNA Evidence, 85
National Human Genome Research Institute (NHGRI), 3, 8, 60, 125, 140–41, 143, 145–46, 149, 153, 211, 222, 227–29
National Indian Child Welfare Association, 207
National Institute of General Medical Sciences (NIGMS), 20
National Institute of Justice, 85
National Institute of Standards and Technology's (NIST), 83
National Institutes of Health (NIH), 8, 21, 32, 53, 141, 169–70, 213
National Institutes of Health (NIH) Revitalization Act, 53–55, 62, 98, 213, 231, 284, 294, 308
National Research Council (NRC), 41, 63, 69, 73–81, 163–64
National Society of Scholars, 6
National Women's Health Network, 213
Native American ancestry, 118, 179–84, 189, 194, 198, 201–7, 237–38
naturalness tropes of diversity, 40–42
Nelkin, Dorothy, 50–51
Neoliberal Order, 105–6
Nephew, Lauren, 308–9
New Deal Order, 105
New England Journal of Medicine, 5–6
NIH Revitalization Act (1994), 11
Nisbet, matt, 305
NitroMed, 136–37
Nixon, Richard, 24
Nobles, Melissa, 63
nonracial general population base, 84–86
"'Nothing to Do with Genes,'" (Kolata), 277

Obama, Barack, 4–5, 166, 305
O'Connor, Reed, 203, 207
O'Connor, Sandra Day, 92, 100–101, 106–7, 113, 134
Office for Protection from Research Risks, 21
Office of Management and Budget (OMB), 28, 31, 53, 63, 199, 229, 294

Index 413

Office of Minority and Women's Business Enterprises (OMWBE), 176–80, 182, 184
Office of Minority Health, 32, 132, 215
Office of Research on Minority Health, 32, 53, 284
Office of the Associate Director for Minority Health, 32
Ogedegbe, Gbenga, 277
Omi, Michael, 3, 63, 84
1000 Genomes Project, 158
Oppenheimer, David, 25–26
Orfield, Gary, 99
Origin of Species, The (Darwin), 19–20
Ortmans, Jonathan, 144
overrepresentation, 282. *See also* representation

Pacific Legal Foundation (PLF), 250–51
Palmer, Scott, 99
Paluck, Levy, 257
pangenome, 227–29
Panofsky, Aaron, 241, 242, 245
Patel, Roshni, 230–31
Pendleton, Clarence, Jr., 32
People v. Barney, 74
People v. Castro (1989), 69
People v. Soto (1999), 89
People v. Wilson, 85
personalized medicine, 1, 62, 154, 162–66, 171–75
Petsko, Gregory, 156
Pew Research Center, 304–5
Pfizer, 287–88, 291, 297, 300, 313
Pharmaceutical Research and Manufacturers of America (PhRMA), 287, 298
Pioneer Fund, 244
Pitkin, Hannah, 232–33
Plotke, David, 231

"Policy Framework for Reconceptualizing the Legal Debate Concerning Affirmative Action in Higher Education, A" (Palmer), 99
political correctness, 95–96, 187
politics of genetic diversity: affirmative action and, 93–96, 185–97, 250–54; biodiversity (biological diversity) and, 261; diversity management and, 55–60; emergence of, 28–30; introduction to, 14–15, 237; Proposition 209 (California), 57–60; racism concerns, 237–40; rationales for, 216–17; rationales for diversity in vaccine trials, 310–13; right-wing politics and, 254–59; *1619 Project*, 257–59; Trump and, 16, 195, 198, 240, 242, 246–48, 254; unnaturalness view of diversity, 259–61; white supremacy and, 240–46
Popejoy, Alice, 215, 237
Popper, Robert, 200
population, defined by HapMap Project, 122
Population Architecture Using Genomics and Epidemiology (PAGE), 158–59, 222–24, 227
population-based research, 46–47, 60–61, 72, 141–42, 155, 159, 176
Post, Robert, 57, 102–3
Powell, Catherine, 266
Powell, Lewis F., 4, 8, 11, 25, 29, 35–36, 40, 58, 105–6, 256
precision medicine, 4, 14, 163–76, 223, 232
Precision Medicine Initiative (PMI), 4–6, 166–68
Pritchard, Jonathan K., 126
Proposition 209 (California), 57–60
prothrombin genetic mutations, 269

Proud Boys, 240
Public Consultation Project (PCP), 142–43, 144, 147–51

race: biological race, 3–4, 20, 189, 190–91, 225–26, 261; biological understandings of, 2, 8–10, 108, 263, 299; defined, 2, 3–4; disentanglement of race and biology, 3, 12, 15, 92–93, 108, 112; genetics/genomic diversity and, 64–66; molecularized, 173–74; self-identified, 5, 46, 83, 179, 229, 300; social understandings of, 2, 14, 64, 184–85; technology vs., 82–83
Race, Evolution, and Behavior (Rushton), 48–49
race-based health disparities, 4, 11, 17, 32, 129–31, 168, 170, 186, 212, 223, 263, 272
race betterment, 50
racial classifications, 19, 25, 34–36, 63–64, 84, 111, 129–30, 179, 183–84, 193–97, 203–7, 249–51, 316
racial discrimination, 11, 33–34, 58, 96, 190, 200, 223, 249
racial exhaustion, 102, 113
racial hierarchy, 1–2, 48, 52, 186, 189–91, 240, 255, 259–61
racial identity, 83, 106, 116, 120, 178–79, 193–95
racial injustice, 1, 9–10, 93, 100–104, 113, 127, 252
racialization: ancestry of slave descendants, 121–22; of crime, 88–89; of disease, 17–18, 265–75; in diverse research cohorts, 174; forensic DNA databases and, 87; of indigeneity, 200–207; individual experiences of, 100; notions of biology, 132; technologies and, 86; of time, 114–16; of trust, 307–10; truths and, 84; of vaccine trials, 283–88
racial justice, 9, 12–13, 186, 195–96, 210, 249, 256, 264, 314
racial prejudice, 11–12, 31, 40–43, 177, 181, 243, 246
racial repair, 241
Raff, Jennifer, 226
Ramaswamy, Vivek, 16, 248
random match probability (RMP), 68, 73, 75–77, 84–86, 89
Reagan, Ronald, 10, 32, 33
Reardon, Jennifer, 20, 45
Regents of the University of California v. Bakke (1978), 4, 8, 11, 25, 29, 35–36, 40, 58, 105, 175
regression in biodiversity, 225–26
Rehnquist, William, 254
Reich, David, 237–39
representation: affirmative action and, 58, 117, 221–22; ancestral genetic diversity, 212, 216–19; bias in, 223–25; biomedical diversity and, 15–16, 53–55, 137–38; citizenship question and, 200; civil rights model, 97–98; conceptualizations of, 15–16, 231–36; DEI and, 4; diversifying clinical trials, 231–36; in drug trials, 137–38, 214, 287–89, 295–302, 310; in human genetic data, 46, 223, 227; minority, 31–32; overrepresentation, 282; political, 28–29, 237; racial, 39; rationale of fairness, 47; underrepresentation, 215–16, 223–25, 231–36
Republican Party, 30, 44, 59, 93, 254, 301, 304–7, 309
Request for Applications (RFA), 159

Index 415

restriction fragment length polymorphism (RFLP), 68, 81
return of research results (ROR), 152
Reverby, Susan, 309
Richmond v. Croson (1989), 38–39, 80, 97–98, 256, 259
Robbin, Alice, 28
Roberts, Dorothy, 89, 186, 187
Roberts, John, 246, 248–49, 251–53
Robertson, Pat, 195
Rufo, Christopher, 4, 255–56
Rural Advancement Foundation International, 48
Rushton, J. Philippe, 48–49

SAGE Working Group of Vaccine Hesitancy, 301–2
Sailer, Steve, 53, 241, 244
Saint Francis College v. Al-Khazraji, 33–35, 59, 80, 127, 195
SARS-CoV-2. *See* COVID-19 pandemic
Satel, Sally, 186
Saving American History Act, 258
Sawyer, Charles, 163–64
Scaife, Richard Mellon, 192
Schmidt, Peter, 26
Schuck, Peter, 6–7, 26, 38–39, 101–11, 113
Schuette v. BAMN (2014), 249
Schuette v. Coalition to Defend Affirmative Action, 185
Schwarzschild, Maimon, 111, 201
scientific rationales for vaccine trials, 288–301
segregation, 19, 27–30, 43, 134, 189, 253–54, 260. *See also* desegregation
self-identified race, 5, 46, 83, 179, 229, 300
SFFA v. Harvard University (2023), 249, 251, 259–60

Shape of the River, The (Bowen, Bok), 98, 104
Shelby County v. Holder, 113, 248–49, 259
Shin, Patrick, 101
Short Tandem Repeats (STRs), 81–83
Shriver, Mark, 114
Siegel, Reva, 249
1619 Project, 257–59
Skrentny, John, 26
Slaughter, Louise, 153
slavery reparations, 13, 96, 118–21, 183, 201
Slovic, Paul, 310
Snowe, Olympia, 153
SNPs (single nucleotide polymorphisms), 83, 122–25, 243–44
social constructionism, 208–10
social diversity, 8, 24, 27, 30, 45, 92, 174–75, 217, 221–25, 239, 250–52
social race, 83–84
social understandings of race, 2, 14, 64, 184–85
Society for Women's Health Research, 213
sociobiology, 23, 50, 245
Sociobiology (Wilson), 23
socioeconomic status (SES), 131, 159
sociolegal diversity: affirmative action and, 185–91, 208–10; ancestry and, 194–97; biodiversity (biological diversity) and, 189, 190–91; DNA testing and, 176–85, 202; Garza and, 188–94; introduction to, 30, 176; racial and ethnic categories, 197–200; racializing indigeneity, 200–207; and Ralph Taylor, 14–15, 176–84, 188, 202, 208–9
Southern Poverty Law Center (SPLC), 239–40

Statement on "Race" and Intelligence (AAA), 49
Stephens, Matthew, 126
Stop WOKE Act, 258
STRUCTURE program, 12, 13, 126–30, 211, 218, 226, 230, 240, 243, 261
Students for Fair Admissions, Inc. v. President and Fellows of Harvard University (2023), 8, 16, 250
Stulberg, Lisa, 26

TallBear, Kim, 201
Taussig, Karen-Sue, 150
taxonomies of diversity, 2, 6–7, 86, 163–64, 205
Taylor, Jared, 226, 244, 247
Taylor, Keeanga-Yamahtta, 312
Taylor, Ralph, 14–15, 176–84, 188, 202, 208–9
technical rationales for diversity, 215–16
technology *vs.* race, 82–83
Thomas, Clarence, 32, 56, 260
Thomas, R. Roosevelt, 10, 39–41, 47, 52, 55, 57, 105–7, 256, 260
Toward Precision Medicine (NRC), 163
Towards a Culturally Competent System of Care (Cross), 42–43
transmembrane serine protease 2 (TMPRSS2) nasal gene expression, 269, 272–73, 281
Troublesome Inheritance, A (Wade), 225, 243
Trump, Donald, 16, 195, 198, 240, 242, 246–48, 254, 258, 266, 312
Tsosie, Krystal, 201
Tuskegee Legacy Project, 303, 308
"Tuskegee Study of Untreated Syphilis in the Negro Male," 21, 308
Tutton, Richard, 149, 151

UK Biobank project, 149
UMAP program, 230
underrepresentation, 215–16, 223–25, 231–36. *See also* representation
United Nations Conference on Environment and Development (UNCED), 44
United Nations Educational, Scientific and Cultural Organization (UNESCO), 20
United States v. Alexander, 282
unnaturalness view of diversity, 259–61
U.S. Census Bureau, 181, 199
U.S. Department of Transportation (USDOT), 177–78, 181–83
U.S. Public Health Service, 21

vaccine trials racialization, 283–88
Vallone, Peter, 82–83
value diversity, 42
Valuing Diversity (1988), 39
Van Arsdale, Adam, 2, 20
variable number of tandem repeats (VNTRs), 68–70, 75, 81–82
Venter, J. Craig, 3, 208
voting rights, 29, 97, 102, 132, 200, 246
Voting Rights Act (1965), 97, 102, 246
Vyas study, 278–79

W. Maurice Young Centre for Applied Ethics, University of British Columbia, 149
Wade, Nicholas, 225, 243
Wages, William, 179
Wailoo, Keith, 265–66
Warren, Elizabeth, 194–95, 207, 237, 280
Warren, Reuben, 308
Washburn, Sherry, 34
Weinstein, Jack, 193

White, Byron, 34–37, 195
white supremacy, 240–46
Whitewashing Race (Brown et al.), 88
Who We Are and How We Got Here (Reich), 237–39
Wilson, Allan, 60, 225
Wilson, E. O., 23, 44, 174, 245
Winant, Howard, 3
Winston, Andrew, 255
wokeism, 95, 248, 254, 257–58
WomenHeart: The National Coalition for Women with Heart Disease, 213
Wood, Peter, 6, 37–38, 103, 110–11

Workforce 2000 (Hudson Institute), 37–39
World Council of Indigenous Peoples, 48
World Health Organization, 290
World Trade Organization (WTO), 312–13
Wyden, Ron, 280

Yancy, Clyde, 275

Zelner, Jon, 278
Ziegler, Mary, 185, 187, 189–90

GPSR Authorized Representative: Easy Access System Europe, Mustamäe tee 50, 10621 Tallinn, Estonia, gpsr.requests@easproject.com